Socio-Affective Computing

Volume 6

Series Editor
Amir Hussain, University of Stirling, Stirling, UK

Co-Editor
Erik Cambria, Nanyang Technological University, Singapore

This exciting Book Series aims to publish state-of-the-art research on socially intelligent, affective and multimodal human-machine interaction and systems. It will emphasize the role of affect in social interactions and the humanistic side of affective computing by promoting publications at the cross-roads between engineering and human sciences (including biological, social and cultural aspects of human life). Three broad domains of social and affective computing will be covered by the book series: (1) social computing, (2) affective computing, and (3) interplay of the first two domains (for example, augmenting social interaction through affective computing). Examples of the first domain will include but not limited to: all types of social interactions that contribute to the meaning, interest and richness of our daily life, for example, information produced by a group of people used to provide or enhance the functioning of a system. Examples of the second domain will include, but not limited to: computational and psychological models of emotions, bodily manifestations of affect (facial expressions, posture, behavior, physiology), and affective interfaces and applications (dialogue systems, games, learning etc.). This series will publish works of the highest quality that advance the understanding and practical application of social and affective computing techniques. Research monographs, introductory and advanced level textbooks, volume editions and proceedings will be considered.

More information about this series at http://www.springer.com/series/13199

Rajiv Shah • Roger Zimmermann

Multimodal Analysis of User-Generated Multimedia Content

Springer

Rajiv Shah
School of Computing
National University of Singapore
Singapore, Singapore

Roger Zimmermann
School of Computing
National University of Singapore
Singapore, Singapore

ISSN 2509-5706 ISSN 2509-5714 (electronic)
Socio-Affective Computing
ISBN 978-3-319-61806-7 ISBN 978-3-319-61807-4 (eBook)
DOI 10.1007/978-3-319-61807-4

Library of Congress Control Number: 2017947053

Printed on acid-free paper

This Springer imprint is published by Springer Nature
The registered company is Springer International Publishing AG
The registered company address is: Gewerbestrasse 11, 6330 Cham, Switzerland

I dedicate this book to my late father,
Ram Dhani Gupta;
my mother, Girija Devi; and my other family members for their
continuous support, motivation, and unconditional love.
I love you all so dearly.

Foreword

We have stepped into an era where every user plays the role of both content provider and content consumer. With many smartphone apps seamlessly converting photographs and videos to social media postings, user-generated multimedia content now becomes the next big data waiting to be turned into useful insights and applications. The book *Multimodal Analysis of User-Generated Multimedia Content* by Rajiv and Roger very carefully selects a few important research topics in analysing big user-generated multimedia data in a multimodal approach pertinent to many novel applications such as content recommendation, content summarization and content uploading. What makes this book stand out among others is the unique focus on multimodal analysis which combines visual, textual and other contextual features of multimedia content to perform better sensemaking.

Rajiv and Roger have made the book a great resource for any reader interested in the above research topics and respective solutions. The literature review chapter gives a very detailed and comprehensive coverage of each topic and comparison of state-of-the-art methods including the ones proposed by the authors. Every chapter that follows is dedicated to a research topic covering the architecture framework of a proposed solution system and its function components. This is accompanied by a fine-grained description of the methods used in the function components. To aid understanding, the description comes with many relevant examples. Beyond describing the methods, the authors also present the performance evaluation of these methods using real-world datasets so as to assess their strengths and weaknesses appropriately.

Despite its deep technical content, the book is surprisingly easy to read. I believe the authors have paid extra attention to organizing the content for easy reading, careful proof editing and good use of figures and examples. The book is clearly written at the level suitable for reading by computer science students in graduate and senior years. It is also a good reference reading for multimedia content analytics researchers in both academia and industry. Whenever appropriate, the authors show their algorithms with clearly defined input, output and steps with comments. This facilitates any further implementation as well as extensions of the methods. This is perhaps the part of the book which will attract the "programming-type" readers most.

I would like to congratulate both Rajiv and Roger for their pioneering work in multimodal analysis for user-generated multimedia content. I believe this book will become widely adopted and referenced in the multimedia community. It is a good guide for anyone who wishes to better understand the challenges and solutions of analysing multimedia data. I wish the authors all the best in their future research endeavour.

School of Information Systems,
Singapore Management University,
Singapore, Singapore
May 2017

Ee-Peng Lim, PhD

Preface

The amount of user-generated multimedia content (UGC) has increased rapidly in recent years due to the ubiquitous availability of smartphones, digital cameras, and affordable network infrastructures. An interesting recent trend is that social media websites such as Flickr and YouTube create opportunities for users to generate multimedia content, instead of creating multimedia content by themselves. Thus, capturing UGC such as user-generated images (UGIs) and user-generated videos (UGVs) anytime and anywhere, and then instantly sharing them on social media platforms such as Instagram and Flickr, have become a very popular activity. Hence, user-generated multimedia content is now an intrinsic part of social media platforms. To benefit users and social media companies from an automatic semantics and sentics understanding of UGC, this book focuses on developing effective algorithms for several significant social media analytics problems. Sentics are common affective patterns associated with natural language concepts exploited for tasks such as emotion recognition from text/speech or sentiment analysis. Knowledge structures derived from the semantics and sentics understanding of user-generated multimedia content are beneficial in an efficient multimedia search, retrieval, and recommendation. However, real-world UGC is complex, and extracting the semantics and sentics from only multimedia content is very difficult because suitable concepts may present in different representations. Moreover, due to the increasing popularity of social media websites and advancements in technology, it is possible now to collect a significant amount of important contextual information (e.g., spatial, temporal, and preference information). Thus, it necessitates analyzing the information of UGC from multiple modalities for a better semantics and sentics understanding. Moreover, the multimodal information is very useful in a social network based news video reporting task (e.g., citizen journalism) which allows people to play active roles in the process of collecting news reports (e.g., CNN iReport). Specifically, we exploit both content and contextual information of UGIs and UGVs to facilitate different multimedia analytics problems.

Further advancements in technology enable mobile devices to collect a significant amount of contextual information in conjunction with captured multimedia content. Since the contextual information greatly helps in the semantics and sentics understanding of user-generated multimedia content, researchers exploit it in their research work related to multimedia analytics problems. Thus, the multimodal information (i.e., both content and contextual information) of UGC benefits several diverse social media analytics problems. For instance, knowledge structures extracted from multiple modalities are useful in an effective multimedia search, retrieval, and recommendation. Specifically, applications related to multimedia summarization, tag ranking and recommendation, preference-aware multimedia recommendation, multimedia-based e-learning, and news video reporting are built by exploiting the multimedia content (e.g., visual content) and associated contextual information (e.g., geo-, temporal, and other sensory data). However, it is very challenging to address these problems efficiently due to the following reasons: (i) difficulty in capturing the semantics of UGC, (ii) the existence of noisy metadata, (iii) difficulty in handling big datasets, (iv) difficulty in learning user preferences, (v) the insufficient accessibility and searchability of video content, and (vi) weak network infrastructures at some locations. Since different information knowledge structures are derived from different sources, it is useful to exploit multimodal information to overcome these challenges.

Exploiting information from multiple sources helps in addressing challenges mentioned above and facilitating different social media analytics applications. Therefore, in this book, we leverage information from multiple modalities and fuse the derived knowledge structures to provide effective solutions for several significant social media analytics problems. Our research focuses on the semantics and sentics understanding of UGC leveraging both content and contextual information. First, for a better understanding of an event from a large collection of UGIs, we present the EventBuilder system. It enables people to automatically generate a summary of the event in real-time by visualizing different social media such as Wikipedia and Flickr. In particular, we exploit Wikipedia as the event background knowledge to obtain more contextual information about the event. This information is very useful in an effective event detection. Next, we solve an optimization problem to produce text summaries for the event. Subsequently, we present the EventSensor system that aims to address sentics understanding and produces a multimedia summary for a given mood. It extracts concepts and mood tags from the visual content and textual metadata of UGCs and exploits them in supporting several significant multimedia analytics problems such as a musical multimedia summary. EventSensor supports sentics-based event summarization by leveraging EventBuilder as its semantics engine component. Moreover, we focus on computing tag relevance for UGIs. Specifically, we leverage personal and social contexts of UGIs and follow a neighbor voting scheme to predict and rank tags. Furthermore, we focus on semantics and sentics understanding from UGVs since they have a significant impact on different areas of a society (e.g., enjoyment, education, and journalism).

Since many outdoor UGVs lack a certain appeal because their soundtracks consist mostly of ambient background noise, we solve the problem of making UGVs more attractive by recommending a matching soundtrack for a UGV by exploiting content and contextual information. In particular, first, we predict scene moods from a real-world video dataset. Users collected this dataset from their daily outdoor activities. Second, we perform heuristic rankings to fuse the predicted confidence scores of multiple models, and, third, we customize the video soundtrack recommendation functionality to make it compatible with mobile devices. Furthermore, we address the problem of knowledge structure extraction from educational UGVs to facilitate e-learning. Specifically, we solve the problem of topic-wise segmentation for lecture videos. To extract the structural knowledge of a multi-topic lecture video and thus make it easily accessible, it is very desirable to divide each video into shorter clips by performing an automatic topic-wise video segmentation. However, the accessibility and searchability of most lecture video content are still insufficient due to the unscripted and spontaneous speech of speakers. We present the ATLAS and TRACE systems to perform the temporal segmentation of lecture videos automatically. In our studies, we construct models from visual, transcript, and Wikipedia features to perform such topic-wise segmentations of lecture videos. Moreover, we investigate the late fusion of video segmentation results derived from state-of-the-art methods by exploiting the multimodal information of lecture videos. Finally, we consider the area of journalism where UGVs have a significant impact on society.

We propose algorithms for news video (UGV) reporting to support journalists. An interesting recent trend, enabled by the ubiquitous availability of mobile devices, is that regular citizens report events which news providers then disseminate, e.g., CNN iReport. Often such news are captured in places with very weak network infrastructure, and it is imperative that a citizen journalist can quickly and reliably upload videos in the face of slow, unstable, and intermittent Internet access. We envision that some middleboxes are deployed to collect these videos over energy-efficient short-range wireless networks. In this study we introduce an adaptive middlebox design, called NEWSMAN, to support citizen journalists. Specifically, the NEWSMAN system jointly considers two aspects under varying network conditions: (i) choosing the optimal transcoding parameters and (ii) determining the uploading schedule for news videos. Finally, since the advances in deep neural network (DNN) technologies enabled significant performance boost in many multimedia analytics problems (e.g., image and video semantic classification, object detection, face matching and retrieval, text detection and recognition in natural scenes, and image and video captioning), we discuss their roles to solve several multimedia analytics problems as part of future directions to readers.

Singapore, Singapore — Rajiv Ratn Shah
Singapore, Singapore — Roger Zimmermann
May 2017

Acknowledgements

Completing this book has been a truly life-changing experience for me, and it would not have been possible to do without the blessing of God. I praise and thank God almighty for giving me strength and wisdom throughout my research work to complete this book. I am grateful to numerous people who have contributed toward shaping this book.

First and foremost, I would like to thank my Ph.D. supervisor Prof. Roger Zimmermann for his great guidance and support throughout my Ph.D. study. I would like to express my deepest gratitude to him for encouraging my research and empowering me to grow as a research scientist. I could not have completed this book without his invaluable motivation and advice. I would like to express my appreciation to the following professors at the National University of Singapore (NUS) for their extremely useful comments: Prof. Mohan S. Kankanhalli, Prof. Wei Tsang Ooi, and Prof. Teck Khim Ng. Furthermore, I would like to thank Prof. Yi Yu, Prof. Suhua Tang, Prof. Shin'ichi Satoh, and Prof. Cheng-Hsin Hsu who have supervised me during my internships at National Tsing Hua University, Taiwan, and National Institute of Informatics, Japan. I am also very grateful to Prof. Ee-Peng Lim and Prof. Jing Jiang for their wonderful guidance and support during my research work in the Living Analytics Research Centre (LARC) at Singapore Management University, Singapore. A special thanks goes to Prof. Ee-Peng Lim for writing the foreword for this book.

I am very much thankful to all my friends who have contributed immensely to my personal and professional time in different universities, cities, and countries during my stay there. Specifically, I would like to thank Yifang Yin, Soujanya Poria, Deepak Lingwal, Vishal Choudhary, Satyendra Yadav, Abhinav Dwivedi, Brahmraj Rawat, Anwar Dilawar Shaikh, Akshay Verma, Anupam Samanta, Deepak Gupta, Jay Prakash Singh, Om Prakash Kaiwartya, Lalit Tulsyan, Manisha Goel, and others. I would also like to acknowledge my debt to my friends and relatives for encouraging throughout my research work. Specifically, I would like to thank Dr. Madhuri Rani, Rajesh Gupta, Priyanka Agrawal, Avinash Singh, Priyavrat Gupta, Santosh Gupta, and others for their unconditional support.

Last but not the least, I would like to express my deepest gratitude to my family. A special love goes to my mother, Girija Devi, who has been a great mentor in my life and had constantly encouraged me to be a better person, and my late father, Ram Dhani Gupta, who has been a great supporter and torchbearer in my life. The struggle and sacrifice of my parents always motivate me to work hard in my research work. The decision to leave my job as a software engineer and pursue higher studies was not easy for me, but I am grateful to my brothers Anoop Ratn and Vikas Ratn for supporting me in the time of need. Without love from my sister Pratiksha Ratn, my sisters-in-law Poonam Gupta and Swati Gupta, my lovely nephews Aahan Ratn and Parin Ratn, and my best friend Rushali Gupta, this book would not have been completed.

Singapore, Singapore
May 2017

Rajiv Ratn Shah

Contents

About the Authors

Rajiv Ratn Shah received his B.Sc. with honors in mathematics from Banaras Hindu University (BHU), India, in 2005. He received his M.Tech. in computer technology and applications from Delhi Technological University (DTU), India, in 2010. Prior to joining Indraprastha Institute of Information Technology Delhi (IIIT Delhi), India, as an assistant professor, Dr. Shah has received his Ph.D. in computer science from the National University of Singapore (NUS), Singapore. Currently, he is also working as a research fellow in Living Analytics Research Centre (LARC) at the Singapore Management University (SMU), Singapore. His research interests include the multimodal analysis of user-generated multimedia content in the support of social media applications, multimodal event detection and recommendation, and multimedia analysis, search, and retrieval. Dr. Shah is the recipient of several awards, including the runner-up in the Grand Challenge competition of ACM International Conference on Multimedia 2015. He is involved in reviewing of many top-tier international conferences and journals. He has published several research works in top-tier conferences and journals such as Springer MultiMedia Modeling, ACM International Conference on Multimedia, IEEE International Symposium on Multimedia, and Elsevier *Knowledge-Based Systems*.

Roger Zimmermann is associate professor of computer science in the *School of Computing* at the *National University of Singapore* (NUS). He is also deputy director with the *Interactive and Digital Media Institute* (IDMI) at NUS and codirector of the *Centre of Social Media Innovations for Communities* (COSMIC), a research institute funded by the National Research Foundation (NRF) of Singapore. Prior to joining NUS, he held the position of research area director with the Integrated Media Systems Center (IMSC) at the University of Southern California (USC). He received his M.S. and Ph.D. degree from the University of Southern California in 1994 and 1998, respectively. Among his research interests are mobile video management, streaming media architectures, distributed and peer-to-peer systems, spatiotemporal data management, and location-based services. Dr. Zimmermann is a senior member of *IEEE* and a member of *ACM*. He has coauthored a book, seven patents, and more than 220 conference publications, journal articles, and book chapters in the areas of multimedia, GIS, and information management. He has received funding from NSF (USA), A*STAR (Singapore), NUS Research Institute (NUSRI), NRF (Singapore), and NSFC (China) as well as several industries such as Fuji Xerox, HP, Intel, and Pratt & Whitney. Dr. Zimmermann is on the editorial boards of the IEEE Multimedia Communications Technical Committee (MMTC) R-Letter and the Springer International Journal of *Multimedia Tools and Applications* (MTAP). He is also an associate editor for the *ACM Transactions on Multimedia Computing, Communications, and Applications* journal (ACM TOMM), and he has been elected to serve as secretary of ACM SIGSPATIAL for the term 1 July 2014 to 30 June 2017. He has served on the conference program committees of many leading conferences and as reviewer of many journals. Recently, he was the general chair of the ACM *Multimedia Systems 2014* and the *IEEE ISM 2015* conferences and TPC cochair of the *ACM TVX 2017* conference.

Abbreviations

UGC	User-generated content
UGI	User-generated image
UGT	User-generated text
UGV	User-generated video
HMM	Hidden Markov model
EventBuilder	Real-time multimedia event summarization by visualizing social media
EventSensor	Leveraging multimodal information for event summarization and concept-level sentiment analysis
DNN	Deep neural network
UTB	User tagging behavior
PD	Photo description
NV	Neighbor voting-based tag ranking system
NVGC	NV corresponding to geo concepts
NVVC	NV corresponding to visual concepts
NVSC	NV corresponding to semantics concepts
NVGVC	NV corresponding to the fusion of geo and visual concepts
NVGSC	NV corresponding to the fusion of geo and semantics concepts
NVVSC	NV corresponding to the fusion of visual and semantics concepts
NVGVSC	NV corresponding to the fusion of geo, visual, and semantics concepts
EF	Early fusion-based tag ranking system
LFE	NV based on late fusion of different modalities with equal weights
LFR	NV based on late fusion with weights determined by the recall of different modalities
DCG	Discounted cumulative gain
NDCG	Normalized discounted cumulative gain
PROMPT	A personalized user tag recommendation for social media photos leveraging personal and social contexts

CRAFT	Concept-level multimodal ranking of Flickr photo tags via recall-based weighting
ADVISOR	A personalized video soundtrack recommendation system
EAT	Emotion annotation tasks
AI	Artificial intelligence
NLP	Natural language processing
E-learning	Electronic learning
ATLAS	Automatic temporal segmentation and annotation of lecture videos based on modeling transition time
TRACE	Linguistic-based approach for automatic lecture video segmentation leveraging Wikipedia texts
NPTEL	National Programme on Technology Enhanced Learning
MIT	Massachusetts Institute of Technology
NUS	National University of Singapore
CNN	Cable News Network
NEWSMAN	Uploading videos over adaptive middleboxes to news servers
PSNR	Peak signal-to-noise ratio
R–D	Rate–distortion
TI	Temporal perceptual information
SI	Spatial perceptual information
Amazon EC2	Amazon Elastic Compute Cloud
EDF	Earlier deadline first
FIFO	First in, first out
SNG	Satellite news gathering
SMS	Short message service
MMS	Multimedia messaging service
FAQ	Frequently asked questions
MKL	Multiple kernel learning

Chapter 1
Introduction

Abstract The amount of user-generated multimedia content (UGC) has increased rapidly in recent years due to the ubiquitous availability of smartphones, digital cameras, and affordable network infrastructures. However, real-world UGC is complex, and extracting the semantics and sentics from only multimedia content is very difficult because suitable concepts may be exhibited in different representations. Since it is possible now to collect a significant amount of relevant contextual information due to advancements in technology, we analyze the information of UGC from multiple modalities to facilitate different social media applications in this book. Specifically, we present our solutions for applications related to multimedia summarization, tag ranking and recommendation, preference-aware multimedia recommendation, multimedia-based e-learning, and news videos uploading by exploiting the multimedia content (*e.g.*, visual content) and associated contextual information (*e.g.*, geo-, temporal, and other sensory data). Moreover, we presented a detailed literature survey and future directions for research on user-generated multimedia content.

Keywords Semantics analysis • Sentics analysis • Multimodal analysis • User-generated multimedia content • Multimedia fusion • Multimedia analysis • Multimedia recommendation • Multimedia uploading

1.1 Background and Motivation

User-generated multimedia content (UGC) has become more prevalent and asynchronous in recent years with the advent of ubiquitous smartphones, digital cameras, affordable network infrastructures, and auto-uploaders. A survey [6] conducted by Ipsos MediaCT, Crowdtap, and the Social Media Advertising Consortium on 839 millennial persons (18–36 years old) indicates that (i) every day, millennials spend a significant amount of time with different types of media, (ii) they spend 30% of the total time with UGC, (iii) millennials prefer social media above all other media types, (iv) they trust information received through UGC 50% more than information from other media sources such as newspapers, magazines, and television advertisement, and (v) UGC is 20% more influential in purchasing decisions of Millennials than other media types. Thus, UGC such as

R. Shah, R. Zimmermann, *Multimodal Analysis of User-Generated Multimedia Content*, Socio-Affective Computing 6, DOI 10.1007/978-3-319-61807-4_1

user-generated texts (UGTs), user-generated images (UGIs), and user-generated videos (UGVs) play a pivotal role in e–commerce, specifically in social commerce. Moreover, instantly sharing UGC anytime and anywhere on social media platforms such as Twitter, Flickr, and NPTEL [3] has become a very popular activity. For instance, in a very popular photo sharing website Instagram, over 1 billion UGIs have been uploaded so far and it has more than 500 million monthly active users [11]. Similarly, over 10 billion UGIs have been uploaded so far in another famous photo sharing website Flickr[1] which has over 112 million users, and an average of 1 million UGIs has uploaded daily [10].

Thus, it is required to extract knowledge structures from UGC on such social media platforms to provide various multimedia-related services and solve several significant multimedia analytics problems. The extracted knowledge structures are very useful in the semantics and sentics understanding of UGC and facilitate several significant social media applications. Sentics are common affective patterns associated with natural language concepts exploited for tasks such as emotion recognition from text/speech or sentiment analysis [19]. Sentics computing is a multidisciplinary approach to natural language processing and understanding at the crossroads between affective computing, information extraction, and commonsense reasoning, which exploits both computer and human sciences to interpret better and process social information on the web [18]. Sentics is also the study of waveforms of touch, emotion, and music, and named by Austrian neuroscientist Manfred Clynes. However, it is a very challenging task to extract such knowledge structures because real-world UGIs and UGVs are complex and noisy, and extracting semantics and sentics from the multimedia content alone is a very difficult problem. Hence, it is desirable to analyze UGC from multiple modalities for a better semantics and sentics understanding. Different modalities uncover different aspects that are useful in determining useful knowledge structures. Such knowledge structures are exploited in solving different multimedia analytics problems.

In this book, we investigate the usage of multimodal information and the fusion of user-generated multimedia content in facilitating different multimedia analytics problems [242, 243]. First, we focus on the semantics and sentics understanding of UGIs to address the multimedia summarization problem. Such summaries are very useful in providing overviews of different events automatically without looking into the vast amount of multimedia content. We particularly address problems related to recommendation and ranking of user tags, summarization of events, and sentics-based multimedia summarization. These problems are very important in providing different significant services to users. For instance, recommendation and ranking of user tags are very beneficial in an effective multimedia search and retrieval. Moreover, multimedia summarization is very useful in providing an overview of a given event. Subsequently, we also focus on the semantics and sentics understanding of UGVs. Similar to the processing of UGIs, we exploit the multimodal

[1] www.flickr.com

information in the semantics and sentics understanding of UGVs, and address several significant multimedia analytics problems such as soundtrack recommendation for UGVs, lecture videos segmentation, and news videos uploading. All such UGVs have a significant impact on a society. For instance, soundtrack recommendation enhances the viewing experience of a UGV, lecture videos segmentation assist in e–learning, and news videos uploading supports citizen journalists.

Capturing UGVs has also become a very popular activity in recent years due to advancements in the manufacturing of mobile devices (*e.g.*, smartphones and tablets) and network engineering (*e.g.*, wireless communications). People now can easily capture UGVs anywhere, anytime, and instantly share their real-life experiences via social websites such as Flickr and YouTube. Enjoying videos has become a very popular entertainment as compared to traditional ways due to its easy access. Thus, besides traditional videos provided by professionals such as movies, music videos, and advertisements, UGVs are also getting higher popularity. UGVs are instantly shareable on social websites. For instance, video hosting services such as YouTube,[2] Vimeo,[3] Dailymotion,[4] and Veoh[5] allow individuals to upload their UGVs and share with others through their mobile devices. In the most popular video sharing website YouTube which has more than 1 billion users, everyday people watch hundreds of millions of hours of UGVs and generate billions of views [21]. Moreover, users upload 300 h of videos every minute on YouTube [21]. Almost 50% of the global viewing time comes from mobile devices, and this is expected to increase rapidly shortly because prices of mobile devices and wireless communications are getting much cheaper. Music videos enhance video watching experience because they do provide not only visual information but also involve music which matches with scenes and locations. However, many outdoor UGVs lack a certain appeal because their soundtracks consist mostly of ambient background noise (*e.g.*, environmental sounds such as cars passing by, *etc.*). Since sound is a very important aspect that contributes greatly to the appeal of a video when it is being viewed, a UGV with a matching soundtrack has more appeal for sharing on social media websites than a normal video without interesting sound.

Considering that a UGV with a matching soundtrack has more appeal for sharing on social media websites (*e.g.*, Flickr, Facebook, and YouTube) and with today's mobile devices that allow immediate sharing of UGC on such social media websites, it is desirable to easily and instantly generate an interesting soundtrack for the UGV before sharing. However, generating soundtracks for UGVs is not easy in the mobile environment due to the following reasons. Firstly, traditionally it is tedious and time-consuming for a user to add a custom soundtrack to a UGV. Secondly, an important aspect is that a good soundtrack should match and enhance the overall mood of the UGV and meet the user's preferences. Lastly, automatically

[2]www.youtube.com

[3]www.vimeo.com

[4]www.dailymotion.com

[5]www.veoh.com

generating a matching soundtrack for the UGV with less user intervention is a challenging task. Thus, it is necessary to construct a music video generation system that enhances the experience of viewing a UGV by adding a soundtrack that matches with both scenes of the UGV and the preferences of a user. In this book, we exploit both multimedia content such as visual features and contextual information such as spatial metadata of UGVs to determine sentics and generate music videos for UGVs. Our study confirms that multimodal information facilitates the understanding of user-generated multimedia content in the support of social media applications. Furthermore, we also consider two more areas where UGVs have a significant impact on a society: (1) education, and (2) journalism.

The number of digital lecture videos has increased dramatically in recent years due to the ubiquitous availability of digital cameras and affordable network infrastructures. Thus, multimedia-based e–learning systems which use electronic educational technologies as a platform for teaching and learning activities have become an important learning environment. It makes distance learning possible by enabling students to learn remotely without being in class. For instance, MIT OpenCourseWare [16] provides open access of virtually all MIT course content using a web-based publication. Now, it is possible to learn from experts in any area through e–learning (*e.g.*, MIT OpenCourseWare [16], and Coursera [12]), without any barriers such as time and distance. Many institutions such as National University of Singapore (NUS) have already started e–learning components in the practice of instructions to prepare themselves for continuing classes even if it is not possible for students to visit the campus due to certain calamities. Thus, e–learning helps in lowering cost, effective learning, faster delivery, and lowering environmental impact in educational learning systems. A long lecture video recording often discusses a specific topic of interest in only a few minutes within the video. Therefore, the requested information may bury within a long video that is stored along with thou-sands of others. It is often relatively easy to find the relevant lecture video in an archive, but the main challenge is to find the proper position within that video. Several websites such as VideoLectures.NET [20] which host lecture videos enable students to access different topics within videos using the annotation of segment boundaries derived from crowd-sourcing. However, the manual annotation of segment boundaries is very time-consuming, subjective, error-prone, and a costly process. Thus, it requires the implementation of a lecture video segmentation system which can automatically segment videos as accurately as possible even if qualities of lecture videos are not sufficiently high. Automatic lecture video segmentation will be very useful in e–learning when it combines with automatic topic modeling, indexing, and recommendation [31]. Subsequently, to facilitate journalists in the area with weak network infrastructures, we propose methods for efficient uploading of news videos.

Citizen journalism allows regular citizens to capture (news) UGVs and report events. Courtney C. Radsch defines citizen journalism as "*an alternative and activist form of newsgathering and reporting that functions outside mainstream media in-stitutions, often as a response to shortcomings in the professional journalistic field, that uses similar journalistic practices but is driven by different objectives and ideals and relies on alternative sources of legitimacy than traditional or mainstream journalism*" [163]. Citizens often can report a breaking news more quickly than traditional news reporters due to the advancement in technology. For instance, on April 4, 2015, Feidin Santana, an American citizen recorded a video that showed a former South Carolina policeman shooting and killing the unarmed Michael Scott [7]. This video has gone viral on social media before it was taken up by any mainstream news channels. This video has helped in revealing the truth about this incident. Thus, the ubiquitous availability of smartphones and cameras has increased the popularity of citizen journalism. However, there is also some incident when any false news is reported by some citizen reporter that causes loss to some organization or person. For instance, Apple suffered a temporary drop in its stock due to a false report which is generated by CNN iReport about Steve Jobs' health in 2008 [1]. CNN allows citizens to report news using modern smartphones, tablets, and websites through its CNN iReport service. This service has more than 1 million citizen journalist users [5], who report news from places where traditional news reporters may not have access. Every month, it garners an average of 15,000 news reports and its content nets 2.6 million views [4]. It is, however, quite challenging for reporters to timely upload news videos, especially from developing countries, where Internet access is slow or even intermittent. Thus, it entails to enable regular citizens to report events quickly and reliably, despite weak network infrastructure at their places.

The presence of contextual information in conjunction with multimedia content has opened up interesting research avenues within the multimedia domain. Thus, the multimodal analysis of UGC is very helpful for an effective information access. It assists in an efficient multimedia analysis, retrieval, and services because UGC is often unstructured and difficult to access in a meaningful way. Moreover, it is difficult to extract relevant content from only one modality because suitable concepts may exhibit in different representations. Furthermore, multimodal information augments knowledge bases by inferring semantics from unstructured multimedia content and contextual information. Therefore, we leverage information from multiple modalities in our solutions to the problems mentioned above. Specifically, we exploit the knowledge structures derived from the fusion of heterogeneous media content to solve different multimedia analytics problems.

1.2 Overview

As illustrated in Fig. 1.1, this book concentrates on the multimodal analysis of user-generated multimedia content (UGC) in the support of social media applications. We determine semantics and sentics knowledge structures from UGC and leverage them in addressing several significant social media problems. Specifically, we present our solutions for five multimedia analytics problems that benefit by leveraging multimodal information such as multimedia content and contextual information (*e.g.*, temporal, geo-, crowdsourced, and other sensory data). First, we solve the

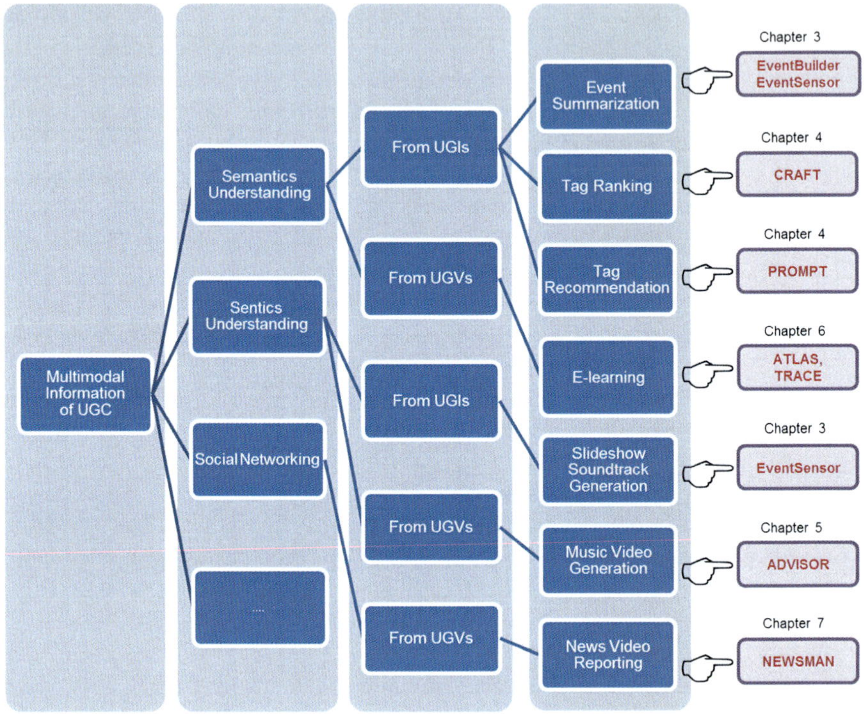

Fig. 1.1 Multimedia applications that benefit from multimodal information

problem of event understanding based on semantics and sentics analysis of UGIs on social media platforms such as Flickr [182, 186]. Subsequently, we address the problem of computing tag relevance for UGIs [181, 185]. Tag relevance scores determine tag recommendation and ranking of UGIs which are subsequently very useful in the searching and retrieval of relevant multimedia content. Next, we answer the problem of soundtrack recommendation for UGVs [187, 188]. A UGV with matching soundtrack enhance video viewing experience. Furthermore, we address research problems in two very important areas (journalism and education) where UGVs have a significant impact on society. Specifically, in the education area, we work out the problem of automatic lecture video segmentation [183, 184]. Finally, in the journalism area, we resolve the problem of user-generated news videos uploading over adaptive middleboxes to news servers in weak network infrastructures [180]. Experimental results have shown that our proposed approaches perform well. Contributions of each work are listed below:

1.2.1 Event Understanding

To efficiently browse multimedia content and obtain a summary of an event from a large collection of UGIs aggregated in social media sharing platforms such as Flickr and Instagram, we present the EventBuilder system. EventBuilder deals with semantics understanding and automatically generates a multimedia summary of a given event in real-time by leveraging different social media such as Wikipedia and Flickr. EventBuilder has two novel characteristics: (i) leveraging Wikipedia as event background knowledge to obtain additional contextual information about an input event, and (ii) visualizing an interesting event in real-time with a diverse set of social media activities. Subsequently, we enable users to obtain a sentics-based multimedia summary from the large collection of UGIs through our proposed sentics engine called, EventSensor. The EventSensor system addresses the sentics understanding from UGIs and produces a multimedia summary for a given mood. It supports sentics-based event summarization by leveraging EventBuilder as its semantics engine component. EventSensor extracts concepts and mood tags from visual content and textual metadata of UGC and exploits them in supporting several significant multimedia-related services such as a musical multimedia summary. Experimental results confirm that both EventBuilder and EventSensor outperform their baselines and effectively summarize knowledge structures on the YFCC100M dataset [201]. The YFCC100M dataset is a collection of 100 million photos and videos from Flickr.

1.2.2 Tag Recommendation and Ranking

Social media platforms such as Flickr allow users to annotate UGIs with descriptive keywords, called, tags which significantly facilitate the effective semantics understanding, search, and retrieval of UGIs. However, manual annotation is very time-consuming and cumbersome for most users, making it difficult to find relevant UGIs. Though there exist some deep neural networks based tag recommendation systems, tags predicted by such systems are limited because most of the available deep neural networks are trained with a few visual concepts. For instance, Yahoo's deep neural network can identify 1756 visual concepts from its publicly available dataset of 100 million UGIs and UGVs. However, the number of concepts that deep neural network can identify is rapidly increasing. For instance, the Google Cloud Vision API [14] can quickly classify photos into thousands of categories such as a *sailboat*, *lion*, and *Eiffel Tower*. Furthermore, Microsoft organized a challenge to recognize faces of 1 million celebrities [65]. Facebook claims to be working on identifying 100,000 objects. However, merely tagging a UGI with the identified objects may not describe the objective aspects of the UGI since often users tag UGIs with some user-defined concepts (*e.g.*, associate objects with some actions, attributes, and locations). Thus, it is very important to learn the tagging behavior of users for tag recommendation. Moreover, recommended tags for a UGI are not necessarily relevant to users' interests. Furthermore, often annotated or predicted tags of a UGI are in a random order and even irrelevant to the visual content. Thus, it necessitates for automatic tag recommendation and ranking systems that consider users' interests and describe objective aspects of the UGI such as visual content and activities. To this end, this book presents a tag recommendation system, called, PROMPT, and a tag ranking system, called, CRAFT. Both systems leverage the multimodal information of a UGI to compute tag relevance. Specifically, for tag recommendation, first, we determine a group of users who have similar interests (tagging behavior) as the user of the UGI. Next, we find candidate tags from visual content and textual metadata leveraging tagging behaviors of users determined in the first step. Particularly, we determine candidate tags from the textual metadata and compute their confidence scores using asymmetric tag co-occurrence scores. Next, we determine candidate user tags from semantically similar neighboring UGIs and compute their scores based on voting counts. Finally, we fuse confidence scores of all candidate tags using a sum method and recommend top five tags to the given UGI. Similar to the neighbor voting based tag recommendation, we propose a tag ranking scheme based on a voting from the UGI neighbors derived from multimodal information. Specifically, we determine the UGI neighbors leveraging geo, visual, and semantics concepts derived from spatial information, visual content, and textual metadata, respectively. Experimental results on a test set from the YFCC100M dataset confirm that the proposed algorithm performs well. In the future, we can exploit our tag recommendation and ranking techniques in SMS/MMS bases FAQ retrieval [189, 190].

1.2.3 Soundtrack Recommendation for UGVs

Most of the outdoor UGVs are captured without much interesting background sounds (*i.e.*, environmental sounds such as cars passing by, *etc.*). Aimed at making outdoor UGVs more attractive, we introduce *ADVISOR*, a personalized video soundtrack recommendation system. We propose a fast and effective heuristic ranking approach based on heterogeneous late fusion by jointly considering three aspects: venue categories, visual scene, and the listening history of a user. Specifically, we combine confidence scores produced by SVM^{hmm} [2, 27, 75] models constructed from geographic, visual, and audio features, to obtain different types of video characteristics. Our contributions are threefold. First, we predict scene moods from a real-world video dataset that was collected from users' daily outdoor activities. Second, we perform heuristic rankings to fuse the predicted confidence scores of multiple models, and third, we customize the video soundtrack recommendation functionality to make it compatible with mobile devices. A series of extensive experiments confirm that our approach performs well and recommends appealing soundtracks for UGVs to enhance the viewing experience.

1.2.4 Automatic Lecture Video Segmentation

The accessibility and searchability of most lecture video content are still insufficient due to the unscripted and spontaneous speech of the speakers. Moreover, this problem becomes even more challenging when the quality of such lecture videos is not sufficiently high. Thus, it is very desirable to enable people to navigate and access specific slides or topics within lecture videos. A huge amount of multimedia data is available due to the ubiquitous availability of cameras and the increasing popularity of e–learning (*i.e.*, electronic learning that leverages multimedia data heavily to facilitate education). Thus, it is very important to have a tool that can align all data available with a lecture video accurately. For instance, the tool can provide a more accurate and detailed alignment of speech transcript, presentation slides, and video content of the lecture video. This tool will help lecture video hosting websites (in fact, it is useful to any video hosting websites) to perform an advanced search, retrieval, and rccommendation at video segments level. That is, a user will not only be recommended a particular lecture video (say, V) but informed that a video segment from 7 to 13 min of the lecture video belong to a particular topic the user is interested. Thus, this problem can be solved in the following two steps: (i) find the temporal segmentation of lecture videos and (ii) determine the annotations for different temporal segments. To this end, we only focus on the first step (*i.e.*, we are interested in performing the temporal segmentation of the lecture video only) because annotations (topic titles) can be determined easily and accurately if the temporal segments are known.

Temporal segments of a lecture video are a coherent text (speech transcript or slide content) block which discusses the same topic. The boundaries of such temporal lecture video segments are known as topic boundaries. We propose the ATLAS and TRACE systems to determine such topic boundaries. ATLAS has two main novelties: (i) an SVM^{hmm} model is proposed to learn temporal transition cues from several modalities and (ii) a fusion scheme is suggested to combine transition cues extracted from the heterogeneous information of lecture videos. Subsequently, we present the TRACE system to automatically determine topic boundaries based on a linguistic approach using Wikipedia texts. TRACE has two main contributions: (i) the extraction of a novel linguistic-based Wikipedia feature to segment lecture videos efficiently and (ii) the investigation of the late fusion of video segmentation results derived from state-of-the-art methods. Specifically for the late fusion, we combine confidence scores produced by models constructed from visual, transcriptional, and Wikipedia features. According to our experiments on lecture videos from VideoLectures.NET [20] and NPTEL [3], proposed algorithms segment topic boundaries (knowledge structures) more accurately compared to existing state-of-the-art algorithms. Evaluation results are very encouraging and thus confirm the effectiveness of our ATLAS and TRACE systems.

1.2.5 Adaptive News Video Uploading

Due to the advent of smartphones and increasing popularity of social media websites, users often capture interesting news videos and share on websites of social media and news channels. However, the captured news videos are often high-qualities and big in sizes. Thus, it is not feasible for users to quickly upload the news videos to websites quickly and reliably. Thus, in order to quickly and reliably upload videos captured by citizen journalists in places with very weak network infrastructure, multiple videos may need to be prioritized, and then optimally transcoded and scheduled. We introduce an adaptive middlebox design, called NEWSMAN, to support citizen journalists. NEWSMAN jointly considers two aspects under varying network conditions: (i) choosing the optimal transcoding parameters, and (ii) determining the uploading schedule for news videos. We design, implement, and evaluate an efficient scheduling algorithm to maximize a user-specified objective function. We conduct a series of experiments using trace-driven simulations, which confirm that our approach is practical and performs well.

1.3 Contributions

The rapid growth amounting to user-generated multimedia content online necessitates for social media companies to automatically extract knowledge structures (concepts) and leverage them in providing diverse multimedia-related services. Due to the ubiquitous availability of smartphones and affordable network infrastructures, the user-generated multimedia content consists of several mobile sensors such as GPS, compass, time, and accelerometer. Moreover, it is possible to obtain several significant contextual information that are very useful in a better understanding of user-generated multimedia content. In this book, we investigate the usage of the multimodal information of the user-generated multimedia content in building improved versions of several significant multimedia systems. Specifically, we build multimedia systems for multimedia summarization, tag recommendation and ranking, soundtrack recommendation for outdoor user-generated videos, segment boundaries detection from lecture videos, and news videos uploading in the in places with very weak network infrastructure. We present novel frameworks for these multimedia systems leveraging multimodal information. Our research confirms that information from multiple modalities (*i.e.*, both multimedia content and contextual information) are very useful and augment knowledge structures. Our proposed systems leverage such knowledge structures and outperform their baselines and state-of-the-arts for above mentioned problems.

Thus, this book determines semantics and sentics knowledge structures from user-generated content leveraging both multimedia content and contextual information. It shows that multimodal information of UGC is very useful in addressing several significant multimedia analytics problems. Subsequently, we build the improved multimedia systems for these multimedia analytics problems that are described in Sects. 1.3.1, 1.3.2, 1.3.3, 1.3.4 and 1.3.5 by exploiting the derived semantics and sentics knowledge structures.

1.3.1 Event Understanding

For event understanding, we presented two real-time multimedia summarization systems: (i) EventBuilder [182] and (ii) EventSensor [186]. We define the problem statement for the EventBuilder system as follows: "For a given event *e* and timestamp *t*, generate an event summary from UGIs on social media websites such as Flickr." Similarly, we define the problem statement for the EventSensor system as follows: "For a given mood tag, generate a multimedia summary from UGIs on social media websites such as Flickr by computing their sentics (affective) details." Experimental results on the YFCC100M dataset confirm that our systems outperform their baselines. Specifically, EventBuilder outperforms its baseline by 11.41% in event detection (see Table 3.7). Moreover, EventBuilder outperforms its baseline for text summaries of events by (i) 19.36% in terms of informative rating, (ii) 27.70% in terms of experience rating, and (ii) 21.58% in terms of acceptance rating (see Tables 3.11 and Fig. 3.9). Our EventSensor system investigated the fusion of multimodal information (*i.e.*, user tags, title, description, and visual concepts) to determine sentics details of UGIs. Experimental results indicate that the feature based on user tags is salient and the most useful in determining sentics details of UGIs (see Fig. 3.10).

1.3.2 Tag Recommendation and Ranking

For tag relevance computation, we presented two systems: (i) PROMPT [181] and (ii) CRAFT [185]. We define the problem statement for the PROMPT system as follows: "For a given social media UGI, automatically recommend *N* tags that describe the objective aspect of the UGI." Our PROMPT system recommends user tags with 76% accuracy, 26% precision, and 20% recall for five predicted tags on the test set with 46,700 photos from Flickr (see Figs. 4.8, 4.9, and 4.10). Thus, there is an improvement of 11.34%, 17.84%, and 17.5% in terms of accuracy, precision, and recall evaluation metrics, respectively, in the performance of the PROMPT system as compared to the best performing state-of-the-art for tag recommendation (*i.e.*, an approach based on random walk, see Sect. 4.2.1).

Next, we present the CRAFT system to work on the problem of ranking tags of a given social media UGI. We define the problem statement for the CRAFT system as follows: "For a given social media photo with *N* tags in random order, automatically rank the *N* tags such that first tags is the most relevant to the UGI and the last tag is the least relevant to the UGI". We compute the final tag relevance for UGIs by performing a late fusion based on weights determined by the recall of modalities. The NDCG score of tags ranked by our CRAFT system is 0.886264, *i.e.*, there is an improvement of 22.24% in the NDCG score for the original order of tags (the baseline). Moreover, there is an improvement of 5.23% and 9.28% in the tag ranking performance (in terms of NDCG scores) of the CRAFT system than the following two most popular state-of-the-arts, respectively: (i) a probabilistic random walk approach (PRW) [109] and (ii) a neighbor voting approach (NVLV) [102] (see Fig. 4.13 and Sect. 4.3.2 for details). Furthermore, our proposed recall based late fusion technique results in 9.23% improvement in terms of the NDCG score than the early fusion technique (see Fig. 4.12). Results from our CRAFT system is consistent with different numbers of neighbors (see Fig. 4.14).

1.3.3 Soundtrack Recommendation for UGVs

We present the ADVISOR system [187, 188] to recommend suitable soundtracks for UGVs. The problem statement for our ADVISOR system is as follows: "For a given outdoor sensor-rich video, recommend a soundtrack that matches with both scenes and user's preferences." We build several learning models to predict scene moods for UGVs. We found that the model M_{GVC} based on the late fusion of learning models M_G and M_F that are build from geo- and visual features, respectively, performed the best. Particularly, M_{GVC} performs 30.83%, 13.93%, and 14.26% better than M_F, M_G, and M_{Cat}, respectively. M_{Cat} is the model build by concatenating geo- and visual features for training. Moreover, the emotion prediction accuracy (70.0%) of the generated soundtrack UGVs from D_{GeoVid} by the ADVISOR system is comparable to the emotion prediction accuracy (68.8%) of soundtrack videos from $D_{Hollywood}$ of the Hollywood movies.

1.3.4 Automatic Lecture Video Segmentation

We present the ATLAS [183] and TRACE [184] systems with the aim to automatically determine segment boundaries for a lecture video for all topic changes within the lecture video. We define the problem statement for this task as follows: "For a given lecture video, we automatically determine segment boundaries within the lecture video content, *i.e.*, a list of timestamps when topic changes within the lecture video". Note that; we only predict segment boundaries, not the topic titles for these boundaries. Determining the topic titles is comparatively an easy problem when the segment boundaries of lecture videos are known. Experimental results confirm that the ATLAS and TRACE systems can effectively segment lecture videos to facilitate the accessibility and traceability within their content despite video qualities are not sufficiently high. Specifically, the segment boundaries derived from the Wikipedia knowledge base outperforms state-of-the-arts regarding precision, *i.e.*, 25.54% and 29.78% better than approaches when only visual content [183] and speech transcript [107] are used in segment boundaries detection from lecture videos, respectively. Moreover, the segment boundaries derived from the Wikipedia knowledge base outperforms state-of-the-arts regarding F1 score, *i.e.*, 48.04% and 12.53% better than approaches when only visual content [183] and speech transcript [107] are used in segment boundaries detection from lecture videos, respectively. Finally, the fusion of segment boundaries derived from visual content, speech transcript, and Wikipedia knowledge base results in the highest recall score.

1.3.5 Adaptive News Video Uploading

We presented the NEWSMAN system [180] to enable citizen journalists in places with very weak network infrastructure to upload news videos. We introduced adaptive middleboxes to between users and news servers to quickly and reliably upload news videos in weak network infrastructures. We presented a novel framework to prioritize multiple news videos, optimally transcode, and then schedule their uploading. NEWSMAN jointly considers two aspects under varying network conditions: (i) choosing the optimal transcoding parameters, and (ii) determining the uploading schedule for news videos. We design, implement, and evaluate an efficient scheduling algorithm to maximize a user-specified objective function. We conduct a series of experiments using trace-driven simulations, which confirm that our approach is practical and performs well. For instance, NEWSMAN outperforms the existing algorithms: (i) by 12 times in terms of system utility (*i.e.*, sum of utilities of all up-loaded videos), and (ii) by 4 times in terms of the number of videos uploaded before their deadline.

1.4 Knowledge Bases and APIs

1.4.1 FourSquare

Foursquare[6] is a well-known company that provides location-based services to different applications. It provides support to application developers through its Foursquare API [13]. It provides API support for the following scenarios. First, the endpoints API that gives information about leaving tips, checking in, seeing what your friends are up to, and venues. Foursquare also provides two real-time push APIs. First, the API for venue push that notifies venue managers when users perform various actions at their venues. Second, the API for user push that notifies

[6]www.foursquare.com

developers when their users check in anywhere. This information is very much useful in increasing profits in business. Another popular API from Foursquare is based on venues service. The venues service allows developers to search for places and access a much useful information such as addresses, tips, popularity, and photos. Foursquare also provides merchant platform that allows developers to write applications that help registered venue owners manage their Foursquare presence. We used the venues service API in our work that maps a geo-location to geo concepts (categories), *i.e.*, it provides the geographic contextual information for the given geo location. For instance, this API also provides distances of geo concepts such as *Theme Park*, *Lake*, *Plaza* and *Beach* with the given GPS point. Thus, geo concepts can serve as an important dimension to represent valuable semantics information of multimedia data with location metadata. Specifically, we can treat each geo concept as a word and exploit the bag-of-words model [93]. Foursquare provides three level hierarchy of geo categories. Level one includes over ten high-level categories such as *Travel and Transport*, *Food*, and *Arts and Entertainment*. Such first level categories are divided into specialized categories further on the second level. For instance, the high-level category, *Arts and Entertainment*, is divided into categories such as *Arcade*, *Casino*, and *Concert Hall*. There are over 1300 categories in the second level. Foursquare categories for a sensor-rich UGV can be corrected by leveraging map matching techniques [244].

1.4.2 Semantics Parser

For a better semantics and sentics analysis, it is important to extract useful information from available text. Since sentiments from the text may not be expressed in the only word, it is required to determine concepts (*i.e.*, multi-word expressions or knowledge structures). Thus, a model based of bag-of-concepts performs better than a model based on bag-of-words in the area of sentiment analysis. Poria *et al.* [143] presented a semantics (concept) parser that extracts multi-word expressions (concepts) from the text for a better sentiment analysis. This concept parser identifies common-sense concepts from a free text without requiring time-consuming phrase structure analysis. For instance, this concept parser determines *rajiv*, *defend_phd*, *defend_from_nus*, *do_job*, and *great_job* concepts from "Rajiv defended his PhD successfully from NUS. He did a great job in his PhD.". The parser leverages linguistic patterns to deconstruct natural language text into meaningful pairs, *e.g.*, ADJ + NOUN, VERB + NOUN, and NOUN + NOUN, and then exploits common-sense knowledge to infer which of such pairs are more relevant in the current context. Later, the derived concepts are exploited in determining the semantics and sentics of user-generated multimedia content.

1.4.3 SenticNet

Poria *et al.* [148] presented enhanced SenticNet with affective labels for concept-based opinion mining. For the sentics analysis of the user-generated multimedia content, we refer to the SenticNet-3 knowledge base. SenticNet-3 is a publicly available resource for concept-level sentiment analysis [41]. It consists of 30,000 common and common-sense concepts such as *food*, *party*, and *accomplish_goal*. The recent version of SenticNet (*i.e.*, SenticNet 4) knowledge base consists of 50,000 common and common-sense concepts [42]. Sentic API[7] provides the semantics and sentics information associated these commonsense concepts [44]. Semantics and sentics provide the denotative and connotative information, respectively. For instance, a given SenticNet concept, *meet_friend*, the SenticNet API provides the following five other related SenticNet concepts as semantics information: *meet person*, *chit chat*, *make friend*, *meet girl*, and *socialize*. Moreover, the sentics associated with the same concept (*i.e.*, *meet person*) are the following: pleasantness: 0.048, attention: -0.08, sensitivity: 0.036, and aptitude: 0. Such sentics information are useful for tasks such as emotion recognition or affective HCI. Furthermore, to provide mood categories for the SenticNet concepts, they followed the Hourglass model of emotions [40]. For instance, the SenticNet API provides *joy* and *surprise* mood categories for the given concept, *meet person*. SenticNet knowledge base documents mood categories following the Hourglass model of emotions into another knowledge base, called EmoSenticNet. EmoSenticNet maps concepts of SenticNet to affective labels such as anger, disgust, joy, sadness, surprise, and fear [155]. It also provides a 100-dimensional vector space for each concept in SenticNet. Furthermore, SenticNet knowledge base also provides polarity information for every concept. It consists of both value (positive or negative) and intensity (a floating number between -1 and $+1$) for polarity. For instance, the Sentic API returns positive polarity with intensity 0.031. Thus, SenticNet knowledge base bridges the conceptual and affective gap between word-level natural language data and the concept-level opinions and sentiments conveyed by them. In other work, Poria *et al.* [154, 156] automatically merged SenticNet and WordNet-Affect emotion lists for sentiment analysis. They merged these two resources by assigning emotion labels to more than 2700 concepts. The above mentioned knowledge bases are very useful in deriving semantics and sentics information from user-generated multimedia content. The derived semantics and sentics information help us in addressing several significant multimedia analytics problems.

[7] http://sentic.net/api/

1.4.4 WordNet

WordNet is a very popular and large lexical database of English [123, 124]. Part of speech such as nouns, verbs, adjectives, and adverbs are grouped into sets of cognitive synonyms (synsets), each expressing a distinct concept. Conceptual-semantic and lexical relations are used to interlink synsets. WordNet is a very useful tool for computational linguistics and natural language processing. WordNet superficially resembles a thesaurus, in that it groups words together based on their meanings. Note that the words in WordNet that are found in close proximity to one another in the network are semantically disambiguated. Moreover, WordNet labels the semantic relations among words, whereas the groupings of words in a thesaurus does not follow any explicit pattern other than meaning similarity. Synonymy is the main relation among words in WordNet. For instance, the word *car* has the following synsets: auto, automobile, machine, and motorcar. Thus, synsets are unordered sets of synonyms words that denote the same concept and are interchangeable in many contexts. Each synset of WordNet is linked to other synsets using a small number of conceptual relations. In our work, we primarily leverage synsets for different words in WordNet.

1.4.5 Stanford POS Tagger

Toutanova *et al.* [204, 205] presented a Part-Of-Speech Tagger (POS Tagger). POS Tagger is a piece of software that reads text in some language and assigns parts of speech (*e.g.*, noun, verb, and adjective) to each word (and other token). For instance, the Stanford Parser provides the following POS Tagging for the sentence, "Rajiv defended his PhD successfully from NUS. He did a great job in his PhD.": "*Rajiv/NNP defended/VBD his/PRP$ PhD/NN successfully/RB from/IN NUS/NNP ./.* He/PRP did/VBD a/DT great/JJ job/NN in/IN his/PRP$ PhD/NN ./.". NNP (Proper noun, singular), VBD (Verb, past tense), PPR$ (Possessive pronoun), NN (Noun, singular or mass), RB (Adverb), IN (Preposition or subordinating conjunction), DT (Deter-miner), and JJ (Adjective) have usual meanings, as described in the Penn Treebank Tagging Guidelines.[8] In our work, we used the Stanford POS Tagger to compute the POS tags. The derived POS tags help us to determine important concepts from a given text, which is subsequently beneficial in our semantics and sentics analysis of user-generated content.

[8]http://www.personal.psu.edu/xxl13/teaching/sp07/apling597e/resources/Tagset.pdf

1.4.6 Wikipedia

Wikipedia is a free online encyclopedia that aims to allow anyone to edit articles. It is the largest and most popular general reference work on the Internet and is ranked among the ten most popular websites. Thus, Wikipedia is considered as one of the most useful and popular resource for knowledge. It provides useful information to understand a given topic quickly and efficiently. In our work, we also exploit information from Wikipedia. We use the Wikipedia API[9] to get text for different Wikipedia articles.

1.5 Roadmap

We organize the rest of this book as follows. Section 1.5 reports important related work to this study. Section 2.5 introduces our solution for event understanding from a large collection of UGIs. In Sect. 3.4, we describe the computation of tag relevance scores for UGIs, which is useful in the recommendation and ranking of user tags. Section 4.4 presents the soundtrack recommendation system for UGVs. Section 5.4 reports an automatic lecture video segmentation system. In Sect. 6.4, we describe the adaptive uploading of news videos (UGVs). Finally, Chap. 8 concludes and suggests potential future work.

References

1. Apple Denies Steve Jobs Heart Attack Report: "It Is Not True". http://www.businessinsider.com/2008/10/apple-s-steve-jobs-rushed-to-er-after-heart-attack-says-cnn-citizen-journalist/. October 2008. Online: Last Accessed Sept 2015.
2. SVM^{hmm}: Sequence Tagging with Structural Support Vector Machines. https://www.cs.cornell.edu/people/tj/svm light/svm hmm.html. August 2008. Online: Last Accessed May 2016.
3. NPTEL. 2009, December. http://www.nptel.ac.in. Online; Accessed Apr 2015.
4. iReport at 5: Nearly 900,000 contributors worldwide. http://www.niemanlab.org/2011/08/ireport-at-5-nearly-900000-contributors-worldwide/. August 2011. Online: Last Accessed Sept 2015.
5. Meet the million: 999,999 iReporters + you! http://www.ireport.cnn.com/blogs/ireport-blog/2012/01/23/meet-the-million-999999-ireporters-you. January 2012. Online: Last Accessed Sept 2015.
6. 5 Surprising Stats about User-generated Content. 2014, April. http://www.smartblogs.com/social-media/2014/04/11/6.-surprising-stats-about-user-generated-content/. Online: Last Accessed Sept 2015.

[9]https://en.wikipedia.org/w/api.php

7. The Citizen Journalist: How Ordinary People are Taking Control of the News. 2015, June. http://www.digitaltrends.com/features/the-citizen-journalist-how-ordinary-people-are-taking-control-of-the-news/. Online: Last Accessed Sept 2015.
8. Wikipedia API. 2015, April. http://tinyurl.com/WikiAPI-AI. API: Last Accessed Apr 2015.
9. Apache Lucene. 2016, June. https://lucene.apache.org/core/. Java API: Last Accessed June 2016.
10. By the Numbers: 14 Interesting Flickr Stats. 2016, May. http://www.expandedramblings.com/index.php/flickr-stats/. Online: Last Accessed May 2016.
11. By the Numbers: 180+ Interesting Instagram Statistics (June 2016). 2016, June. http://www.expandedramblings.com/index.php/important-instagram-stats/. Online: Last Accessed July 2016.
12. Coursera. 2016, May. https://www.coursera.org/. Online: Last Accessed May 2016.
13. FourSquare API. 2016, June. https://developer.foursquare.com/. Last Accessed June 2016.
14. Google Cloud Vision API. 2016, December. https://cloud.google.com/vision/. Online: Last Accessed Dec 2016.
15. Google Forms. 2016, May. https://docs.google.com/forms/. Online: Last Accessed May 2016.
16. MIT Open Course Ware. 2016, May. http://www.ocw.mit.edu/. Online: Last Accessed May 2016.
17. Porter Stemmer. 2016, May. https://tartarus.org/martin/PorterStemmer/. Online: Last Accessed May 2016.
18. SenticNet. 2016, May. http://www.sentic.net/computing/. Online: Last Accessed May 2016.
19. Sentics. 2016, May. https://en.wiktionary.org/wiki/sentics. Online: Last Accessed May 2016.
20. VideoLectures.Net. 2016, May. http://www.videolectures.net/. Online: Last Accessed May, 2016.
21. YouTube Statistics. 2016, July. http://www.youtube.com/yt/press/statistics.html. Online: Last Accessed July, 2016.
22. Abba, H.A., S.N.M. Shah, N.B. Zakaria, and A.J. Pal. 2012. Deadline based performance evalu-ation of job scheduling algorithms. In *Proceedings of the IEEE International Conference on Cyber-Enabled Distributed Computing and Knowledge Discovery*, 106–110.
23. Achanta, R.S., W.-Q. Yan, and M.S. Kankanhalli. (2006). Modeling Intent for Home Video Repurposing. In *Proceedings of the IEEE MultiMedia,* (1):46–55.
24. Adcock, J., M. Cooper, A. Girgensohn, and L. Wilcox. 2005. Interactive Video Search using Multilevel Indexing. In *Proceedings of the Springer Image and Video Retrieval*, 205–214.
25. Agarwal, B., S. Poria, N. Mittal, A. Gelbukh, and A. Hussain. 2015. Concept-level Sentiment Analysis with Dependency-based Semantic Parsing: A Novel Approach. In *Proceedings of the Springer Cognitive Computation,* 1–13.
26. Aizawa, K., D. Tancharoen, S. Kawasaki, and T. Yamasaki. 2004. Efficient Retrieval of Life Log based on Context and Content. In *Proceedings of the ACM Workshop on Continuous Archival and Retrieval of Personal Experiences*, 22–31.
27. Altun, Y., I. Tsochantaridis, and T. Hofmann. 2003. Hidden Markov Support Vector Machines. In *Proceedings of the International Conference on Machine Learning*, 3–10.
28. Anderson, A., K. Ranghunathan, and A. Vogel. 2008. Tagez: Flickr Tag Recommendation. In *Proceedings of the Association for the Advancement of Artificial Intelligence.*
29. Atrey, P.K., A. El Saddik, and M.S. Kankanhalli. 2011. Effective Multimedia Surveillance using a Human-centric Approach. *Proceedings of the Springer Multimedia Tools and Applications* 51(2): 697–721.
30. Barnard, K., P. Duygulu, D. Forsyth, N. De Freitas, D.M. Blei, and M.I. Jordan. 2003. Matching Words and Pictures. *Proceedings of the Journal of Machine Learning Research* 3: 1107–1135.
31. Basu, S., Y. Yu, V.K. Singh, and R. Zimmermann. 2016. Videopedia: Lecture Video Recommendation for Educational Blogs Using Topic Modeling. In *Proceedings of the Springer International Conference on Multimedia Modeling*, 238–250.

32. Basu, S., Y. Yu, and R. Zimmermann. 2016. Fuzzy Clustering of Lecture Videos based on Topic Modeling. In *Proceedings of the IEEE International Workshop on Content-Based Multimedia Indexing*, 1–6.
33. Basu, S., R. Zimmermann, K.L. OHalloran, S. Tan, and K. Marissa. 2015. Performance Evaluation of Students Using Multimodal Learning Systems. In *Proceedings of the Springer International Conference on Multimedia Modeling*, 135–147.
34. Beeferman, D., A. Berger, and J. Lafferty. 1999. Statistical Models for Text Segmentation. *Proceedings of the Springer Machine Learning* 34(1–3): 177–210.
35. Bernd, J., D. Borth, C. Carrano, J. Choi, B. Elizalde, G. Friedland, L. Gottlieb, K. Ni, R. Pearce, D. Poland, et al. 2015. Kickstarting the Commons: The YFCC100M and the YLI Corpora. In *Proceedings of the ACM Workshop on Community-Organized Multimodal Mining: Opportunities for Novel Solutions*, 1–6.
36. Bhatt, C.A., and M.S. Kankanhalli. 2011. Multimedia Data Mining: State of the Art and Challenges. *Proceedings of the Multimedia Tools and Applications* 51(1): 35–76.
37. Bhatt, C.A., A. Popescu-Belis, M. Habibi, S. Ingram, S. Masneri, F. McInnes, N. Pappas, and O. Schreer. 2013. Multi-factor Segmentation for Topic Visualization and Recommendation: the MUST-VIS System. In *Proceedings of the ACM International Conference on Multimedia*, 365–368.
38. Bhattacharjee, S., W.C. Cheng, C.-F. Chou, L. Golubchik, and S. Khuller. 2000. BISTRO: A Frame-work for Building Scalable Wide-area Upload Applications. *Proceedings of the ACM SIGMETRICS Performance Evaluation Review* 28(2): 29–35.
39. Cambria, E., J. Fu, F. Bisio, and S. Poria. 2015. AffectiveSpace 2: Enabling Affective Intuition for Concept-level Sentiment Analysis. In *Proceedings of the AAAI Conference on Artificial Intelligence*, 508–514.
40. Cambria, E., A. Livingstone, and A. Hussain. 2012. The Hourglass of Emotions. In *Proceedings of the Springer Cognitive Behavioural Systems*, 144–157.
41. Cambria, E., D. Olsher, and D. Rajagopal. 2014. SenticNet 3: A Common and Common-sense Knowledge Base for Cognition-driven Sentiment Analysis. In *Proceedings of the AAAI Conference on Artificial Intelligence*, 1515–1521.
42. Cambria, E., S. Poria, R. Bajpai, and B. Schuller. 2016. SenticNet 4: A Semantic Resource for Sentiment Analysis based on Conceptual Primitives. In *Proceedings of the International Conference on Computational Linguistics (COLING)*, 2666–2677.
43. Cambria, E., S. Poria, F. Bisio, R. Bajpai, and I. Chaturvedi. 2015. The CLSA Model: A Novel Framework for Concept-Level Sentiment Analysis. In *Proceedings of the Springer Computational Linguistics and Intelligent Text Processing*, 3–22.
44. Cambria, E., S. Poria, A. Gelbukh, and K. Kwok. 2014. Sentic API: A Common-sense based API for Concept-level Sentiment Analysis. *CEUR Workshop Proceedings* 144: 19–24.
45. Cao, J., Z. Huang, and Y. Yang. 2015. Spatial-aware Multimodal Location Estimation for Social Images. In *Proceedings of the ACM Conference on Multimedia Conference*, 119–128.
46. Chakraborty, I., H. Cheng, and O. Javed. 2014. Entity Centric Feature Pooling for Complex Event Detection. In *Proceedings of the Workshop on HuEvent at the ACM International Conference on Multimedia*, 1–5.
47. Che, X., H. Yang, and C. Meinel. 2013. Lecture Video Segmentation by Automatically Analyzing the Synchronized Slides. In *Proceedings of the ACM International Conference on Multimedia*, 345–348.
48. Chen, B., J. Wang, Q. Huang, and T. Mei. 2012. Personalized Video Recommendation through Tripartite Graph Propagation. In *Proceedings of the ACM International Conference on Multimedia*, 1133–1136.
49. Chen, S., L. Tong, and T. He. 2011. Optimal Deadline Scheduling with Commitment. In *Proceedings of the IEEE Annual Allerton Conference on Communication, Control, and Computing*, 111–118.
50. Chen, W.-B., C. Zhang, and S. Gao. 2012. Segmentation Tree based Multiple Object Image Retrieval. In *Proceedings of the IEEE International Symposium on Multimedia*, 214–221.

51. Chen, Y., and W.J. Heng. 2003. Automatic Synchronization of Speech Transcript and Slides in Presentation. *Proceedings of the IEEE International Symposium on Circuits and Systems* 2: 568–571.
52. Cohen, J. 1960. A Coefficient of Agreement for Nominal Scales. *Proceedings of the Durham Educational and Psychological Measurement* 20(1): 37–46.
53. Cristani, M., A. Pesarin, C. Drioli, V. Murino, A. Rodà,M. Grapulin, and N. Sebe. 2010. Toward an Automatically Generated Soundtrack from Low-level Cross-modal Correlations for Automotive Scenarios. In *Proceedings of the ACM International Conference on Multimedia*, 551–560.
54. Dang-Nguyen, D.-T., L. Piras, G. Giacinto, G. Boato, and F.G. De Natale. 2015. A Hybrid Approach for Retrieving Diverse Social Images of Landmarks. In *Proceedings of the IEEE International Conference on Multimedia and Expo (ICME)*, 1–6.
55. Fabro, M. Del, A. Sobe, and L. Böszörmenyi. 2012. Summarization of Real-life Events based on Community-contributed Content. In *Proceedings of the International Conferences on Advances in Multimedia*, 119–126.
56. Du, L., W.L. Buntine, and M. Johnson. 2013. Topic Segmentation with a Structured Topic Model. In *Proceedings of the North American Chapter of the Association for Computational Linguistics: Human Language Technologies*, 190–200.
57. Fan, Q., K. Barnard, A. Amir, A. Efrat, and M. Lin. 2006. Matching Slides to Presentation Videos using SIFT and Scene Background Matching. In *Proceedings of the ACM International Conference on Multimedia*, 239–248.
58. Filatova, E. and V. Hatzivassiloglou. 2004. Event-based Extractive Summarization. In *Proceedings of the ACL Workshop on Summarization*, 104–111.
59. Firan, C.S., M. Georgescu, W. Nejdl, and R. Paiu. 2010. Bringing Order to Your Photos: Event-driven Classification of Flickr Images based on Social Knowledge. In *Proceedings of the ACM International Conference on Information and Knowledge Management*, 189–198.
60. Gao, S., C. Zhang, and W.-B. Chen. 2012. An Improvement of Color Image Segmentation through Projective Clustering. In *Proceedings of the IEEE International Conference on Information Reuse and Integration*, 152–158.
61. Garg, N. and I. Weber. 2008. Personalized, Interactive Tag Recommendation for Flickr. In *Proceedings of the ACM Conference on Recommender Systems*, 67–74.
62. Ghias, A., J. Logan, D. Chamberlin, and B.C. Smith. 1995. Query by Humming: Musical Information Retrieval in an Audio Database. In *Proceedings of the ACM International Conference on Multimedia*, 231–236.
63. Golder, S.A., and B.A. Huberman. 2006. Usage Patterns of Collaborative Tagging Systems. *Proceedings of the Journal of Information Science* 32(2): 198–208.
64. Gozali, J.P., M.-Y. Kan, and H. Sundaram. 2012. Hidden Markov Model for Event Photo Stream Segmentation. In *Proceedings of the IEEE International Conference on Multimedia and Expo Workshops*, 25–30.
65. Guo, Y., L. Zhang, Y. Hu, X. He, and J. Gao. 2016. Ms-celeb-1m: Challenge of recognizing one million celebrities in the real world. *Proceedings of the Society for Imaging Science and Technology Electronic Imaging* 2016(11): 1–6.
66. Hanjalic, A., and L.-Q. Xu. 2005. Affective Video Content Representation and Modeling. *Proceedings of the IEEE Transactions on Multimedia* 7(1): 143–154.
67. Haubold, A. and J.R. Kender. 2005. Augmented Segmentation and Visualization for Presentation Videos. In *Proceedings of the ACM International Conference on Multimedia*, 51–60.
68. Healey, J.A., and R.W. Picard. 2005. Detecting Stress during Real-world Driving Tasks using Physiological Sensors. *Proceedings of the IEEE Transactions on Intelligent Transportation Systems* 6(2): 156–166.
69. Hefeeda, M., and C.-H. Hsu. 2010. On Burst Transmission Scheduling in Mobile TV Broadcast Networks. *Proceedings of the IEEE/ACM Transactions on Networking* 18(2): 610–623.

70. Hevner, K. 1936. Experimental Studies of the Elements of Expression in Music. *Proceedings of the American Journal of Psychology* 48: 246–268.
71. Hochbaum, D.S.. 1996. Approximating Covering and Packing Problems: Set Cover, Vertex Cover, Independent Set, and related Problems. In *Proceedings of the PWS Approximation algorithms for NP-hard problems*, 94–143.
72. Hong, R., J. Tang, H.-K. Tan, S. Yan, C. Ngo, and T.-S. Chua. 2009. Event Driven Summarization for Web Videos. In *Proceedings of the ACM SIGMM Workshop on Social Media*, 43–48.
73. P. ITU-T Recommendation. 1999. Subjective Video Quality Assessment Methods for Multimedia Applications.
74. Jiang, L., A.G. Hauptmann, and G. Xiang. 2012. Leveraging High-level and Low-level Features for Multimedia Event Detection. In *Proceedings of the ACM International Conference on Multimedia*, 449–458.
75. Joachims, T., T. Finley, and C.-N. Yu. 2009. Cutting-plane Training of Structural SVMs. *Proceedings of the Machine Learning Journal* 77(1): 27–59.
76. Johnson, J., L. Ballan, and L. Fei-Fei. 2015. Love Thy Neighbors: Image Annotation by Exploiting Image Metadata. In *Proceedings of the IEEE International Conference on Computer Vision*, 4624–4632.
77. Johnson, J., A. Karpathy, and L. Fei-Fei. 2015. Densecap: Fully Convolutional Localization Networks for Dense Captioning. *In Proceedings of the arXiv preprint arXiv:1511.07571*.
78. Jokhio, F., A. Ashraf, S. Lafond, I. Porres, and J. Lilius. 2013. Prediction-based dynamic resource allocation for video transcoding in cloud computing. In *Proceedings of the IEEE International Conference on Parallel, Distributed and Network-Based Processing*, 254–261.
79. Kaminskas, M., I. Fernández-Tobías, F. Ricci, and I. Cantador. 2014. Knowledge-based Identification of Music Suited for Places of Interest. *Proceedings of the Springer Information Technology & Tourism* 14(1): 73–95.
80. Kaminskas, M. and F. Ricci. 2011. Location-adapted Music Recommendation using Tags. In *Proceedings of the Springer User Modeling, Adaption and Personalization*, 183–194.
81. Kan, M.-Y. 2001. Combining Visual Layout and Lexical Cohesion Features for Text Segmentation. *In Proceedings of the Citeseer*.
82. Kan, M.-Y. 2003. *Automatic Text Summarization as Applied to Information Retrieval*. PhD thesis, Columbia University.
83. Kan, M.-Y., J.L. Klavans, and K.R. McKeown. 1998. Linear Segmentation and Segment Significance. *In Proceedings of the arXiv preprint cs/9809020*.
84. Kan, M.-Y., K.R. McKeown, and J.L. Klavans. 2001. Applying Natural Language Generation to Indicative Summarization. *Proceedings of the ACL European Workshop on Natural Language Generation* 8: 1–9.
85. Kang, H.B.. 2003. Affective Content Detection using HMMs. In *Proceedings of the ACM International Conference on Multimedia*, 259–262.
86. Kang, Y.-L., J.-H. Lim, M.S. Kankanhalli, C.-S. Xu, and Q. Tian. 2004. Goal Detection in Soccer Video using Audio/Visual Keywords. *Proceedings of the IEEE International Conference on Image Processing* 3: 1629–1632.
87. Kang, Y.-L., J.-H. Lim, Q. Tian, and M.S. Kankanhalli. 2003. Soccer Video Event Detection with Visual Keywords. In *Proceedings of the Joint Conference of International Conference on Information, Communications and Signal Processing, and Pacific Rim Conference on Multimedia*, 3:1796–1800.
88. Kankanhalli, M.S., and T.-S. Chua. 2000. Video Modeling using Strata-based Annotation. *Proceedings of the IEEE MultiMedia* 7(1): 68–74.
89. Kennedy, L., M. Naaman, S. Ahern, R. Nair, and T. Rattenbury. 2007. How Flickr Helps us Make Sense of the World: Context and Content in Community-contributed Media Collections. In *Proceedings of the ACM International Conference on Multimedia*, 631–640.

90. Kennedy, L.S., S.-F. Chang, and I.V. Kozintsev. 2006. To Search or to Label?: Predicting the Performance of Search-based Automatic Image Classifiers. In *Proceedings of the ACM International Workshop on Multimedia Information Retrieval*, 249–258.
91. Kim, Y.E., E.M. Schmidt, R. Migneco, B.G. Morton, P. Richardson, J. Scott, J.A. Speck, and D. Turnbull. 2010. Music Emotion Recognition: A State of the Art Review. In *Proceedings of the International Society for Music Information Retrieval*, 255–266.
92. Klavans, J.L., K.R. McKeown, M.-Y. Kan, and S. Lee. 1998. Resources for Evaluation of Summarization Techniques. *In Proceedings of the arXiv preprint cs/9810014*.
93. Ko, Y.. 2012. A Study of Term Weighting Schemes using Class Information for Text Classification. In *Proceedings of the ACM Special Interest Group on Information Retrieval*, 1029–1030.
94. Kort, B., R. Reilly, and R.W. Picard. 2001. An Affective Model of Interplay between Emotions and Learning: Reengineering Educational Pedagogy-Building a Learning Companion. *Proceedings of the IEEE International Conference on Advanced Learning Technologies* 1: 43–47.
95. Kucuktunc, O., U. Gudukbay, and O. Ulusoy. 2010. Fuzzy Color Histogram-based Video Segmentation. *Proceedings of the Computer Vision and Image Understanding* 114(1): 125–134.
96. Kuo, F.-F., M.-F. Chiang, M.-K. Shan, and S.-Y. Lee. 2005. Emotion-based Music Recommendation by Association Discovery from Film Music. In *Proceedings of the ACM International Conference on Multimedia*, 507–510.
97. Lacy, S., T. Atwater, X. Qin, and A. Powers. 1988. Cost and Competition in the Adoption of Satellite News Gathering Technology. *Proceedings of the Taylor & Francis Journal of Media Economics* 1(1): 51–59.
98. Lambert, P., W. De Neve, P. De Neve, I. Moerman, P. Demeester, and R. Van de Walle. 2006. Rate-distortion performance of H. 264/AVC compared to state-of-the-art video codecs. *Proceedings of the IEEE Transactions on Circuits and Systems for Video Technology* 16(1): 134–140.
99. Laurier, C., M. Sordo, J. Serrà, and P. Herrera. 2009. Music Mood Representations from Social Tags. In *Proceedings of the International Society for Music Information Retrieval*, 381–386.
100. Li, C.T. and M.K. Shan. 2007. Emotion-based Impressionism Slideshow with Automatic Music Accompaniment. In *Proceedings of the ACM International Conference on Multimedia*, 839–842.
101. Li, J., and J.Z. Wang. 2008. Real-time Computerized Annotation of Pictures. *Proceedings of the IEEE Transactions on Pattern Analysis and Machine Intelligence* 30(6): 985–1002.
102. Li, X., C.G. Snoek, and M. Worring. 2009. Learning Social Tag Relevance by Neighbor Voting. *Proceedings of the IEEE Transactions on Multimedia* 11(7): 1310–1322.
103. Li, X., T. Uricchio, L. Ballan, M. Bertini, C.G. Snoek, and A.D. Bimbo. 2016. Socializing the Semantic Gap: A Comparative Survey on Image Tag Assignment, Refinement, and Retrieval. *Proceedings of the ACM Computing Surveys (CSUR)* 49(1): 14.
104. Li, Z., Y. Huang, G. Liu, F. Wang, Z.-L. Zhang, and Y. Dai. 2012. Cloud Transcoder: Bridging the Format and Resolution Gap between Internet Videos and Mobile Devices. In *Proceedings of the ACM International Workshop on Network and Operating System Support for Digital Audio and Video*, 33–38.
105. Liang, C., Y. Guo, and Y. Liu. 2008. Is Random Scheduling Sufficient in P2P Video Streaming? In *Proceedings of the IEEE International Conference on Distributed Computing Systems*, 53–60. IEEE.
106. Lim, J.-H., Q. Tian, and P. Mulhem. 2003. Home Photo Content Modeling for Personalized Event-based Retrieval. *Proceedings of the IEEE MultiMedia* 4: 28–37.
107. Lin, M., M. Chau, J. Cao, and J.F. Nunamaker Jr. 2005. Automated Video Segmentation for Lecture Videos: A Linguistics-based Approach. *Proceedings of the IGI Global International Journal of Technology and Human Interaction* 1(2): 27–45.

108. Liu, C.L., and J.W. Layland. 1973. Scheduling Algorithms for Multiprogramming in a Hard-real-time Environment. *Proceedings of the ACM Journal of the ACM* 20(1): 46–61.
109. Liu, D., X.-S. Hua, L. Yang, M. Wang, and H.-J. Zhang. 2009. Tag Ranking. In *Proceedings of the ACM World Wide Web Conference*, 351–360.
110. Liu, T., C. Rosenberg, and H.A. Rowley. 2007. Clustering Billions of Images with Large Scale Nearest Neighbor Search. In *Proceedings of the IEEE Winter Conference on Applications of Computer Vision*, 28–28.
111. Liu, X. and B. Huet. 2013. Event Representation and Visualization from Social Media. In *Proceedings of the Springer Pacific-Rim Conference on Multimedia*, 740–749.
112. Liu, Y., D. Zhang, G. Lu, and W.-Y. Ma. 2007. A Survey of Content-based Image Retrieval with High-level Semantics. *Proceedings of the Elsevier Pattern Recognition* 40(1): 262–282.
113. Livingston, S., and D.A.V. BELLE. 2005. The Effects of Satellite Technology on Newsgathering from Remote Locations. *Proceedings of the Taylor & Francis Political Communication* 22(1): 45–62.
114. Long, R., H. Wang, Y. Chen, O. Jin, and Y. Yu. 2011. Towards Effective Event Detection, Tracking and Summarization on Microblog Data. In *Proceedings of the Springer Web-Age Information Management*, 652–663.
115. L. Lu, H. You, and H. Zhang. 2001. A New Approach to Query by Humming in Music Retrieval. In *Proceedings of the IEEE International Conference on Multimedia and Expo*, 22–25.
116. Lu, Y., H. To, A. Alfarrarjeh, S.H. Kim, Y. Yin, R. Zimmermann, and C. Shahabi. 2016. GeoUGV: User-generated Mobile Video Dataset with Fine Granularity Spatial Metadata. In *Proceedings of the ACM International Conference on Multimedia Systems*, 43.
117. Mao, J., W. Xu, Y. Yang, J. Wang, Z. Huang, and A. Yuille. 2014. Deep Captioning with Multimodal Recurrent Neural Networks (M-RNN). *In Proceedings of the arXiv preprint arXiv:1412.6632*.
118. Matusiak, K.K. 2006. Towards User-centered Indexing in Digital Image Collections. *Proceedings of the OCLC Systems & Services: International Digital Library Perspectives* 22(4): 283–298.
119. McDuff, D., R. El Kaliouby, E. Kodra, and R. Picard. 2013. Measuring Voter's Candidate Preference Based on Affective Responses to Election Debates. In *Proceedings of the IEEE Humaine Association Conference on Affective Computing and Intelligent Interaction*, 369–374.
120. McKeown, K.R., J.L. Klavans, and M.-Y. Kan. Method and System for Topical Segmentation, Segment Significance and Segment Function, 29 2002. US Patent 6,473,730.
121. Mezaris, V., A. Scherp, R. Jain, M. Kankanhalli, H. Zhou, J. Zhang, L. Wang, and Z. Zhang. 2011. Modeling and Rrepresenting Events in Multimedia. In *Proceedings of the ACM International Conference on Multimedia*, 613–614.
122. Mezaris, V., A. Scherp, R. Jain, and M.S. Kankanhalli. 2014. Real-life Events in Multimedia: Detection, Representation, Retrieval, and Applications. *Proceedings of the Springer Multimedia Tools and Applications* 70(1): 1–6.
123. Miller, G., and C. Fellbaum. 1998. *Wordnet: An Electronic Lexical Database*. Cambridge, MA: MIT Press.
124. Miller, G.A. 1995. WordNet: A Lexical Database for English. *Proceedings of the Communications of the ACM* 38(11): 39–41.
125. Moxley, E., J. Kleban, J. Xu, and B. Manjunath. 2009. Not All Tags are Created Equal: Learning Flickr Tag Semantics for Global Annotation. In *Proceedings of the IEEE International Conference on Multimedia and Expo*, 1452–1455.
126. Mulhem, P., M.S. Kankanhalli, J. Yi, and H. Hassan. 2003. Pivot Vector Space Approach for Audio-Video Mixing. *Proceedings of the IEEE MultiMedia* 2: 28–40.
127. Naaman, M. 2012. Social Multimedia: Highlighting Opportunities for Search and Mining of Multimedia Data in Social Media Applications. *Proceedings of the Springer Multimedia Tools and Applications* 56(1): 9–34.

128. Natarajan, P., P.K. Atrey, and M. Kankanhalli. 2015. Multi-Camera Coordination and Control in Surveillance Systems: A Survey. *Proceedings of the ACM Transactions on Multimedia Computing, Communications, and Applications* 11(4): 57.
129. Nayak, M.G. 2004. *Music Synthesis for Home Videos*. PhD thesis.
130. Neo, S.-Y., J. Zhao, M.-Y. Kan, and T.-S. Chua. 2006. Video Retrieval using High Level Features: Exploiting Query Matching and Confidence-based Weighting. In *Proceedings of the Springer International Conference on Image and Video Retrieval*, 143–152.
131. Ngo, C.-W., F. Wang, and T.-C. Pong. 2003. Structuring Lecture Videos for Distance Learning Applications. In *Proceedings of the IEEE International Symposium on Multimedia Software Engineering*, 215–222.
132. Nguyen, V.-A., J. Boyd-Graber, and P. Resnik. 2012. SITS: A Hierarchical Nonparametric Model using Speaker Identity for Topic Segmentation in Multiparty Conversations. In *Proceedings of the Annual Meeting of the Association for Computational Linguistics*, 78–87.
133. Nwana, A.O. and T. Chen. 2016. Who Ordered This?: Exploiting Implicit User Tag Order Preferences for Personalized Image Tagging. *In Proceedings of the arXiv preprint arXiv:1601.06439*.
134. Papagiannopoulou, C. and V. Mezaris. 2014. Concept-based Image Clustering and Summarization of Event-related Image Collections. In *Proceedings of the Workshop on HuEvent at the ACM International Conference on Multimedia*, 23–28.
135. Park, M.H., J.H. Hong, and S.B. Cho. 2007. Location-based Recommendation System using Bayesian User's Preference Model in Mobile Devices. In *Proceedings of the Springer Ubiquitous Intelligence and Computing*, 1130–1139.
136. Petkos, G., S. Papadopoulos, V. Mezaris, R. Troncy, P. Cimiano, T. Reuter, and Y. Kompatsiaris. 2014. Social Event Detection at MediaEval: a Three-Year Retrospect of Tasks and Results. In *Proceedings of the Workshop on Social Events in Web Multimedia at ACM International Conference on Multimedia Retrieval*.
137. Pevzner, L., and M.A. Hearst. 2002. A Critique and Improvement of an Evaluation Metric for Text Segmentation. *Proceedings of the Computational Linguistics* 28(1): 19–36.
138. Picard, R.W., and J. Klein. 2002. Computers that Recognise and Respond to User Emotion: Theoretical and Practical Implications. *Proceedings of the Interacting with Computers* 14(2): 141–169.
139. Picard, R.W., E. Vyzas, and J. Healey. 2001. Toward machine emotional intelligence: Analysis of affective physiological state. *Proceedings of the IEEE Transactions on Pattern Analysis and Machine Intelligence* 23(10): 1175–1191.
140. Poisson, S.D. and C.H. Schnuse. 1841. Recherches Sur La Pprobabilité Des Jugements En Mmatieré Criminelle Et En Matieré Civile. Meyer.
141. Poria, S., E. Cambria, R. Bajpai, and A. Hussain. 2017. A Review of Affective Computing: From Unimodal Analysis to Multimodal Fusion. *Proceedings of the Elsevier Information Fusion* 37: 98–125.
142. Poria, S., E. Cambria, and A. Gelbukh. 2016. Aspect Extraction for Opinion Mining with a Deep Convolutional Neural Network. *Proceedings of the Elsevier Knowledge-Based Systems* 108: 42–49.
143. Poria, S., E. Cambria, A. Gelbukh, F. Bisio, and A. Hussain. 2015. Sentiment Data Flow Analysis by Means of Dynamic Linguistic Patterns. *Proceedings of the IEEE Computational Intelligence Magazine* 10(4): 26–36.
144. Poria, S., E. Cambria, and A.F. Gelbukh. 2015. Deep Convolutional Neural Network Textual Features and Multiple Kernel Learning for Utterance-level Multimodal Sentiment Analysis. In *Proceedings of the EMNLP*, 2539–2544.
145. Poria, S., E. Cambria, D. Hazarika, N. Mazumder, A. Zadeh, and L.-P. Morency. 2017. Context-Dependent Sentiment Analysis in User-Generated Videos. *In Proceedings of the Association for Computational Linguistics*.

146. Poria, S., E. Cambria, D. Hazarika, and P. Vij. 2016. A Deeper Look into Sarcastic Tweets using Deep Convolutional Neural Networks. In *Proceedings of the International Conference on Computational Linguistics (COLING)*.
147. Poria, S., E. Cambria, N. Howard, G.-B. Huang, and A. Hussain. 2016. Fusing Audio Visual and Textual Clues for Sentiment Analysis from Multimodal Content. *Proceedings of the Elsevier Neurocomputing* 174: 50–59.
148. Poria, S., E. Cambria, N. Howard, and A. Hussain. 2015. Enhanced SenticNet with Affective Labels for Concept-based Opinion Mining: Extended Abstract. In *Proceedings of the International Joint Conference on Artificial Intelligence*.
149. Poria, S., E. Cambria, A. Hussain, and G.-B. Huang. 2015. Towards an Intelligent Framework for Multimodal Affective Data Analysis. *Proceedings of the Elsevier Neural Networks* 63: 104–116.
150. Poria, S., E. Cambria, L.-W. Ku, C. Gui, and A. Gelbukh. 2014. A Rule-based Approach to Aspect Extraction from Product Reviews. In *Proceedings of the Second Workshop on Natural Language Processing for Social Media (SocialNLP)*, 28–37.
151. Poria, S., I. Chaturvedi, E. Cambria, and F. Bisio. 2016. Sentic LDA: Improving on LDA with Semantic Similarity for Aspect-based Sentiment Analysis. In *Proceedings of the IEEE International Joint Conference on Neural Networks (IJCNN)*, 4465–4473.
152. Poria, S., I. Chaturvedi, E. Cambria, and A. Hussain. 2016. Convolutional MKL based Multimodal Emotion Recognition and Sentiment Analysis. In *Proceedings of the IEEE International Conference on Data Mining (ICDM)*, 439–448.
153. Poria, S., A. Gelbukh, B. Agarwal, E. Cambria, and N. Howard. 2014. Sentic Demo: A Hybrid Concept-level Aspect-based Sentiment Analysis Toolkit. *In Proceedings of the ESWC*.
154. Poria, S., A. Gelbukh, E. Cambria, D. Das, and S. Bandyopadhyay. 2012. Enriching SenticNet Polarity Scores Through Semi-Supervised Fuzzy Clustering. In *Proceedings of the IEEE International Conference on Data Mining Workshops (ICDMW)*, 709–716.
155. Poria, S., A. Gelbukh, E. Cambria, A. Hussain, and G.-B. Huang. 2014. EmoSenticSpace: A Novel Framework for Affective Common-sense Reasoning. *Proceedings of the Elsevier Knowledge-Based Systems* 69: 108–123.
156. Poria, S., A. Gelbukh, E. Cambria, P. Yang, A. Hussain, and T. Durrani. 2012. Merging SenticNet and WordNet-Affect Emotion Lists for Sentiment Analysis. *Proceedings of the IEEE International Conference on Signal Processing (ICSP)* 2: 1251–1255.
157. Poria, S., A. Gelbukh, A. Hussain, S. Bandyopadhyay, and N. Howard. 2013. Music Genre Classification: A Semi-Supervised Approach. In *Proceedings of the Springer Mexican Conference on Pattern Recognition*, 254–263.
158. Poria, S., N. Ofek, A. Gelbukh, A. Hussain, and L. Rokach. 2014. Dependency Tree-based Rules for Concept-level Aspect-based Sentiment Analysis. In *Proceedings of the Springer Semantic Web Evaluation Challenge*, 41–47.
159. Poria, S., H. Peng, A. Hussain, N. Howard, and E. Cambria. 2017. Ensemble Application of Convolutional Neural Networks and Multiple Kernel Learning for Multimodal Sentiment Analysis. *In Proceedings of the Elsevier Neurocomputing*.
160. Pye, D., N.J. Hollinghurst, T.J. Mills, and K.R. Wood. 1998. Audio-visual Segmentation for Content-based Retrieval. In *Proceedings of the International Conference on Spoken Language Processing*.
161. Qiao, Z., P. Zhang, C. Zhou, Y. Cao, L. Guo, and Y. Zhang. 2014. Event Recommendation in Event-based Social Networks.
162. Raad, E.J. and R. Chbeir. 2014. Foto2Events: From Photos to Event Discovery and Linking in Online Social Networks. In *Proceedings of the IEEE Big Data and Cloud Computing*, 508–515, .
163. Radsch, C.C.. 2013. The Revolutions will be Blogged: Cyberactivism and the 4th Estate in Egypt. Doctoral Disseration. American University.

164. Rae, A., B. Sigurbjörnssön, and R. van Zwol. 2010. Improving Tag Recommendation using Social Networks. In *Proceedings of the Adaptivity, Personalization and Fusion of Heterogeneous Information*, 92–99.
165. Rahmani, H., B. Piccart, D. Fierens, and H. Blockeel. 2010. Three Complementary Approaches to Context Aware Movie Recommendation. In *Proceedings of the ACM Workshop on Context-Aware Movie Recommendation*, 57–60.
166. Rattenbury, T., N. Good, and M. Naaman. 2007. Towards Automatic Extraction of Event and Place Semantics from Flickr Tags. In *Proceedings of the ACM Special Interest Group on Information Retrieval.*
167. Rawat, Y. and M. S. Kankanhalli. 2016. ConTagNet: Exploiting User Context for Image Tag Recommendation. In *Proceedings of the ACM International Conference on Multimedia*, 1102–1106.
168. Repp, S., A. Groß, and C. Meinel. 2008. Browsing within Lecture Videos based on the Chain Index of Speech Transcription. *Proceedings of the IEEE Transactions on Learning Technologies* 1(3): 145–156.
169. Repp, S. and C. Meinel. 2006. Semantic Indexing for Recorded Educational Lecture Videos. In *Proceedings of the IEEE International Conference on Pervasive Computing and Communications Workshops*, 5.
170. Repp, S., J. Waitelonis, H. Sack, and C. Meinel. 2007. Segmentation and Annotation of Audiovisual Recordings based on Automated Speech Recognition. In *Proceedings of the Springer Intelligent Data Engineering and Automated Learning*, 620–629.
171. Russell, J.A. 1980. A Circumplex Model of Affect. *Proceedings of the Journal of Personality and Social Psychology* 39: 1161–1178.
172. Sahidullah, M., and G. Saha. 2012. Design, Analysis and Experimental Evaluation of Block based Transformation in MFCC Computation for Speaker Recognition. *Proceedings of the Speech Communication* 54: 543–565.
173. J. Salamon, J. Serra, and E. Gomez´. Tonal Representations for Music Retrieval: From Version Identification to Query-by-Humming. *In* Proceedings of the Springer International Journal of Multimedia Information Retrieval, 2(1):45–58, 2013.
174. Schedl, M. and D. Schnitzer. 2014. Location-Aware Music Artist Recommendation. In *Proceedings of the Springer MultiMedia Modeling*, 205–213.
175. M. Schedl and F. Zhou. 2016. Fusing Web and Audio Predictors to Localize the Origin of Music Pieces for Geospatial Retrieval. In *Proceedings of the Springer European Conference on Information Retrieval*, 322–334.
176. Scherp, A., and V. Mezaris. 2014. Survey on Modeling and Indexing Events in Multimedia. *Proceedings of the Springer Multimedia Tools and Applications* 70(1): 7–23.
177. Scherp, A., V. Mezaris, B. Ionescu, and F. De Natale. 2014. HuEvent '14: Workshop on Human-Centered Event Understanding from Multimedia. In *Proceedings of the ACM International Conference on Multimedia*, 1253–1254, .
178. Schmitz, P.. 2006. Inducing Ontology from Flickr Tags. In *Proceedings of the Collaborative Web Tagging Workshop at ACM World Wide Web Conference*, volume 50.
179. Schuller, B., C. Hage, D. Schuller, and G. Rigoll. 2010. Mister DJ, Cheer Me Up!: Musical and Textual Features for Automatic Mood Classification. *Proceedings of the Journal of New Music Research* 39(1): 13–34.
180. Shah, R.R., M. Hefeeda, R. Zimmermann, K. Harras, C.-H. Hsu, and Y. Yu. 2016. NEWSMAN: Uploading Videos over Adaptive Middleboxes to News Servers In Weak Network Infrastructures. In *Proceedings of the Springer International Conference on Multimedia Modeling*, 100–113.
181. Shah, R.R., A. Samanta, D. Gupta, Y. Yu, S. Tang, and R. Zimmermann. 2016. PROMPT: Personalized User Tag Recommendation for Social Media Photos Leveraging Multimodal Information. In *Proceedings of the ACM International Conference on Multimedia*, 486–492.

182. Shah, R.R., A.D. Shaikh, Y. Yu, W. Geng, R. Zimmermann, and G. Wu. 2015. EventBuilder: Real-time Multimedia Event Summarization by Visualizing Social Media. In *Proceedings of the ACM International Conference on Multimedia*, 185–188.
183. Shah, R.R., Y. Yu, A.D. Shaikh, S. Tang, and R. Zimmermann. 2014. ATLAS: Automatic Temporal Segmentation and Annotation of Lecture Videos Based on Modelling Transition Time. In *Proceedings of the ACM International Conference on Multimedia*, 209–212.
184. Shah, R.R., Y. Yu, A.D. Shaikh, and R. Zimmermann. 2015. TRACE: A Linguistic-based Approach for Automatic Lecture Video Segmentation Leveraging Wikipedia Texts. In *Proceedings of the IEEE International Symposium on Multimedia*, 217–220.
185. Shah, R.R., Y. Yu, S. Tang, S. Satoh, A. Verma, and R. Zimmermann. 2016. Concept-Level Multimodal Ranking of Flickr Photo Tags via Recall Based Weighting. In *Proceedings of the MMCommon's Workshop at ACM International Conference on Multimedia*, 19–26.
186. Shah, R.R., Y. Yu, A. Verma, S. Tang, A.D. Shaikh, and R. Zimmermann. 2016. Leveraging Multimodal Information for Event Summarization and Concept-level Sentiment Analysis. In *Proceedings of the Elsevier Knowledge-Based Systems*, 102–109.
187. Shah, R.R., Y. Yu, and R. Zimmermann. 2014. ADVISOR: Personalized Video Soundtrack Recommendation by Late Fusion with Heuristic Rankings. In *Proceedings of the ACM International Conference on Multimedia*, 607–616.
188. Shah, R.R., Y. Yu, and R. Zimmermann. 2014. User Preference-Aware Music Video Generation Based on Modeling Scene Moods. In *Proceedings of the ACM International Conference on Multimedia Systems*, 156–159.
189. Shaikh, A.D., M. Jain, M. Rawat, R.R. Shah, and M. Kumar. 2013. Improving Accuracy of SMS Based FAQ Retrieval System. In *Proceedings of the Springer Multilingual Information Access in South Asian Languages*, 142–156.
190. Shaikh, A.D., R.R. Shah, and R. Shaikh. 2013. SMS based FAQ Retrieval for Hindi, English and Malayalam. In *Proceedings of the ACM Forum on Information Retrieval Evaluation*, 9.
191. Shamma, D.A., R. Shaw, P.L. Shafton, and Y. Liu. 2007. Watch What I Watch: Using Community Activity to Understand Content. In *Proceedings of the ACM International Workshop on Multimedia Information Retrieval*, 275–284.
192. Shaw, B., J. Shea, S. Sinha, and A. Hogue. 2013. Learning to Rank for Spatiotemporal Search. In *Proceedings of the ACM International Conference on Web Search and Data Mining*, 717–726.
193. Sigurbjörnsson, B. and R. Van Zwol. 2008. Flickr Tag Recommendation based on Collective Knowledge. In *Proceedings of the ACM World Wide Web Conference*, 327–336.
194. Snoek, C.G., M. Worring, and A.W.Smeulders. 2005. Early versus Late Fusion in Semantic Video Analysis. In *Proceedings of the ACM International Conference on Multimedia*, 399–402.
195. Snoek, C.G., M. Worring, J.C. Van Gemert, J.-M. Geusebroek, and A.W. Smeulders. 2006. The Challenge Problem for Automated Detection of 101 Semantic Concepts in Multimedia. In *Proceedings of the ACM International Conference on Multimedia*, 421–430.
196. Soleymani, M., J.J.M. Kierkels, G. Chanel, and T. Pun. 2009. A Bayesian Framework for Video Affective Representation. In *Proceedings of the IEEE International Conference on Affective Computing and Intelligent Interaction and Workshops*, 1–7.
197. Stober, S., and A. . Nürnberger. 2013. Adaptive Music Retrieval–a State of the Art. *Proceedings of the Springer Multimedia Tools and Applications* 65(3): 467–494.
198. Stoyanov, V., N. Gilbert, C. Cardie, and E. Riloff. 2009. Conundrums in Noun Phrase Coreference Resolution: Making Sense of the State-of-the-art. In *Proceedings of the ACL International Joint Conference on Natural Language Processing of the Asian Federation of Natural Language Processing*, 656–664.
199. Stupar, A. and S. Michel. 2011. Picasso: Automated Soundtrack Suggestion for Multi-modal Data. In *Proceedings of the ACM Conference on Information and Knowledge Management*, 2589–2592.

200. Thayer, R.E. 1989. *The Biopsychology of Mood and Arousal*. New York: Oxford University Press.
201. Thomee, B., B. Elizalde, D.A. Shamma, K. Ni, G. Friedland, D. Poland, D. Borth, and L.-J. Li. 2016. YFCC100M: The New Data in Multimedia Research. *Proceedings of the Communications of the ACM* 59(2): 64–73.
202. Tirumala, A., F. Qin, J. Dugan, J. Ferguson, and K. Gibbs. 2005. Iperf: The TCP/UDP Bandwidth Measurement Tool. http://dast.nlanr.net/Projects/Iperf/
203. Torralba, A., R. Fergus, and W.T. Freeman. 2008. 80 Million Tiny Images: A Large Data set for Nonparametric Object and Scene Recognition. *Proceedings of the IEEE Transactions on Pattern Analysis and Machine Intelligence* 30(11): 1958–1970.
204. Toutanova, K., D. Klein, C.D. Manning, and Y. Singer. 2003. Feature-Rich Part-of-Speech Tagging with a Cyclic Dependency Network. In *Proceedings of the North American Chapter of the Association for Computational Linguistics on Human Language Technology*, 173–180.
205. Toutanova, K. and C.D. Manning. 2000. Enriching the Knowledge Sources used in a Maximum Entropy Part-of-Speech Tagger. In *Proceedings of the Annual Meeting of the Association for Computational Linguistics*, 63–70.
206. Utiyama, M. and H. Isahara. 2001. A Statistical Model for Domain-Independent Text Segmentation. In *Proceedings of the Annual Meeting on Association for Computational Linguistics*, 499–506.
207. Vishal, K., C. Jawahar, and V. Chari. 2015. Accurate Localization by Fusing Images and GPS Signals. In *Proceedings of the IEEE Computer Vision and Pattern Recognition Workshops*, 17–24.
208. Wang, C., F. Jing, L. Zhang, and H.-J. Zhang. 2008. Scalable Search-based Image Annotation. *Proceedings of the Springer Multimedia Systems* 14(4): 205–220.
209. Wang, H.L., and L.F. Cheong. 2006. Affective Understanding in Film. *Proceedings of the IEEE Transactions on Circuits and Systems for Video Technology* 16(6): 689–704.
210. Wang, J., J. Zhou, H. Xu, T. Mei, X.-S. Hua, and S. Li. 2014. Image Tag Refinement by Regularized Latent Dirichlet Allocation. *Proceedings of the Elsevier Computer Vision and Image Understanding* 124: 61–70.
211. Wang, P., H. Wang, M. Liu, and W. Wang. 2010. An Algorithmic Approach to Event Summarization. In *Proceedings of the ACM Special Interest Group on Management of Data*, 183–194.
212. Wang, X., Y. Jia, R. Chen, and B. Zhou. 2015. Ranking User Tags in Micro-Blogging Website. In *Proceedings of the IEEE ICISCE*, 400–403.
213. Wang, X., L. Tang, H. Gao, and H. Liu. 2010. Discovering Overlapping Groups in Social Media. In *Proceedings of the IEEE International Conference on Data Mining*, 569–578.
214. Wang, Y. and M.S. Kankanhalli. 2015. Tweeting Cameras for Event Detection. In *Proceedings of the IW3C2 International Conference on World Wide Web*, 1231–1241.
215. Webster, A.A., C.T. Jones, M.H. Pinson, S.D. Voran, and S. Wolf. 1993. Objective Video Quality Assessment System based on Human Perception. In *Proceedings of the IS&T/SPIE's Symposium on Electronic Imaging: Science and Technology*, 15–26. International Society for Optics and Photonics.
216. Wei, C.Y., N. Dimitrova, and S.-F. Chang. 2004. Color-mood Analysis of Films based on Syntactic and Psychological Models. In *Proceedings of the IEEE International Conference on Multimedia and Expo*, 831–834.
217. Whissel, C. 1989. The Dictionary of Affect in Language. In *Emotion: Theory, Research and Experience. Vol. 4. The Measurement of Emotions*, ed. R. Plutchik and H. Kellerman, 113–131. New York: Academic.
218. Wu, L., L. Yang, N. Yu, and X.-S. Hua. 2009. Learning to Tag. In *Proceedings of the ACM World Wide Web Conference*, 361–370.

219. Xiao, J., W. Zhou, X. Li, M. Wang, and Q. Tian. 2012. Image Tag Re-ranking by Coupled Probability Transition. In *Proceedings of the ACM International Conference on Multimedia*, 849–852.
220. Xie, D., B. Qian, Y. Peng, and T. Chen. 2009. A Model of Job Scheduling with Deadline for Video-on-Demand System. In *Proceedings of the IEEE International Conference on Web Information Systems and Mining*, 661–668.
221. Xu, M., L.-Y. Duan, C. Xu, M. Kankanhalli, and Q. Tian. 2003. Event Detection in Basketball Video using Multiple Modalities. *Proceedings of the IEEE Joint Conference of the Fourth International Conference on Information, Communications and Signal Processing, and Fourth Pacific Rim Conference on Multimedia* 3: 1526–1530.
222. Xu, M., N.C. Maddage, C. Xu, M. Kankanhalli, and Q. Tian. 2003. Creating Audio Keywords for Event Detection in Soccer Video. In *Proceedings of the IEEE International Conference on Multimedia and Expo*, 2:II–281.
223. Yamamoto, N., J. Ogata, and Y. Ariki. 2003. Topic Segmentation and Retrieval System for Lecture Videos based on Spontaneous Speech Recognition. In *Proceedings of the INTERSPEECH*, 961–964.
224. Yang, H., M. Siebert, P. Luhne, H. Sack, and C. Meinel. 2011. Automatic Lecture Video Indexing using Video OCR Technology. In *Proceedings of the IEEE International Symposium on Multimedia*, 111–116.
225. Yang, Y.H., Y.C. Lin, Y.F. Su, and H.H. Chen. 2008. A Regression Approach to Music Emotion Recognition. *Proceedings of the IEEE Transactions on Audio, Speech, and Language Processing* 16(2): 448–457.
226. Ye, G., D. Liu, I.-H. Jhuo, and S.-F. Chang. 2012. Robust Late Fusion with Rank Minimization. In *Proceedings of the IEEE Conference on Computer Vision and Pattern Recognition*, 3021–3028.
227. Ye, Q., Q. Huang, W. Gao, and D. Zhao. 2005. Fast and Robust Text Detection in Images and Video Frames. *Proceedings of the Elsevier Image and Vision Computing* 23(6): 565–576.
228. Yin, Y., Z. Shen, L. Zhang, and R. Zimmermann. 2015. Spatial-temporal Tag Mining for Automatic Geospatial Video Annotation. *Proceedings of the ACM Transactions on Multimedia Computing, Communications, and Applications* 11(2): 29.
229. Yoon, S. and V. Pavlovic. 2014. Sentiment Flow for Video Interestingness Prediction. In *Proceedings of the Workshop on HuEvent at the ACM International Conference on Multimedia*, 29–34.
230. Yu, Y., K. Joe, V. Oria, F. Moerchen, J.S. Downie, and L. Chen. 2009. Multi-version Music Search using Acoustic Feature Union and Exact Soft Mapping. *Proceedings of the World Scientific International Journal of Semantic Computing* 3(02): 209–234.
231. Yu, Y., Z. Shen, and R. Zimmermann. 2012. Automatic Music Soundtrack Generation for Out-door Videos from Contextual Sensor Information. In *Proceedings of the ACM International Conference on Multimedia*, 1377–1378.
232. Zaharieva, M., M. Zeppelzauer, and C. Breiteneder. 2013. Automated Social Event Detection in Large Photo Collections. In *Proceedings of the ACM International Conference on Multimedia Retrieval*, 167–174.
233. Zhang, J., X. Liu, L. Zhuo, and C. Wang. 2015. Social Images Tag Ranking based on Visual Words in Compressed Domain. *Proceedings of the Elsevier Neurocomputing* 153: 278–285.
234. Zhang, J., S. Wang, and Q. Huang. 2015. Location-Based Parallel Tag Completion for Geo-tagged Social Image Retrieval. In *Proceedings of the ACM International Conference on Multimedia Retrieval*, 355–362.
235. Zhang, M., J. Wong, W. Tavanapong, J. Oh, and P. de Groen. 2004. Media Uploading Systems with Hard Deadlines. In *Proceedings of the Citeseer International Conference on Internet and Multimedia Systems and Applications*, 305–310.
236. Zhang, M., J. Wong, W. Tavanapong, J. Oh, and P. de Groen. 2008. Deadline-constrained Media Uploading Systems. *Proceedings of the Springer Multimedia Tools and Applications* 38(1): 51–74.

237. Zhang, W., J. Lin, X. Chen, Q. Huang, and Y. Liu. 2006. Video Shot Detection using Hidden Markov Models with Complementary Features. *Proceedings of the IEEE International Conference on Innovative Computing, Information and Control* 3: 593–596.
238. Zheng, L., V. Noroozi, and P.S. Yu. 2017. Joint Deep Modeling of Users and Items using Reviews for Recommendation. In *Proceedings of the ACM International Conference on Web Search and Data Mining*, 425–434.
239. Zhou, X.S. and T.S. Huang. 2000. CBIR: from Low-level Features to High-level Semantics. In *Proceedings of the International Society for Optics and Photonics Electronic Imaging*, 426–431.
240. Zhuang, J. and S.C. Hoi. 2011. A Two-view Learning Approach for Image Tag Ranking. In *Proceedings of the ACM International Conference on Web Search and Data Mining*, 625–634.
241. Zimmermann, R. and Y. Yu. 2013. Social Interactions over Geographic-aware Multimedia Systems. In *Proceedings of the ACM International Conference on Multimedia*, 1115–1116.
242. Shah, R.R. 2016. Multimodal-based Multimedia Analysis, Retrieval, and Services in Support of Social Media Applications. In *Proceedings of the ACM International Conference on Multimedia*, 1425–1429.
243. Shah, R.R. 2016. Multimodal Analysis of User-Generated Content in Support of Social Media Applications. In *Proceedings of the ACM International Conference in Multimedia Retrieval*, 423–426.
244. Yin, Y., R.R. Shah, and R. Zimmermann. 2016. A General Feature-based Map Matching Framework with Trajectory Simplification. In *Proceedings of the ACM SIGSPATIAL International Workshop on GeoStreaming*, 7.

Chapter 2
Literature Review

Abstract In this chapter we cover a detailed literature survey for five multimedia analytics problems which we have addressed in this book. First, we present a literature review for event understanding in Sect. 2.1. Next, we cover the literature review for tag recommendation and ranking in Sect. 2.2. Subsequently, Sect. 2.3 describes the literature review for soundtrack recommendation. Next, we present the literature review for lecture videos segmentation in Sect. 2.4. Finally, we describes the literature review for the adaptive news videos uploading in Sect. 2.5.

Keywords Literature review • Semantics analysis • Sentics analysis • Multimodal analysis • User-generated multimedia content • Multimedia fusion • Multimedia analysis • Multimedia recommendation • Multimedia uploading

2.1 Event Understanding

In event understanding, our purpose is to produce summaries for multimedia content from social media automatically. We describe the steps of such a process as follows: (i) identifying events and sentiments from all UGIs, (ii) producing the summary for a given event based on semantics analysis, and (iii) generating the summary based on sentics analysis. In this section, we briefly provide some recent progress on event detection and summarization, semantics and sentiments analysis, and soundtrack recommendation for multimedia content [242, 243].

The area of event modeling, detection, and understanding from multimedia content observes significant work [122, 136, 176, 177] over the past few years. Earlier methods [59, 166, 178, 232] leveraged multimodal information such as user tags, spatial and temporal information and multimedia content to detect events automatically from a large collection of UGC such as Flickr. Rattenbury et al. [166] extracted place and event semantics for tags using Flickr metadata. Kan [82] presented his thesis on automatic text summarization as applied to information retrieval using indicative and informative summaries. Raad et al. [162] presented a clustering algorithm to automatically detect personal events from photos shared online on the social network of a specific user. They defined an event model that captures event triggers and relationships that can exist between detected events.

R. Shah, R. Zimmermann, *Multimodal Analysis of User-Generated Multimedia Content*, Socio-Affective Computing 6, DOI 10.1007/978-3-319-61807-4_2

Furthermore, they leveraged event features (*Who*, *Where*, and *When*) to refine clustering results using defined rules. Moreover, they used appropriate time-space granularities to detect multi-location, multi-day, and multi-person events. Fabro et al. [55] presented an algorithm for the summarization of real-life events based on community-contributed multimedia content using photos from Flickr and videos from YouTube. They evaluated the coverage of the produced summaries by comparing them with Wikipedia articles that report on the corresponding events (see Sect. 1.4.6 for details on Wikipedia API). They also found that the composed summaries show a good coverage of interesting situations that happened during the selected events on the event summarization. We leverage Wikipedia in our event summarization system since it is one of the comprehensive sources of knowledge. Long et al. [114] presented a unified workflow of event detection, tracking, and summarization of microblog data such as Twitter. They selected topical words from the microblog data leveraging its characteristics for event detection. Moreover, Naaman [127] presented an approach for social media applications for the searching and mining of multimedia data.

Lim et al. [106] addressed the semantic gap between feature-based indices computed automatically and human query by focusing on the notion of an event in home photos. They employed visual keywords indexing which derived from a visual content domain with relevant semantics labels. To detect complex events in videos on YouTube, Chakraborty et al. [46] proposed an entity-centric region of interest detection and visual-semantic pooling scheme. Events can ubiquitously find in multimedia content (*e.g.*, UGT, UGI, UGV) that are created, shared, or encountered on social media websites such as Twitter, Flickr, and YouTube [121]. A significant research work has been carried out in detecting events from a video. Kang et al. [86, 87] presented the detection of events such as goals and corner kicks from soccer videos by using audio/visual keywords. Similarly, Xu et al. [221] leveraged multiple modalities to detect basketball events from videos by using audio/visual keywords. Xu et al. [222] presented a framework to detect events in a soccer video using audio keywords derived from low-level audio features by using support vector machine learning. Multi-camera surveillance systems are being increasingly used in public and prohibited places such as banks, airports, and military premises. Natarajan et al. [128] presented a research survey on the state-of-the-art overview of various techniques for multi-camera coordination and control that have adopted in surveillance systems. Atrey et al. [29] presented the detection of surveillance events such as human movements and abandoned objects, by exploiting visual and aural information. Wang et al. [214] leveraged visual sensors to tweet semantic concepts for event detection and proposed a novel multi-layer tweeting cameras framework. They also described an approach to infer high-level semantics from the fused information of physical sensors and social media sensors.

Low-level visual features often use for event detections or the selection of representative images from a collection of images/videos [136]. Papagiannopoulou and Mezaris [134] presented a clustering approach to producing an event-related image collection summarization using trained visual concept detectors based on

Table 2.1 A comparison with the previous work on the semantics understanding of an event

Approach	Visual	Textual	Spatial	Temporal	Social
Semantics understanding of an event [166]			✓	✓	
Semantics understanding of an event based on social interactions [55, 127, 162]		✓			✓
Event understanding and summarization [58, 114, 211]		✓			
Event detection from videos [86, 87]	✓				
The Event Builder system [182, 186]	✓	✓	✓	✓	

image features such as SIFT, RGB-SIFT, and OpponentSIFT. Wang et al. [211] summarized events based on minimum description length principle. They achieved summaries through learning an HMM from event data. Liu and Huet [111] attempted to retrieve and summarize events on a given topic and proposed a framework to extract and illustrate social events automatically on any given query by leveraging social media data. Filatova and Hatzivassiloglou [58] proposed a set of event-based features based on *TF-IDF* scores to produce event summaries. We leveraged these event-based features [58] to produce text summaries for given events. Moxley et al. [125] explored tag uses in geo-referenced image collections crawled from Flickr, with the aim of improving an automatic annotation system. Hong et al. [72] proposed a framework to produce multi-video event summarization for web videos. Yoon and Pavlovic [229] presented a video interestingness prediction framework that includes a mid-level representation of sentiment sequence as an interestingness determinant. As illustrated in Table 2.1, we leveraged information from multiple modalities for an efficient event understanding. Moreover, our EventBuilder system utilized information from existing knowledge bases such as Wikipedia. However, the earlier work [166] leveraged temporal and spatial metadata, and the work [55, 127, 162] exploited social interactions for event understanding. Moreover, the work [58, 114, 211] performed event understanding and summarization based on textual data.

Due to unstructured, heterogeneous nature, and sheer volume of multimedia data, it is required to discover important features from raw data during pre-processing [36]. Data cleaning, normalization, and transformation are also required during pre-processing to remove noises from data and normalize the huge difference between maximum and minimum values of data. Next, various data mining techniques can be applied to discover interesting patterns in data that are not ordinarily accessible by basic queries. First, we review the area of affective computing and emotion recognition. Picard et al. [139] proposed that machine intelligence needs to include emotional intelligence. They analyzed four physiological signals that exhibit problematic day-to-day variations and found that the technique of seeding a Fisher Projection with the results of Sequential Floating Forward Search improves the performance of the Fisher Projection, and provided the highest recognition rates for classification of affect from physiology. Kort et al. [94] build a model to the interplay of emotions upon learning with the aim that learning will proceed at an optimal pace, *i.e.*, the model can recognize a learner's affective state and respond appropriately to it. Picard and Klein [138] discussed a high-level process to begin to directly address the human emotional component in human-computer interaction (HCI). They broadly discussed the following two issues: (i) the consideration of human needs beyond efficiency and productivity, and (ii) what kinds of emotional needs do human tend to have one a day-to-day basis that, if unmet, can significantly degrade the quality of life. Healey and Picard [68] presented methods for collecting and analyzing physiological data during real-world driving tasks to determine a driver's relative stress level. Such methods can also be employed to people in activities that involve much attention such as learning and gaming. McDuff et al. [119] presented an analysis of naturalistic and spontaneous responses to video segments of electoral debates. They showed that it is possible to measure significantly different responses to the candidates using automated facial expression analysis. Moreover, such different responses can predict self-report candidate preferences. They were also able to identify moments within the video clips at which initially similar expressions are seen, but the temporal evolution of the expressions leads to very different political associations.

Next, we review the area of sentiment analysis which attempts to determine the sentics details of multimedia content based on the concepts exhibited from their visual content and metadata. Over the past few years, we witness the significant contributions [25, 43, 153, 158] in the area of sentiment analysis. Sentiments are very useful in personalized search, retrieval, and recommendation systems. Cambria et al. [41] presented SenticNet-3 that bridges the conceptual and affective gap between word-level natural language data and the concept-level opinions and sentiments conveyed by them (see Sect. 1.4.3 for details). They also presented AffectiveSpace-2 to determine affective intuitions for concepts [39]. Poria et al. [149] presented an intelligent framework for multimodal affective data analysis. Leveraging the above knowledge bases, we determine sentics details from multimedia content. Recent advances in deep neural networks help Google Cloud Vision API [14] to analyze emotional facial attributes in photos such as joy, sorrow, and anger. Thus, the results of sentiment analysis can be improved significantly leveraging deep learning technologies. In our proposed EventSensor system, we perform the sentiment analysis to determine moods associated with UGIs, and subsequently provide a sentics-based multimedia summary. We add a matching soundtrack to the slideshow of UGIs based on the determined moods.

Next, we review the area of soundtrack recommendation for multimedia content. The area of music recommendation for multimedia content is largely unexplored. Earlier approaches [100, 199] added soundtracks to the slideshow of UGIs. However, they largely focused on low-level visual features. There are a few approaches [66, 196, 209] to recognizing emotions from videos but the field of soundtrack recommendation for UGVs [53, 231] is largely unexplored. Rahmani et al. [165] proposed context-aware movie recommendation techniques based on background information such as users' preferences, movie reviews, actors and directors of movies. Since the main contribution of our work is to determine sentics details (mood tag) of the multimedia content, we randomly select soundtracks

Table 2.2 A comparison with the previous work on the sentics understanding of social media content

Approach	Visual	Textual	Audio	Knowledge bases	Spatial
Sentics understanding from UGIs [100, 199] and UGVs [66, 196, 209]	✓				
Sentics understanding from UGVs [53, 231]	✓		✓		✓
The EventSensor system [186]	✓	✓		✓	

corresponding to the determined mood tag from an existing mood-tagged music dataset [187]. As illustrated in Table 2.2, our EventSensor system leveraged information from multiple modalities such as visual content and textual metadata for an efficient sentics understanding of UGIs. We also exploited information from existing knowledge bases such as SenticNet [41], EmoSenticNet, and WordNet (see Sect. 1.4 for details on knowledge bases).

2.2 Tag Recommendation and Ranking

First, we describe recent progress on tag recommendation, and subsequently, we discuss earlier work on tag ranking. Li et al. [102] proposed a neighbor voting algorithm that learns tag relevance by accumulating votes from visual neighbors of an input photo. They showed the effectiveness of their approach by ranking tags of the photo. Sigurbjörnsson and Van Zwol [193] presented a tag recommendation system to predict tags based on tag co-occurrence for each user input tag and merge them into a single candidate list using the proposed aggregate (Vote or Sum) and promote (descriptive, stability, and rank promotion) methods. Rae et al. [164] proposed an extendable framework that can recommend additional tags to partially annotated images using a combination of different personalized and collective contexts. For instance, they leveraged the following information: (i) all photos in the system, (ii) a user's own photos, (iii) photos of the user's social contacts, and (iv) photos posted in groups of which the user is a member. These approaches are not fully automatic since they expect a user to input (annotate) a few initial tags.

Anderson et al. [28] presented a tag prediction system for Flickr photos, which combines both linguistic and visual features of a photo. Nwana and Chen [133] proposed a novel way of measuring tag preferences, and also proposed a new personalized tagging objective function that explicitly considers a user's preferred tag orderings using a (partially) greedy algorithm. Wu et al. [218] proposed a multi-modality recommendation based on both tag and visual correlation, and formulated the tag recommendation as a learning problem. Each modality is used to generate a ranking feature, and Rankboost algorithm is applied to learn an optimal combination of these ranking features from different modalities. Liu et al. [109] proposed a tag ranking scheme, aiming to automatically rank tags associated with a given photo according to their relevance to the photo content. They estimated initial relevance scores for the tags based on probability density estimation, and then performed a random walk over a tag similarity graph to refine the relevance scores. Wang et al. [213] proposed a novel co-clustering framework, which takes advantage of networking information between users and tags in social media, to discover these overlapping communities. They clustered edges instead of nodes to determine overlapping clusters (*i.e.*, a single user belongs to multiple social groups). Recent work [133, 167] exploit user context for photo tag recommendation. Garg and Weber [61] proposed a system that suggests related tags to a user, based on the tags that she or other people have used in the past along with (some of) the tags already entered. The suggested tags are dynamically updated with every additional tag entered/selected. Image captioning is an active area and seems to have subsumed image captioning. A recent work on image captioning is presented by Johnson et al. [77] that addressed the localization and description task jointly using a Fully Convolutional Localization Network (FCLN) architecture. FCLN processes an image with a single, efficient forward pass, requires no external regions proposals, and can be trained end-to-end with a single round of optimization. As illustrated in Table 2.3, our PROMPT system leveraged information from personal and social contexts to recommend personalized user tags for social media photos. First, we determine a group of users who have similar tagging behavior for a given user. Next, we find candidate tags from visual content, textual metadata, and tags of neighboring photos to leverage information from social context. We initialize scores of the candidate tags using asymmetric tag co-occurrence probabilities and normalized scores of tags after neighbor voting, and later perform random walk to promote the tags that have many close neighbors and weaken isolated tags. Finally, we recommend the top five user tags to the given photo.

There exists significant prior work to perform tag ranking for a UGI [102, 109, 233]. Liu et al. [109] proposed a tag ranking scheme by first estimating the initial relevance scores for tags based on probability density estimation and then performing a random walk over a tag similarity graph to refine relevance scores. However, such a process incurs high online computation cost for a tag to tag relevance and iterative update for the tag to image relevance. Li et al. [102] proposed a neighbor voting scheme for tag ranking based on the intuition that if different people annotate visually similar photos using the same tags then these tags are likely to describe objective aspects of the visual content. They computed neighbors using low-level visual features. Zhang et al. [233] also leveraged the neighbor voting model for tag ranking based on visual words in a compressed domain. They computed tag ranking for photos in three steps: (i) low-resolution photos are constructed, (ii) visual words are created using SIFT descriptors of the low-resolution photos, and (iii) tags are ranked according to voting from neighbors derived based on visual words similarity. Computing low-level features from the visual content of photos and videos is a very costly and time-consuming process since it requires information at every pixel levels. Despite the increasing popularity in GPUs uses enables multimedia systems to analyze photos from pixels much more quickly than before, it may not be possible to employ GPUs on a very large scale

Table 2.3 A comparison with the previous work on tag recommendation

Approach	Neighbour voting	Tag concurrence	Random Walk	High-level features	Low-level features
[102, 233]	✓				✓
[185]	✓			✓	
[193]		✓			✓
[109, 219]			✓		✓
PROMPT [181]	✓	✓	✓	✓	✓

and continuously growing multimedia databases. However, computing high-level features from available concepts leveraging the bag-of-words model is relatively fast. Thus, we leverage high-level features to compute tag relevance for photos.

Moxley et al. [125] explored the ability to learn tag semantics by mining geo-referenced photos, and categorizing tags as places, landmarks, and visual descriptors automatically. There exist some earlier work [219, 240] which leverage both textual and visual content to computing tag relevance for photos. Zhuang and Hoi [240] proposed a two-way learning approach by exploiting both the textual and visual content of photos to discover the relationship between tags and photos. They formulated the two-view tag weighting problem as an optimization task and solved using a stochastic coordinate descent algorithm. Xiao et al. [219] proposed a coupled probability transition algorithm to estimate the text-visual group relevance and next utilized it in inferring tag relevance for a new photo. Wang et al. [210] presented the regularized Latent Dirichlet Allocation (rLDA) model for tag refinement and estimated both tag similarity and tag relevance. Neo et al. [130] presented a high-level feature indexing on shots or frames for news video retrieval. First, they utilized extensive query analysis to relate various high-level features and query terms by matching the textual description and context in a time-dependent manner. Second, they introduced a framework to effectively fuse the relation weights with the detectors' confidence scores. This results in individual high-level features that are weighted on a per-query basis. Such work motivated us to leverage high-level features from different modalities and employ fusion techniques to compute tag relevance since suitable concepts may present in different representations.

In a recent work, Zhang et al. [234] proposed a framework to learn the relation between a geo-tagged photo and a tag within different Points of Interests (POI). Moreover, Wang et al. [212] proposed a user tag ranking scheme in the microblogging website Sina Weibo based on the relations between users. They derived such relations from re-tweeting or notification to other users. Furthermore, Xiao et al. [219] utilized the Latent Semantic Indexing (LSI) model in their tag ranking algorithm. First, they computed the tag relevance using LSI, and next performed a random walk to discover the final significance of each tag. Similar to earlier work [102, 233] we follow neighbor voting scheme to compute tag relevance for photos. However, we employ high-level features instead of low-level features in determining neighbors for photos. As illustrated in Table 2.4, our CRAFT system that ranks

Table 2.4 A comparison with the previous work on tag ranking

Approach	Visual	Textual	Spatial
Neighbor voting based tag ranking [102, 233]	✓		
Tag ranking based on users' interaction [212]		✓	
Random walk based tag ranking [109, 219]	✓	✓	
The proposed CRAFT system [185]	✓	✓	✓

tags of social media photos leveraged information from all three modalities (*i.e.*, visual, textual and spatial content). However, earlier work ignored spatial domain to compute tag relevance for UGIs. Moreover, our CRAFT system used high-level features instead of traditional low-level features in state-of-the-arts. Since the performance of SMS and MMS based FAQ retrieval can be improved by leveraging important keywords (i.e., tags) [189, 190], we would like to leverage our tag ranking method for an efficient SMS/MMS based FAQ retrieval.

2.3 Soundtrack Recommendation for UGVs

Our purpose is to support real-time user preference-aware video soundtrack recommendations via mobile devices. We describe the steps of such a process as follows. (i) A user captures a video on a smartphone. (ii) An emotion-cognition model predicts video scene moods based on a heterogeneous late fusion of geo and visual features. (iii) A list of songs is recommended for the video matching with the user's listening history automatically. (iv) The system sets the most appropriate song from the recommended list as the soundtrack for the video by leveraging the experience of professional mood-based associations between music and movie content. In this section, we briefly provide some recent progress on emotion recognition and music recommendation systems and techniques.

Despite significant efforts that have focused on music recommendation techniques [80, 96, 135, 174] in recent years, researchers have paid little attention to music recommendation for sets of images or UGVs. Kuo et al. [96] investigated the association discovery between emotions and music features of film music and proposed an emotion-based music recommendation system. As of now, the music recommendation area for a set of images has been largely unexplored and consists of only a few state-of-the-art approaches such as an emotion-based impressionism slideshow system from images of paintings by Li et al. [100]. This method extracts features such as the dominant color, the color coherence vector, and the color moment for color and light. It also extracts some statistical measures from the gray level co-occurrence matrix for textures and computes the primitive length of textures. Furthermore, Wei et al. [216] tried to establish an association between color and mood by exploiting the color-related features using an SVM classifier. Mulhem et al. [126] proposed an audiovideo mixing technique which uses a method based on pivot vector space mapping for home videos. Since the process of manual audio-video mixing is very tedious, time-consuming, and costly, they matched video shots with music segments based on aesthetic cinematographic heuristics to perform this task automatically. Nayak [129] presented a novel approach to accomplish audio-video mixing based on generating content-related music by translating primitive elements of video to audio features. To synchronize music with events happening in a video, Nayak [129] used sequence comparison to synthesize new pitch sequence and varied the tempo of music according to the motion of video sequences.

There exist a few approaches [66, 196, 209] to recognize emotions from videos but the field of video soundtrack recommendation for UGVs [188, 231] is largely unexplored. Hanjalic et al. [66] proposed a computational framework for affective video content representation and modeling based on the dimensional approach to affect. They developed models for arousal and valence time curves using low-level features extracted from video content, which maps the affective video content onto a 2D emotion space characterized by arousal and valence. Soleymani et al. [196] introduced a Bayesian classification framework for affective video tagging which takes contextual information into account since emotions that are elicited in response to a video scene contains valuable information for multimedia indexing and tagging. Based on this, they proposed an affective indexing and retrieval system which extracts features from different modalities of a movie, such as a video, audio, and others. To understand the affective content of general Hollywood movies, Wang et al. [209] formulated a few effective audiovisual cues to help bridge the affective gap between emotions and low-level features. They introduced a method to extract affective information from multifaceted audio streams and classified every scene of Hollywood domain movies probabilistically into affective categories. They further processed the visual and audio signals separately for each scene to find the audio-visual cues and then concatenated them to form scene vectors which were sent to an SVM to obtain probabilistic membership vectors. Audio cues at the scene level were obtained using the SVM and the visual cues were computed for each scene by using the segmented shots and keyframes. Since amateur home videos often fail to convey the desired intent due to several reasons such as the limitations of traditional consumer-quality video cameras, it necessities a better approach for video-intent delivery. Achanta et al. [23] presented a general approach based on offline cinematography and automated continuity editing concepts for video-intent delivery. Moreover, they demonstrated the use of the video-intent delivery for four basic emotions such as cheer, serenity, gloom, and excitement.

Cristani et al. [53] introduced a music recommendation policy for a video sequence taken by a camera mounted on board a car. They established the association between audio and video features from low-level cross-modal correlations. Yu et al. [231] presented a system to automatically generate soundtracks for UGVs based on their concurrently captured contextual sensor information. The proposed system correlates viewable scene information from sensors with geographic contextual tags from OpenStreetMap[1] to investigate the relationship between geo-categories and mood tags. Since the video soundtrack generation system by

[1] www.openstreetmap.org

Yu et al. [231] does not consider the visual content of the video or the contextual information other than geo-categories, soundtracks recommended by this system are very subjective. Furthermore, the system used a pre-defined mapping between geo-categories and mood tags, and hence the system is not adaptive in nature. In our earlier work [188], we recommend soundtracks for a UGV based on modeling scene moods using a SVM^{hmm} model. In particular, first, the SVM^{hmm} model predicts scene moods based on the sequence of concatenated geo- and visual features. Next, a list of matching songs corresponding to the predicted scene moods are retrieved.

Currently, sensor-rich media content is receiving increasing attention because sensors provide additional external information such as location from GPS, viewing direction from a compass unit, and so on. Sensor-based media can be useful for applications (*e.g.*, life log recording, and location-based queries and recommendations) [26]. Map matching techniques along with Foursquare categories can be used to accurately determine knowledge structures from sensor-rich videos [244]. Kim et al. [91] discussed the use of textual information such as web documents, social tags and lyrics to derive an emotion of a music sample. Rahmani et al. [165] proposed context-aware movie recommendation techniques based on background information such as users' preferences, movie reviews, actors and directors of the movie, and others. Chen et al. [48] proposed an approach by leveraging a tri-partite graph (user, video, query) to recommend personalized videos. Kaminskas et al. [80] proposed a location-aware music recommendation system using tags, which recommends songs that suit a place of interest. Park et al. [135] proposed a location-based recommendation system based on location, time, the mood of a user and other contextual information in mobile environments. In a recent work, Schedl et al. [174] proposed a few hybrid music recommendation algorithms that integrate information of the music content, the music context, and the user context, to build a music retrieval system. For the ADVISOR system, these earlier work inspired us to mainly focus on sensor-annotated videos that contain additional information provided by sensors and other contextual information such as a user's listening history, music genre information, and others. Preferred music genre from the user's listening history can be automatically determined using a semi-supervised approach [157].

Multi-feature late fusion techniques are very useful for various applications such as video event detection and object recognition [226]. Snoek et al. [194, 195] performed early and late fusion schemes for semantic video analysis and found that the late fusion scheme performs better than the early fusion scheme. Ghias et al. [62] and Lu et al. [115] used heuristic approaches for querying desired songs from a music database by humming a tune. These earlier work inspired us to build the ADVISOR system by performing heterogeneous late fusion to recognize moods from videos and retrieve a ranked list of songs using a heuristic approach. To the best of our knowledge, this is the first work that correlates preference-aware activities from different behavioral signals of individual users, *e.g.*, online listening activities and physical activities. As illustrated in Table 2.5, we exploited information from multiple modalities such as visual, audio, and spatial information to recommend soundtrack for outdoor user-generated videos. Earlier work mostly ignored the spatial information while determining sentics details for scenes in an outdoor UGV.

Table 2.5 A comparison with the previous work on emotion discover and recommending soundtracks for UGIs and UGVs

Approach	Visual content	Audio content	Spatial content	Machine learning model
Soundtrack recommendation for a group of photos [216]	✓			✓
Soundtrack recommendation for a group of photos [100]	✓			
Emotion discovery from a video [126, 129]	✓	✓		
Emotion discovery from a video [66, 196]	✓			✓
Emotion discovery from a video [209]	✓	✓		✓
Soundtrack recommendation for an outdoor video [231]		✓	✓	
The proposed ADVISOR system [187]	✓	✓	✓	✓

2.4 Lecture Video Segmentation

Our purpose is to perform temporal segmentation of lecture videos to assist efficient topic-wise browsing within videos. We describe such a process in the following three steps. First, computing segment boundaries from content information such as video content and SRT. Second, deriving segment boundaries by leveraging information from existing knowledge bases such as Wikipedia. Third, studying the effect of the late fusion of segment boundaries derived from different modalities. In this section, we briefly provide some recent progress on segment boundaries detection, topic modeling, and e–learning for lecture videos.

The rapid growth in the number of digital lecture videos makes distance learning very easy [131, 169]. Traditional video retrieval based on a feature extraction can not be efficiently applied to e-learning applications due to the unstructured and linear features of lecture videos [168]. For the effective content–based retrieval of the appropriate information in such e-learning applications, it is desirable to have a systematic indexing which can be achieved by an efficient video segmentation algorithm. The manual segmentation of a lecture video into smaller cohesive units is an accepted approach to finding appropriate information [131, 223]. However, an automatic temporal segmentation and annotation of lecture videos is a challenging task, since it depends on many factors such as speaker presentation style, characteristic of a camera (*i.e.*, video quality, static or dynamic position/view, *etc.*), and others. Moreover, it is a cross-disciplinary area which requires knowledge of text analysis, visual analysis, speech analysis, machine learning, and others. Additionally, it is not feasible due to the high cost of manual segmentation and rapid growth in the size of a large lecture video database. In the last a few years, several researchers attempted to solve this problem. Adcock et al. [24] employed a story segmentation system to develop a video search system. They evaluated their system on news videos. However, this system is not directly applicable to the topic–wise segmentation of lecture videos because topics of a lecture video are related and not as independent as different news segments in a news video.

Earlier work [67, 107, 160, 170, 223] attempted to segment videos automatically by exploiting visual, audio, and linguistic features. Lin et al. [107] proposed a lecture video segmentation method based on natural language processing (NLP) techniques. Haubold and Kender [67] investigated methods of segmenting, visualizing, and indexing presentation videos by separately considering audio and visual data. Pye et al. [160] performed the segmentation of an audio/video content by the fusion of segmentation achieved by audio and video analysis in the context of television news retrieval. Yamamoto et al. [223] proposed a segmentation method of a continuous lecture speech into topics by associating the lecture speech with the lecture textbook. They performed the association by computing the similarity between topic vectors and a sequence of lecture vectors obtained through spontaneous speech recognition. Moreover, they determined segment boundaries from videos using visual content based on the video shot detection [95]. Most of the state-of-the-arts on the lecture video segmentation by exploiting the visual content are based on a color histogram. Zhang et al. [237] presented a video shot detection method using Hidden Markov Models (HMM) with complementary features such as HSV color histogram difference and statistical corner change ratio (SCCR).

However, not all features from a color space, such as RGB, HSV, or Lab from a particular color image are equally effective in describing the visual characteristics of segments. Therefore, Gao et al. [60] proposed a projective clustering algorithm to improve color image segmentation, which can be used for a better lecture video segmentation. Since a video consists of a number of frames/images, the MRIA algorithm [50] which performs image segmentation and hierarchical tree construction for multiple object image retrieval, can be used for the lecture video segmentation. There exist earlier work [224] on the lecture video segmentation based on an optical character recognition (OCR). Ye et al. [227] presented a fast and robust text detection in images and video frames. However, the video OCR technology is not useful in many cases since the video quality of most of the videos in existing lecture-video databases are not sufficiently high for OCR. Moreover, the image analysis of lecture videos fails even if they are of high quality since the most of the time, a speaker is in focus and the presenting topic is not visible. Fan et al. [57] tried to match slides with presentation videos by exploiting visual content features. Chen et al. [51] attempted to synchronize presentation slides with the speaker video automatically.

Machine learning models [37, 47, 183] were used to perform the segmentation of lecture videos based on the different events such as slide transitions, visibility of speaker only, and visibility of both speaker and slide. Research on video retrieval in the past has focused on either low- or high-level features, but the retrieval effectiveness is either limited or applicable to a few domains. Thus, Kankanhalli and Chua [88] proposed a strata-based annotation method for digital video modeling to achieve efficient browsing and retrieval. Strata-based annotation methods provide a middle ground that model video content as the overlapping strata of concepts. As illustrated in Table 2.6, we exploited information from multiple modalities such as

Table 2.6 A comparison with the previous work on lecture video segmentation

Approach	Visual	SRT	Wikipedia	Speech
Audio and linguistic based video segmentation [223]		✓		✓
Visual features based video segmentation [37, 57, 88]	✓			
Linguistic based video segmentation [107]		✓		
Visual and linguistic based video segmentation [183, 224]	✓	✓		
The TRACE system [184]	✓	✓	✓	

visual, transcript, and Wikipedia to perform the topic-wise segmentation of lecture videos. However, earlier work on lecture video segmentation mostly ignored existing knowledge bases.

There exist several work in the literature [81, 83, 120] that segment documents. Kan et al. [83] presented a method for discovering a segmental discourse structure of a document while categorizing segment function. They demonstrated how retrieval of noun phrases and pronominal forms, along with a zero-sum weighting scheme, determined topicalized segmentation. Furthermore, they used term distribution to aid in identifying the role that the segment performs in the document. Kan [81] proposed integrating features from lexical cohesion with elements from layout recognition to build a composite framework. He used supervised machine learning on this composite feature set to derive discourse structure on the topic level. Utiyama and Isahara [206] presented a statistical model for domain-independent text segmentation. However, these document segmentation approaches may not work very well in the case of lecture video segmentation since often the speech transcript is noisy and may consists repetitions and breaks (pauses). Speech transcripts are more personalized as compared to normal documents. Thus, it is more difficult to segments speech transcripts compared to normal documents.

2.5 Adaptive News Video Uploading

The NEWSMAN scheduling process is described as follows [180], (i) reporters directly upload news videos to the news organizations if the Internet connectivity is good, otherwise (ii) reporters upload news videos to a middlebox, and (iii) the scheduler at the middlebox determines an uploading schedule and optimal bitrates for transcoding. In this section, we survey some recent related work.

In addition to traditional news reporting systems such as satellite news networks, the use of satellite news gathering (SNG) by local stations has also increased during recent years. However, SNG has not been adopted as widely as satellite news networks due to reasons such as: (i) the high setup and maintenance costs of SNG, (ii) the non-portability of SNG equipment to many locations due to its big size [97, 113], and (iii) the unspontaneous occurrence of events which can be captured by reporters (or people) nearby that location. These constraints have popularized news reporting by regular citizens through services such as CNN iReport[2]. CNN iReport is a very popular service provided by CNN to support common man to work as a reporter. It helps citizens to raise their voice on any important issues, and helps news providers to get breaking news quickly where they have no access.

Unlike significant efforts that have focused on systems supporting downloading applications such as video streaming and file sharing [105, 220], little attention has been paid to systems that support uploading applications [38, 236]. News videos from citizens are of not much use if they are not delivered to news providers within some time when any events happened. Thus, it is required to quickly upload the news videos to news providers's websites quickly in the highest possible quality. Therefore, media uploading with hard deadlines require an optimal deadline scheduling algorithm [22, 49, 235]. Abba et al. [22] proposed a prioritized deadline based scheduling algorithm using a project management technique for an efficient job execution with deadline constraints of jobs. Chen et al. [49] proposed an online preemptive scheduling of jobs with deadlines arriving sporadically. The scheduler either accepts or declines a job immediately upon arrival based on a contract where the scheduler looses the profit of the job and pays a penalty if the accepted job is not finished within its deadline. The objective of the online scheduler is to maximize the overall profit, *i.e.*, the total profit of completed jobs before their deadlines is more than the penalty paid for the jobs that missed their deadlines. Online scheduling algorithms such as earliest deadline first (EDF) [108] are often used for applications with deadlines. Since we consider jobs with diverse deadlines, we leverage the EDF concept in our system to determine the uploading schedule that will maximize the system utility.

Recent years have seen significant progress in the area of rate-distortion (R–D) optimized image and video coding [69, 98]. In lossy compression, there is a tradeoff between the bitrate and the distortion. R–D models are functions that describe the relationship between the bitrate and expected level of distortion in the reconstructed video. Our aim is to upload the news video in the highest possible quality with less distortion. In NEWSMAN, R–D models enable the optimization of the received video quality under different network conditions. To avoid unnecessary complexity of deriving R–D models of individual news videos, NEWSMAN categorizes news videos into a few classes using *temporal perceptual information* (TI) and *spatial perceptual information* (SI), which are the measures of temporal changes and spatial details, respectively [73, 215]. Due to limited storage space, less powerful CPU, and constrained battery capacity, earlier works [78, 104] suggested to perform transcoding at resourceful clouds (middleboxes in our case) instead of at mobile devices. In our work we follow this model, *i.e.*, we transcode videos at middleboxes based on the bitrate determined from our proposed algorithm. In recent years, advances in deep neural network (DNN) technologies have yielded immense success in computer vision, natural language processing (NLP), and

[2]http://ireport.cnn.com/

speech processing. In future, we would like to exploit multimodal information leveraging DNN technologies to optimally determine the transcoding bitrate.

References

1. Apple Denies Steve Jobs Heart Attack Report: "It Is Not True". http://www.businessinsider.com/2008/10/apple-s-steve-jobs-rushed-to-er-after-heart-attack-says-cnn-citizen-journalist/. October 2008. Online: Last Accessed Sept 2015.
2. SVM^{hmm}: Sequence Tagging with Structural Support Vector Machines. https://www.cs.cornell.edu/people/tj/svm light/svm hmm.html. August 2008. Online: Last Accessed May 2016.
3. NPTEL. 2009, December. http://www.nptel.ac.in. Online; Accessed Apr 2015.
4. iReport at 5: Nearly 900,000 contributors worldwide. http://www.niemanlab.org/2011/08/ireport-at-5-nearly-900000-contributors-worldwide/. August 2011. Online: Last Accessed Sept 2015.
5. Meet the million: 999,999 iReporters + you! http://www.ireport.cnn.com/blogs/ireport-blog/2012/01/23/meet-the-million-999999-ireporters-you. January 2012. Online: Last Accessed Sept 2015.
6. 5 Surprising Stats about User-generated Content. 2014, April. http://www.smartblogs.com/social-media/2014/04/11/6.-surprising-stats-about-user-generated-content/. Online: Last Accessed Sept 2015.
7. The Citizen Journalist: How Ordinary People are Taking Control of the News. 2015, June. http://www.digitaltrends.com/features/the-citizen-journalist-how-ordinary-people-are-taking-control-of-the-news/. Online: Last Accessed Sept 2015.
8. Wikipedia API. 2015, April. http://tinyurl.com/WikiAPI-AI. API: Last Accessed Apr 2015.
9. Apache Lucene. 2016, June. https://lucene.apache.org/core/. Java API: Last Accessed June 2016.
10. By the Numbers: 14 Interesting Flickr Stats. 2016, May. http://www.expandedramblings.com/index.php/flickr-stats/. Online: Last Accessed May 2016.
11. By the Numbers: 180+ Interesting Instagram Statistics (June 2016). 2016, June. http://www.expandedramblings.com/index.php/important-instagram-stats/. Online: Last Accessed July 2016.
12. Coursera. 2016, May. https://www.coursera.org/. Online: Last Accessed May 2016.
13. FourSquare API. 2016, June. https://developer.foursquare.com/. Last Accessed June 2016.
14. Google Cloud Vision API. 2016, December. https://cloud.google.com/vision/. Online: Last Accessed Dec 2016.
15. Google Forms. 2016, May. https://docs.google.com/forms/. Online: Last Accessed May 2016.
16. MIT Open Course Ware. 2016, May. http://www.ocw.mit.edu/. Online: Last Accessed May 2016.
17. Porter Stemmer. 2016, May. https://tartarus.org/martin/PorterStemmer/. Online: Last Accessed May 2016.
18. SenticNet. 2016, May. http://www.sentic.net/computing/. Online: Last Accessed May 2016.
19. Sentics. 2016, May. https://en.wiktionary.org/wiki/sentics. Online: Last Accessed May 2016.
20. VideoLectures.Net. 2016, May. http://www.videolectures.net/. Online: Last Accessed May, 2016.
21. YouTube Statistics. 2016, July. http://www.youtube.com/yt/press/statistics.html. Online: Last Accessed July, 2016.

22. Abba, H.A., S.N.M. Shah, N.B. Zakaria, and A.J. Pal. 2012. Deadline based performance evalu-ation of job scheduling algorithms. In *Proceedings of the IEEE International Conference on Cyber-Enabled Distributed Computing and Knowledge Discovery*, 106–110.
23. Achanta, R.S., W.-Q. Yan, and M.S. Kankanhalli. (2006). Modeling Intent for Home Video Repurposing. In *Proceedings of the IEEE MultiMedia*, (1):46–55.
24. Adcock, J., M. Cooper, A. Girgensohn, and L. Wilcox. 2005. Interactive Video Search using Multilevel Indexing. In *Proceedings of the Springer Image and Video Retrieval*, 205–214.
25. Agarwal, B., S. Poria, N. Mittal, A. Gelbukh, and A. Hussain. 2015. Concept-level Sentiment Analysis with Dependency-based Semantic Parsing: A Novel Approach. In *Proceedings of the Springer Cognitive Computation*, 1–13.
26. Aizawa, K., D. Tancharoen, S. Kawasaki, and T. Yamasaki. 2004. Efficient Retrieval of Life Log based on Context and Content. In *Proceedings of the ACM Workshop on Continuous Archival and Retrieval of Personal Experiences*, 22–31.
27. Altun, Y., I. Tsochantaridis, and T. Hofmann. 2003. Hidden Markov Support Vector Machines. In *Proceedings of the International Conference on Machine Learning*, 3–10.
28. Anderson, A., K. Ranghunathan, and A. Vogel. 2008. Tagez: Flickr Tag Recommendation. In *Proceedings of the Association for the Advancement of Artificial Intelligence*.
29. Atrey, P.K., A. El Saddik, and M.S. Kankanhalli. 2011. Effective Multimedia Surveillance using a Human-centric Approach. *Proceedings of the Springer Multimedia Tools and Applications* 51(2): 697–721.
30. Barnard, K., P. Duygulu, D. Forsyth, N. De Freitas, D.M. Blei, and M.I. Jordan. 2003. Matching Words and Pictures. *Proceedings of the Journal of Machine Learning Research* 3: 1107–1135.
31. Basu, S., Y. Yu, V.K. Singh, and R. Zimmermann. 2016. Videopedia: Lecture Video Recommendation for Educational Blogs Using Topic Modeling. In *Proceedings of the Springer International Conference on Multimedia Modeling*, 238–250.
32. Basu, S., Y. Yu, and R. Zimmermann. 2016. Fuzzy Clustering of Lecture Videos based on Topic Modeling. In *Proceedings of the IEEE International Workshop on Content-Based Multimedia Indexing*, 1–6.
33. Basu, S., R. Zimmermann, K.L. OHalloran, S. Tan, and K. Marissa. 2015. Performance Evaluation of Students Using Multimodal Learning Systems. In *Proceedings of the Springer International Conference on Multimedia Modeling*, 135–147.
34. Beeferman, D., A. Berger, and J. Lafferty. 1999. Statistical Models for Text Segmentation. *Proceedings of the Springer Machine Learning* 34(1–3): 177–210.
35. Bernd, J., D. Borth, C. Carrano, J. Choi, B. Elizalde, G. Friedland, L. Gottlieb, K. Ni, R. Pearce, D. Poland, et al. 2015. Kickstarting the Commons: The YFCC100M and the YLI Corpora. In *Proceedings of the ACM Workshop on Community-Organized Multimodal Mining: Opportunities for Novel Solutions*, 1–6.
36. Bhatt, C.A., and M.S. Kankanhalli. 2011. Multimedia Data Mining: State of the Art and Challenges. *Proceedings of the Multimedia Tools and Applications* 51(1): 35–76.
37. Bhatt, C.A., A. Popescu-Belis, M. Habibi, S. Ingram, S. Masneri, F. McInnes, N. Pappas, and O. Schreer. 2013. Multi-factor Segmentation for Topic Visualization and Recommendation: the MUST-VIS System. In *Proceedings of the ACM International Conference on Multimedia*, 365–368.
38. Bhattacharjee, S., W.C. Cheng, C.-F. Chou, L. Golubchik, and S. Khuller. 2000. BISTRO: A Frame-work for Building Scalable Wide-area Upload Applications. *Proceedings of the ACM SIGMETRICS Performance Evaluation Review* 28(2): 29–35.
39. Cambria, E., J. Fu, F. Bisio, and S. Poria. 2015. AffectiveSpace 2: Enabling Affective Intuition for Concept-level Sentiment Analysis. In *Proceedings of the AAAI Conference on Artificial Intelligence*, 508–514.
40. Cambria, E., A. Livingstone, and A. Hussain. 2012. The Hourglass of Emotions. In *Proceedings of the Springer Cognitive Behavioural Systems*, 144–157.

41. Cambria, E., D. Olsher, and D. Rajagopal. 2014. SenticNet 3: A Common and Common-sense Knowledge Base for Cognition-driven Sentiment Analysis. In *Proceedings of the AAAI Conference on Artificial Intelligence*, 1515–1521.
42. Cambria, E., S. Poria, R. Bajpai, and B. Schuller. 2016. SenticNet 4: A Semantic Resource for Sentiment Analysis based on Conceptual Primitives. In *Proceedings of the International Conference on Computational Linguistics (COLING)*, 2666–2677.
43. Cambria, E., S. Poria, F. Bisio, R. Bajpai, and I. Chaturvedi. 2015. The CLSA Model: A Novel Framework for Concept-Level Sentiment Analysis. In *Proceedings of the Springer Computational Linguistics and Intelligent Text Processing*, 3–22.
44. Cambria, E., S. Poria, A. Gelbukh, and K. Kwok. 2014. Sentic API: A Common-sense based API for Concept-level Sentiment Analysis. *CEUR Workshop Proceedings* 144: 19–24.
45. Cao, J., Z. Huang, and Y. Yang. 2015. Spatial-aware Multimodal Location Estimation for Social Images. In *Proceedings of the ACM Conference on Multimedia Conference*, 119–128.
46. Chakraborty, I., H. Cheng, and O. Javed. 2014. Entity Centric Feature Pooling for Complex Event Detection. In *Proceedings of the Workshop on HuEvent at the ACM International Conference on Multimedia*, 1–5.
47. Che, X., H. Yang, and C. Meinel. 2013. Lecture Video Segmentation by Automatically Analyzing the Synchronized Slides. In *Proceedings of the ACM International Conference on Multimedia*, 345–348.
48. Chen, B., J. Wang, Q. Huang, and T. Mei. 2012. Personalized Video Recommendation through Tripartite Graph Propagation. In *Proceedings of the ACM International Conference on Multimedia*, 1133–1136.
49. Chen, S., L. Tong, and T. He. 2011. Optimal Deadline Scheduling with Commitment. In *Proceedings of the IEEE Annual Allerton Conference on Communication, Control, and Computing*, 111–118.
50. Chen, W.-B., C. Zhang, and S. Gao. 2012. Segmentation Tree based Multiple Object Image Retrieval. In *Proceedings of the IEEE International Symposium on Multimedia*, 214–221.
51. Chen, Y., and W.J. Heng. 2003. Automatic Synchronization of Speech Transcript and Slides in Presentation. *Proceedings of the IEEE International Symposium on Circuits and Systems* 2: 568–571.
52. Cohen, J. 1960. A Coefficient of Agreement for Nominal Scales. *Proceedings of the Durham Educational and Psychological Measurement* 20(1): 37–46.
53. Cristani, M., A. Pesarin, C. Drioli, V. Murino, A. Rodà,M. Grapulin, and N. Sebe. 2010. Toward an Automatically Generated Soundtrack from Low-level Cross-modal Correlations for Automotive Scenarios. In *Proceedings of the ACM International Conference on Multimedia*, 551–560.
54. Dang-Nguyen, D.-T., L. Piras, G. Giacinto, G. Boato, and F.G. De Natale. 2015. A Hybrid Approach for Retrieving Diverse Social Images of Landmarks. In *Proceedings of the IEEE International Conference on Multimedia and Expo (ICME)*, 1–6.
55. Fabro, M. Del, A. Sobe, and L. Böszörmenyi. 2012. Summarization of Real-life Events based on Community-contributed Content. In *Proceedings of the International Conferences on Advances in Multimedia*, 119–126.
56. Du, L., W.L. Buntine, and M. Johnson. 2013. Topic Segmentation with a Structured Topic Model. In *Proceedings of the North American Chapter of the Association for Computational Linguistics: Human Language Technologies*, 190–200.
57. Fan, Q., K. Barnard, A. Amir, A. Efrat, and M. Lin. 2006. Matching Slides to Presentation Videos using SIFT and Scene Background Matching. In *Proceedings of the ACM International Conference on Multimedia*, 239–248.
58. Filatova, E. and V. Hatzivassiloglou. 2004. Event-based Extractive Summarization. In *Proceedings of the ACL Workshop on Summarization*, 104–111.
59. Firan, C.S., M. Georgescu, W. Nejdl, and R. Paiu. 2010. Bringing Order to Your Photos: Event-driven Classification of Flickr Images based on Social Knowledge. In *Proceedings of the ACM International Conference on Information and Knowledge Management*, 189–198.

60. Gao, S., C. Zhang, and W.-B. Chen. 2012. An Improvement of Color Image Segmentation through Projective Clustering. In *Proceedings of the IEEE International Conference on Information Reuse and Integration*, 152–158.
61. Garg, N. and I. Weber. 2008. Personalized, Interactive Tag Recommendation for Flickr. In *Proceedings of the ACM Conference on Recommender Systems*, 67–74.
62. Ghias, A., J. Logan, D. Chamberlin, and B.C. Smith. 1995. Query by Humming: Musical Information Retrieval in an Audio Database. In *Proceedings of the ACM International Conference on Multimedia*, 231–236.
63. Golder, S.A., and B.A. Huberman. 2006. Usage Patterns of Collaborative Tagging Systems. *Proceedings of the Journal of Information Science* 32(2): 198–208.
64. Gozali, J.P., M.-Y. Kan, and H. Sundaram. 2012. Hidden Markov Model for Event Photo Stream Segmentation. In *Proceedings of the IEEE International Conference on Multimedia and Expo Workshops*, 25–30.
65. Guo, Y., L. Zhang, Y. Hu, X. He, and J. Gao. 2016. Ms-celeb-1m: Challenge of recognizing one million celebrities in the real world. *Proceedings of the Society for Imaging Science and Technology Electronic Imaging* 2016(11): 1–6.
66. Hanjalic, A., and L.-Q. Xu. 2005. Affective Video Content Representation and Modeling. *Proceedings of the IEEE Transactions on Multimedia* 7(1): 143–154.
67. Haubold, A. and J.R. Kender. 2005. Augmented Segmentation and Visualization for Presentation Videos. In *Proceedings of the ACM International Conference on Multimedia*, 51–60.
68. Healey, J.A., and R.W. Picard. 2005. Detecting Stress during Real-world Driving Tasks using Physiological Sensors. *Proceedings of the IEEE Transactions on Intelligent Transportation Systems* 6(2): 156–166.
69. Hefeeda, M., and C.-H. Hsu. 2010. On Burst Transmission Scheduling in Mobile TV Broadcast Networks. *Proceedings of the IEEE/ACM Transactions on Networking* 18(2): 610–623.
70. Hevner, K. 1936. Experimental Studies of the Elements of Expression in Music. *Proceedings of the American Journal of Psychology* 48: 246–268.
71. Hochbaum, D.S. 1996. Approximating Covering and Packing Problems: Set Cover, Vertex Cover, Independent Set, and related Problems. In *Proceedings of the PWS Approximation algorithms for NP-hard problems*, 94–143.
72. Hong, R., J. Tang, H.-K. Tan, S. Yan, C. Ngo, and T.-S. Chua. 2009. Event Driven Summarization for Web Videos. In *Proceedings of the ACM SIGMM Workshop on Social Media*, 43–48.
73. P. ITU-T Recommendation. 1999. Subjective Video Quality Assessment Methods for Multimedia Applications.
74. Jiang, L., A.G. Hauptmann, and G. Xiang. 2012. Leveraging High-level and Low-level Features for Multimedia Event Detection. In *Proceedings of the ACM International Conference on Multimedia*, 449–458.
75. Joachims, T., T. Finley, and C.-N. Yu. 2009. Cutting-plane Training of Structural SVMs. *Proceedings of the Machine Learning Journal* 77(1): 27–59.
76. Johnson, J., L. Ballan, and L. Fei-Fei. 2015. Love Thy Neighbors: Image Annotation by Exploiting Image Metadata. In *Proceedings of the IEEE International Conference on Computer Vision*, 4624–4632.
77. Johnson, J., A. Karpathy, and L. Fei-Fei. 2015. Densecap: Fully Convolutional Localization Networks for Dense Captioning. *In Proceedings of the arXiv preprint arXiv:1511.07571*.
78. Jokhio, F., A. Ashraf, S. Lafond, I. Porres, and J. Lilius. 2013. Prediction-based dynamic resource allocation for video transcoding in cloud computing. In *Proceedings of the IEEE International Conference on Parallel, Distributed and Network-Based Processing*, 254–261.
79. Kaminskas, M., I. Fernández-Tobías, F. Ricci, and I. Cantador. 2014. Knowledge-based Identification of Music Suited for Places of Interest. *Proceedings of the Springer Information Technology & Tourism* 14(1): 73–95.

80. Kaminskas, M. and F. Ricci. 2011. Location-adapted Music Recommendation using Tags. In *Proceedings of the Springer User Modeling, Adaption and Personalization*, 183–194.
81. Kan, M.-Y. 2001. Combining Visual Layout and Lexical Cohesion Features for Text Segmentation. *In Proceedings of the Citeseer*.
82. Kan, M.-Y. 2003. *Automatic Text Summarization as Applied to Information Retrieval*. PhD thesis, Columbia University.
83. Kan, M.-Y., J.L. Klavans, and K.R. McKeown.1998. Linear Segmentation and Segment Significance. *In Proceedings of the arXiv preprint cs/9809020*.
84. Kan, M.-Y., K.R. McKeown, and J.L. Klavans. 2001. Applying Natural Language Generation to Indicative Summarization. *Proceedings of the ACL European Workshop on Natural Language Generation* 8: 1–9.
85. Kang, H.B. 2003. Affective Content Detection using HMMs. In *Proceedings of the ACM International Conference on Multimedia*, 259–262.
86. Kang, Y.-L., J.-H. Lim, M.S. Kankanhalli, C.-S. Xu, and Q. Tian. 2004. Goal Detection in Soccer Video using Audio/Visual Keywords. *Proceedings of the IEEE International Conference on Image Processing* 3: 1629–1632.
87. Kang, Y.-L., J.-H. Lim, Q. Tian, and M.S. Kankanhalli. 2003. Soccer Video Event Detection with Visual Keywords. In *Proceedings of the Joint Conference of International Conference on Information, Communications and Signal Processing, and Pacific Rim Conference on Multimedia*, 3:1796–1800.
88. Kankanhalli, M.S., and T.-S. Chua. 2000. Video Modeling using Strata-based Annotation. *Proceedings of the IEEE MultiMedia* 7(1): 68–74.
89. Kennedy, L., M. Naaman, S. Ahern, R. Nair, and T. Rattenbury. 2007. How Flickr Helps us Make Sense of the World: Context and Content in Community-contributed Media Collections. In *Proceedings of the ACM International Conference on Multimedia*, 631–640.
90. Kennedy, L.S., S.-F. Chang, and I.V. Kozintsev. 2006. To Search or to Label?: Predicting the Performance of Search-based Automatic Image Classifiers. In *Proceedings of the ACM International Workshop on Multimedia Information Retrieval*, 249–258.
91. Kim, Y.E., E.M. Schmidt, R. Migneco, B.G. Morton, P. Richardson, J. Scott, J.A. Speck, and D. Turnbull. 2010. Music Emotion Recognition: A State of the Art Review. In *Proceedings of the International Society for Music Information Retrieval*, 255–266.
92. Klavans, J.L., K.R. McKeown, M.-Y. Kan, and S. Lee. 1998. Resources for Evaluation of Summarization Techniques. *In Proceedings of the arXiv preprint cs/9810014*.
93. Ko, Y. 2012. A Study of Term Weighting Schemes using Class Information for Text Classification. In *Proceedings of the ACM Special Interest Group on Information Retrieval*, 1029–1030.
94. Kort, B., R. Reilly, and R.W. Picard. 2001. An Affective Model of Interplay between Emotions and Learning: Reengineering Educational Pedagogy-Building a Learning Companion. *Proceedings of the IEEE International Conference on Advanced Learning Technologies* 1: 43–47.
95. Kucuktunc, O., U. Gudukbay, and O. Ulusoy. 2010. Fuzzy Color Histogram-based Video Segmentation. *Proceedings of the Computer Vision and Image Understanding* 114(1): 125–134.
96. Kuo, F.-F., M.-F. Chiang, M.-K. Shan, and S.-Y. Lee. 2005. Emotion-based Music Recommendation by Association Discovery from Film Music. In *Proceedings of the ACM International Conference on Multimedia*, 507–510.
97. Lacy, S., T. Atwater, X. Qin, and A. Powers. 1988. Cost and Competition in the Adoption of Satellite News Gathering Technology. *Proceedings of the Taylor & Francis Journal of Media Economics* 1(1): 51–59.
98. Lambert, P., W. De Neve, P. De Neve, I. Moerman, P. Demeester, and R. Van de Walle. 2006. Rate-distortion performance of H. 264/AVC compared to state-of-the-art video codecs. *Proceedings of the IEEE Transactions on Circuits and Systems for Video Technology* 16(1): 134–140.

99. Laurier, C., M. Sordo, J. Serrà, and P. Herrera. 2009. Music Mood Representations from Social Tags. In *Proceedings of the International Society for Music Information Retrieval*, 381–386.
100. Li, C.T. and M.K. Shan. 2007. Emotion-based Impressionism Slideshow with Automatic Music Accompaniment. In *Proceedings of the ACM International Conference on Multimedia*, 839–842.
101. Li, J., and J.Z. Wang. 2008. Real-time Computerized Annotation of Pictures. *Proceedings of the IEEE Transactions on Pattern Analysis and Machine Intelligence* 30(6): 985–1002.
102. Li, X., C.G. Snoek, and M. Worring. 2009. Learning Social Tag Relevance by Neighbor Voting. *Proceedings of the IEEE Transactions on Multimedia* 11(7): 1310–1322.
103. Li, X., T. Uricchio, L. Ballan, M. Bertini, C.G. Snoek, and A.D. Bimbo. 2016. Socializing the Semantic Gap: A Comparative Survey on Image Tag Assignment, Refinement, and Retrieval. *Proceedings of the ACM Computing Surveys (CSUR)* 49(1): 14.
104. Li, Z., Y. Huang, G. Liu, F. Wang, Z.-L. Zhang, and Y. Dai. 2012. Cloud Transcoder: Bridging the Format and Resolution Gap between Internet Videos and Mobile Devices. In *Proceedings of the ACM International Workshop on Network and Operating System Support for Digital Audio and Video*, 33–38.
105. Liang, C., Y. Guo, and Y. Liu. 2008. Is Random Scheduling Sufficient in P2P Video Streaming? In *Proceedings of the IEEE International Conference on Distributed Computing Systems*, 53–60. IEEE.
106. Lim, J.-H., Q. Tian, and P. Mulhem. 2003. Home Photo Content Modeling for Personalized Event-based Retrieval. *Proceedings of the IEEE MultiMedia* 4: 28–37.
107. Lin, M., M. Chau, J. Cao, and J.F. Nunamaker Jr. 2005. Automated Video Segmentation for Lecture Videos: A Linguistics-based Approach. *Proceedings of the IGI Global International Journal of Technology and Human Interaction* 1(2): 27–45.
108. Liu, C.L., and J.W. Layland. 1973. Scheduling Algorithms for Multiprogramming in a Hard-real-time Environment. *Proceedings of the ACM Journal of the ACM* 20(1): 46–61.
109. Liu, D., X.-S. Hua, L. Yang, M. Wang, and H.-J. Zhang. 2009. Tag Ranking. In *Proceedings of the ACM World Wide Web Conference*, 351–360.
110. Liu, T., C. Rosenberg, and H.A. Rowley. 2007. Clustering Billions of Images with Large Scale Nearest Neighbor Search. In *Proceedings of the IEEE Winter Conference on Applications of Computer Vision*, 28–28.
111. Liu, X. and B. Huet. 2013. Event Representation and Visualization from Social Media. In *Proceedings of the Springer Pacific-Rim Conference on Multimedia*, 740–749.
112. Liu, Y., D. Zhang, G. Lu, and W.-Y. Ma. 2007. A Survey of Content-based Image Retrieval with High-level Semantics. *Proceedings of the Elsevier Pattern Recognition* 40(1): 262–282.
113. Livingston, S., and D.A.V. BELLE. 2005. The Effects of Satellite Technology on Newsgathering from Remote Locations. *Proceedings of the Taylor & Francis Political Communication* 22(1): 45–62.
114. Long, R., H. Wang, Y. Chen, O. Jin, and Y. Yu. 2011. Towards Effective Event Detection, Tracking and Summarization on Microblog Data. In *Proceedings of the Springer Web-Age Information Management*, 652–663.
115. L. Lu, H. You, and H. Zhang. 2001. A New Approach to Query by Humming in Music Retrieval. In *Proceedings of the IEEE International Conference on Multimedia and Expo*, 22–25.
116. Lu, Y., H. To, A. Alfarrarjeh, S.H. Kim, Y. Yin, R. Zimmermann, and C. Shahabi. 2016. GeoUGV: User-generated Mobile Video Dataset with Fine Granularity Spatial Metadata. In *Proceedings of the ACM International Conference on Multimedia Systems*, 43.
117. Mao, J., W. Xu, Y. Yang, J. Wang, Z. Huang, and A. Yuille. 2014. Deep Captioning with Multimodal Recurrent Neural Networks (M-RNN). *In Proceedings of the arXiv preprint arXiv:1412.6632*.

118. Matusiak, K.K. 2006. Towards User-centered Indexing in Digital Image Collections. *Proceedings of the OCLC Systems & Services: International Digital Library Perspectives* 22(4): 283–298.
119. McDuff, D., R. El Kaliouby, E. Kodra, and R. Picard. 2013. Measuring Voter's Candidate Preference Based on Affective Responses to Election Debates. In *Proceedings of the IEEE Humaine Association Conference on Affective Computing and Intelligent Interaction*, 369–374.
120. McKeown, K.R., J.L. Klavans, and M.-Y. Kan. Method and System for Topical Segmentation, Segment Significance and Segment Function, 29 2002. US Patent 6,473,730.
121. Mezaris, V., A. Scherp, R. Jain, M. Kankanhalli, H. Zhou, J. Zhang, L. Wang, and Z. Zhang. 2011. Modeling and Rrepresenting Events in Multimedia. In *Proceedings of the ACM International Conference on Multimedia*, 613–614.
122. Mezaris, V., A. Scherp, R. Jain, and M.S. Kankanhalli. 2014. Real-life Events in Multimedia: Detection, Representation, Retrieval, and Applications. *Proceedings of the Springer Multimedia Tools and Applications* 70(1): 1–6.
123. Miller, G., and C. Fellbaum. 1998. *Wordnet: An Electronic Lexical Database*. Cambridge, MA: MIT Press.
124. Miller, G.A. 1995. WordNet: A Lexical Database for English. *Proceedings of the Communications of the ACM* 38(11): 39–41.
125. Moxley, E., J. Kleban, J. Xu, and B. Manjunath. 2009. Not All Tags are Created Equal: Learning Flickr Tag Semantics for Global Annotation. In *Proceedings of the IEEE International Conference on Multimedia and Expo*, 1452–1455.
126. Mulhem, P., M.S. Kankanhalli, J. Yi, and H. Hassan. 2003. Pivot Vector Space Approach for Audio-Video Mixing. *Proceedings of the IEEE MultiMedia* 2: 28–40.
127. Naaman, M. 2012. Social Multimedia: Highlighting Opportunities for Search and Mining of Multimedia Data in Social Media Applications. *Proceedings of the Springer Multimedia Tools and Applications* 56(1): 9–34.
128. Natarajan, P., P.K. Atrey, and M. Kankanhalli. 2015. Multi-Camera Coordination and Control in Surveillance Systems: A Survey. *Proceedings of the ACM Transactions on Multimedia Computing, Communications, and Applications* 11(4): 57.
129. Nayak, M.G. 2004. *Music Synthesis for Home Videos*. PhD thesis.
130. Neo, S.-Y., J. Zhao, M.-Y. Kan, and T.-S. Chua. 2006. Video Retrieval using High Level Features: Exploiting Query Matching and Confidence-based Weighting. In *Proceedings of the Springer International Conference on Image and Video Retrieval*, 143–152.
131. Ngo, C.-W., F. Wang, and T.-C. Pong. 2003. Structuring Lecture Videos for Distance Learning Applications. In *Proceedings of the IEEE International Symposium on Multimedia Software Engineering*, 215–222.
132. Nguyen, V.-A., J. Boyd-Graber, and P. Resnik. 2012. SITS: A Hierarchical Nonparametric Model using Speaker Identity for Topic Segmentation in Multiparty Conversations. In *Proceedings of the Annual Meeting of the Association for Computational Linguistics*, 78–87.
133. Nwana, A.O. and T. Chen. 2016. Who Ordered This?: Exploiting Implicit User Tag Order Preferences for Personalized Image Tagging. *In Proceedings of the arXiv preprint arXiv:1601.06439*.
134. Papagiannopoulou, C. and V. Mezaris. 2014. Concept-based Image Clustering and Summarization of Event-related Image Collections. In *Proceedings of the Workshop on HuEvent at the ACM International Conference on Multimedia*, 23–28.
135. Park, M.H., J.H. Hong, and S.B. Cho. 2007. Location-based Recommendation System using Bayesian User's Preference Model in Mobile Devices. In *Proceedings of the Springer Ubiquitous Intelligence and Computing*, 1130–1139.
136. Petkos, G., S. Papadopoulos, V. Mezaris, R. Troncy, P. Cimiano, T. Reuter, and Y. Kompatsiaris. 2014. Social Event Detection at MediaEval: a Three-Year Retrospect of Tasks and Results. In *Proceedings of the Workshop on Social Events in Web Multimedia at ACM International Conference on Multimedia Retrieval*.

137. Pevzner, L., and M.A. Hearst. 2002. A Critique and Improvement of an Evaluation Metric for Text Segmentation. *Proceedings of the Computational Linguistics* 28(1): 19–36.
138. Picard, R.W., and J. Klein. 2002. Computers that Recognise and Respond to User Emotion: Theoretical and Practical Implications. *Proceedings of the Interacting with Computers* 14(2): 141–169.
139. Picard, R.W., E. Vyzas, and J. Healey. 2001. Toward machine emotional intelligence: Analysis of affective physiological state. *Proceedings of the IEEE Transactions on Pattern Analysis and Machine Intelligence* 23(10): 1175–1191.
140. Poisson, S.D. and C.H. Schnuse. 1841. Recherches Sur La Pprobabilité Des Jugements En Mmatieré Criminelle Et En Matieré Civile. Meyer.
141. Poria, S., E. Cambria, R. Bajpai, and A. Hussain. 2017. A Review of Affective Computing: From Unimodal Analysis to Multimodal Fusion. *Proceedings of the Elsevier Information Fusion* 37: 98–125.
142. Poria, S., E. Cambria, and A. Gelbukh. 2016. Aspect Extraction for Opinion Mining with a Deep Convolutional Neural Network. *Proceedings of the Elsevier Knowledge-Based Systems* 108: 42–49.
143. Poria, S., E. Cambria, A. Gelbukh, F. Bisio, and A. Hussain. 2015. Sentiment Data Flow Analysis by Means of Dynamic Linguistic Patterns. *Proceedings of the IEEE Computational Intelligence Magazine* 10(4): 26–36.
144. Poria, S., E. Cambria, and A.F. Gelbukh. 2015. Deep Convolutional Neural Network Textual Features and Multiple Kernel Learning for Utterance-level Multimodal Sentiment Analysis. In *Proceedings of the EMNLP*, 2539–2544.
145. Poria, S., E. Cambria, D. Hazarika, N. Mazumder, A. Zadeh, and L.-P. Morency. 2017. Context-Dependent Sentiment Analysis in User-Generated Videos. *In Proceedings of the Association for Computational Linguistics.*
146. Poria, S., E. Cambria, D. Hazarika, and P. Vij. 2016. A Deeper Look into Sarcastic Tweets using Deep Convolutional Neural Networks. In *Proceedings of the International Conference on Computational Linguistics (COLING).*
147. Poria, S., E. Cambria, N. Howard, G.-B. Huang, and A. Hussain. 2016. Fusing Audio Visual and Textual Clues for Sentiment Analysis from Multimodal Content. *Proceedings of the Elsevier Neurocomputing* 174: 50–59.
148. Poria, S., E. Cambria, N. Howard, and A. Hussain. 2015. Enhanced SenticNet with Affective Labels for Concept-based Opinion Mining: Extended Abstract. In *Proceedings of the International Joint Conference on Artificial Intelligence.*
149. Poria, S., E. Cambria, A. Hussain, and G.-B. Huang. 2015. Towards an Intelligent Framework for Multimodal Affective Data Analysis. *Proceedings of the Elsevier Neural Networks* 63: 104–116.
150. Poria, S., E. Cambria, L.-W. Ku, C. Gui, and A. Gelbukh. 2014. A Rule-based Approach to Aspect Extraction from Product Reviews. In *Proceedings of the Second Workshop on Natural Language Processing for Social Media (SocialNLP)*, 28–37.
151. Poria, S., I. Chaturvedi, E. Cambria, and F. Bisio. 2016. Sentic LDA: Improving on LDA with Semantic Similarity for Aspect-based Sentiment Analysis. In *Proceedings of the IEEE International Joint Conference on Neural Networks (IJCNN)*, 4465–4473.
152. Poria, S., I. Chaturvedi, E. Cambria, and A. Hussain. 2016. Convolutional MKL based Multimodal Emotion Recognition and Sentiment Analysis. In *Proceedings of the IEEE International Conference on Data Mining (ICDM)*, 439–448.
153. Poria, S., A. Gelbukh, B. Agarwal, E. Cambria, and N. Howard. 2014. Sentic Demo: A Hybrid Concept-level Aspect-based Sentiment Analysis Toolkit. *In Proceedings of the ESWC.*
154. Poria, S., A. Gelbukh, E. Cambria, D. Das, and S. Bandyopadhyay. 2012. Enriching SenticNet Polarity Scores Through Semi-Supervised Fuzzy Clustering. In *Proceedings of the IEEE International Conference on Data Mining Workshops (ICDMW)*, 709–716.

155. Poria, S., A. Gelbukh, E. Cambria, A. Hussain, and G.-B. Huang. 2014. EmoSenticSpace: A Novel Framework for Affective Common-sense Reasoning. *Proceedings of the Elsevier Knowledge-Based Systems* 69: 108–123.
156. Poria, S., A. Gelbukh, E. Cambria, P. Yang, A. Hussain, and T. Durrani. 2012. Merging SenticNet and WordNet-Affect Emotion Lists for Sentiment Analysis. *Proceedings of the IEEE International Conference on Signal Processing (ICSP)* 2: 1251–1255.
157. Poria, S., A. Gelbukh, A. Hussain, S. Bandyopadhyay, and N. Howard. 2013. Music Genre Classification: A Semi-Supervised Approach. In *Proceedings of the Springer Mexican Conference on Pattern Recognition*, 254–263.
158. Poria, S., N. Ofek, A. Gelbukh, A. Hussain, and L. Rokach. 2014. Dependency Tree-based Rules for Concept-level Aspect-based Sentiment Analysis. In *Proceedings of the Springer Semantic Web Evaluation Challenge*, 41–47.
159. Poria, S., H. Peng, A. Hussain, N. Howard, and E. Cambria. 2017. Ensemble Application of Convolutional Neural Networks and Multiple Kernel Learning for Multimodal Sentiment Analysis. *In Proceedings of the Elsevier Neurocomputing*.
160. Pye, D., N.J. Hollinghurst, T.J. Mills, and K.R. Wood. 1998. Audio-visual Segmentation for Content-based Retrieval. In *Proceedings of the International Conference on Spoken Language Processing*.
161. Qiao, Z., P. Zhang, C. Zhou, Y. Cao, L. Guo, and Y. Zhang. 2014. Event Recommendation in Event-based Social Networks.
162. Raad, E.J. and R. Chbeir. 2014. Foto2Events: From Photos to Event Discovery and Linking in Online Social Networks. In *Proceedings of the IEEE Big Data and Cloud Computing*, 508–515, .
163. Radsch, C.C. 2013. The Revolutions will be Blogged: Cyberactivism and the 4th Estate in Egypt. Doctoral Disseration. American University.
164. Rae, A., B. Sigurbjörnssön, and R. van Zwol. 2010. Improving Tag Recommendation using Social Networks. In *Proceedings of the Adaptivity, Personalization and Fusion of Heterogeneous Information*, 92–99.
165. Rahmani, H., B. Piccart, D. Fierens, and H. Blockeel. 2010. Three Complementary Approaches to Context Aware Movie Recommendation. In *Proceedings of the ACM Workshop on Context-Aware Movie Recommendation*, 57–60.
166. Rattenbury, T., N. Good, and M. Naaman. 2007. Towards Automatic Extraction of Event and Place Semantics from Flickr Tags. In *Proceedings of the ACM Special Interest Group on Information Retrieval*.
167. Rawat, Y. and M. S. Kankanhalli. 2016. ConTagNet: Exploiting User Context for Image Tag Recommendation. In *Proceedings of the ACM International Conference on Multimedia*, 1102–1106.
168. Repp, S., A. Groß, and C. Meinel. 2008. Browsing within Lecture Videos based on the Chain Index of Speech Transcription. *Proceedings of the IEEE Transactions on Learning Technologies* 1(3): 145–156.
169. Repp, S. and C. Meinel. 2006. Semantic Indexing for Recorded Educational Lecture Videos. In *Proceedings of the IEEE International Conference on Pervasive Computing and Communications Workshops*, 5.
170. Repp, S., J. Waitelonis, H. Sack, and C. Meinel. 2007. Segmentation and Annotation of Audiovisual Recordings based on Automated Speech Recognition. In *Proceedings of the Springer Intelligent Data Engineering and Automated Learning*, 620–629.
171. Russell, J.A. 1980. A Circumplex Model of Affect. *Proceedings of the Journal of Personality and Social Psychology* 39: 1161–1178.
172. Sahidullah, M., and G. Saha. 2012. Design, Analysis and Experimental Evaluation of Block based Transformation in MFCC Computation for Speaker Recognition. *Proceedings of the Speech Communication* 54: 543–565.

173. J. Salamon, J. Serra, and E. Gomez´. Tonal Representations for Music Retrieval: From Version Identification to Query-by-Humming. *In* Proceedings of the Springer International Journal of Multimedia Information Retrieval, 2(1):45–58, 2013.
174. Schedl, M. and D. Schnitzer. 2014. Location-Aware Music Artist Recommendation. In *Proceedings of the Springer MultiMedia Modeling*, 205–213.
175. M. Schedl and F. Zhou. 2016. Fusing Web and Audio Predictors to Localize the Origin of Music Pieces for Geospatial Retrieval. In *Proceedings of the Springer European Conference on Information Retrieval*, 322–334.
176. Scherp, A., and V. Mezaris. 2014. Survey on Modeling and Indexing Events in Multimedia. *Proceedings of the Springer Multimedia Tools and Applications* 70(1): 7–23.
177. Scherp, A., V. Mezaris, B. Ionescu, and F. De Natale. 2014. HuEvent '14: Workshop on Human-Centered Event Understanding from Multimedia. In *Proceedings of the ACM International Conference on Multimedia*, 1253–1254, .
178. Schmitz, P. 2006. Inducing Ontology from Flickr Tags. In *Proceedings of the Collaborative Web Tagging Workshop at ACM World Wide Web Conference*, volume 50.
179. Schuller, B., C. Hage, D. Schuller, and G. Rigoll. 2010. Mister DJ, Cheer Me Up!: Musical and Textual Features for Automatic Mood Classification. *Proceedings of the Journal of New Music Research* 39(1): 13–34.
180. Shah, R.R., M. Hefeeda, R. Zimmermann, K. Harras, C.-H. Hsu, and Y. Yu. 2016. NEWSMAN: Uploading Videos over Adaptive Middleboxes to News Servers In Weak Network Infrastructures. In *Proceedings of the Springer International Conference on Multimedia Modeling*, 100–113.
181. Shah, R.R., A. Samanta, D. Gupta, Y. Yu, S. Tang, and R. Zimmermann. 2016. PROMPT: Personalized User Tag Recommendation for Social Media Photos Leveraging Multimodal Information. In *Proceedings of the ACM International Conference on Multimedia*, 486–492.
182. Shah, R.R., A.D. Shaikh, Y. Yu, W. Geng, R. Zimmermann, and G. Wu. 2015. EventBuilder: Real-time Multimedia Event Summarization by Visualizing Social Media. In *Proceedings of the ACM International Conference on Multimedia*, 185–188.
183. Shah, R.R., Y. Yu, A.D. Shaikh, S. Tang, and R. Zimmermann. 2014. ATLAS: Automatic Temporal Segmentation and Annotation of Lecture Videos Based on Modelling Transition Time. In *Proceedings of the ACM International Conference on Multimedia*, 209–212.
184. Shah, R.R., Y. Yu, A.D. Shaikh, and R. Zimmermann. 2015. TRACE: A Linguistic-based Approach for Automatic Lecture Video Segmentation Leveraging Wikipedia Texts. In *Proceedings of the IEEE International Symposium on Multimedia*, 217–220.
185. Shah, R.R., Y. Yu, S. Tang, S. Satoh, A. Verma, and R. Zimmermann. 2016. Concept-Level Multimodal Ranking of Flickr Photo Tags via Recall Based Weighting. In *Proceedings of the MMCommon's Workshop at ACM International Conference on Multimedia*, 19–26.
186. Shah, R.R., Y. Yu, A. Verma, S. Tang, A.D. Shaikh, and R. Zimmermann. 2016. Leveraging Multimodal Information for Event Summarization and Concept-level Sentiment Analysis. In *Proceedings of the Elsevier Knowledge-Based Systems*, 102–109.
187. Shah, R.R., Y. Yu, and R. Zimmermann. 2014. ADVISOR: Personalized Video Soundtrack Recommendation by Late Fusion with Heuristic Rankings. In *Proceedings of the ACM International Conference on Multimedia*, 607–616.
188. Shah, R.R., Y. Yu, and R. Zimmermann. 2014. User Preference-Aware Music Video Generation Based on Modeling Scene Moods. In *Proceedings of the ACM International Conference on Multimedia Systems*, 156–159.
189. Shaikh, A.D., M. Jain, M. Rawat, R.R. Shah, and M. Kumar. 2013. Improving Accuracy of SMS Based FAQ Retrieval System. In *Proceedings of the Springer Multilingual Information Access in South Asian Languages*, 142–156.
190. Shaikh, A.D., R.R. Shah, and R. Shaikh. 2013. SMS based FAQ Retrieval for Hindi, English and Malayalam. In *Proceedings of the ACM Forum on Information Retrieval Evaluation*, 9.

191. Shamma, D.A., R. Shaw, P.L. Shafton, and Y. Liu. 2007. Watch What I Watch: Using Community Activity to Understand Content. In *Proceedings of the ACM International Workshop on Multimedia Information Retrieval*, 275–284.
192. Shaw, B., J. Shea, S. Sinha, and A. Hogue. 2013. Learning to Rank for Spatiotemporal Search. In *Proceedings of the ACM International Conference on Web Search and Data Mining*, 717–726.
193. Sigurbjörnsson, B. and R. Van Zwol. 2008. Flickr Tag Recommendation based on Collective Knowledge. In *Proceedings of the ACM World Wide Web Conference*, 327–336.
194. Snoek, C.G., M. Worring, and A.W.Smeulders. 2005. Early versus Late Fusion in Semantic Video Analysis. In *Proceedings of the ACM International Conference on Multimedia*, 399–402.
195. Snoek, C.G., M. Worring, J.C. Van Gemert, J.-M. Geusebroek, and A.W. Smeulders. 2006. The Challenge Problem for Automated Detection of 101 Semantic Concepts in Multimedia. In *Proceedings of the ACM International Conference on Multimedia*, 421–430.
196. Soleymani, M., J.J.M. Kierkels, G. Chanel, and T. Pun. 2009. A Bayesian Framework for Video Affective Representation. In *Proceedings of the IEEE International Conference on Affective Computing and Intelligent Interaction and Workshops*, 1–7.
197. Stober, S., and A. . Nürnberger. 2013. Adaptive Music Retrieval–a State of the Art. *Proceedings of the Springer Multimedia Tools and Applications* 65(3): 467–494.
198. Stoyanov, V., N. Gilbert, C. Cardie, and E. Riloff. 2009. Conundrums in Noun Phrase Coreference Resolution: Making Sense of the State-of-the-art. In *Proceedings of the ACL International Joint Conference on Natural Language Processing of the Asian Federation of Natural Language Processing*, 656–664.
199. Stupar, A. and S. Michel. 2011. Picasso: Automated Soundtrack Suggestion for Multi-modal Data. In *Proceedings of the ACM Conference on Information and Knowledge Management*, 2589–2592.
200. Thayer, R.E. 1989. *The Biopsychology of Mood and Arousal*. New York: Oxford University Press.
201. Thomee, B., B. Elizalde, D.A. Shamma, K. Ni, G. Friedland, D. Poland, D. Borth, and L.-J. Li. 2016. YFCC100M: The New Data in Multimedia Research. *Proceedings of the Communications of the ACM* 59(2): 64–73.
202. Tirumala, A., F. Qin, J. Dugan, J. Ferguson, and K. Gibbs. 2005. Iperf: The TCP/UDP Bandwidth Measurement Tool. http://dast.nlanr.net/Projects/Iperf/
203. Torralba, A., R. Fergus, and W.T. Freeman. 2008. 80 Million Tiny Images: A Large Data set for Nonparametric Object and Scene Recognition. *Proceedings of the IEEE Transactions on Pattern Analysis and Machine Intelligence* 30(11): 1958–1970.
204. Toutanova, K., D. Klein, C.D. Manning, and Y. Singer. 2003. Feature-Rich Part-of-Speech Tagging with a Cyclic Dependency Network. In *Proceedings of the North American Chapter of the Association for Computational Linguistics on Human Language Technology*, 173–180.
205. Toutanova, K. and C.D. Manning. 2000. Enriching the Knowledge Sources used in a Maximum Entropy Part-of-Speech Tagger. In *Proceedings of the Annual Meeting of the Association for Computational Linguistics*, 63–70.
206. Utiyama, M. and H. Isahara. 2001. A Statistical Model for Domain-Independent Text Segmentation. In *Proceedings of the Annual Meeting on Association for Computational Linguistics*, 499–506.
207. Vishal, K., C. Jawahar, and V. Chari. 2015. Accurate Localization by Fusing Images and GPS Signals. In *Proceedings of the IEEE Computer Vision and Pattern Recognition Workshops*, 17–24.
208. Wang, C., F. Jing, L. Zhang, and H.-J. Zhang. 2008. Scalable Search-based Image Annotation. *Proceedings of the Springer Multimedia Systems* 14(4): 205–220.
209. Wang, H.L., and L.F. Cheong. 2006. Affective Understanding in Film. *Proceedings of the IEEE Transactions on Circuits and Systems for Video Technology* 16(6): 689–704.

210. Wa
Re
Im
211. Wa
Su
Da
212. W
W
213. W
M
214. W
Pr
215. W
Qu
Sy
Op
216. W
Sy
or
217. W
E
11
218. W
W
219. X
Pr
84
220. X
V
Ir
221. X
B
th
P
222. X
fo
o
223. Y
L
I
224. Y
I
s
225. Y
E
8
226. Y
z
3
227. Y
Y

Chapter 3
Event Understanding

Abstract The rapid growth in the amount of photos/videos online necessitates for social media companies to automatically extract knowledge structures (concepts) from photos and videos to provide diverse multimedia-related services such as event detection and summarization. However, real-world photos and videos aggregated in social media sharing platforms (*e.g.*, Flickr and Instagram) are complex and noisy, and extracting semantics and sentics from the multimedia content alone is a very difficult task because suitable concepts may be exhibited in different representations. Since semantics and sentics knowledge structures are very useful in multimedia search, retrieval, and recommendation, it is desirable to analyze UGCs from multiple modalities for a better understanding. To this end, we first present the EventBuilder system that deals with semantics understanding and automatically generates a multimedia summary for a given event in real-time by leveraging different social media such as Wikipedia and Flickr. Subsequently, we present the EventSensor system that aims to address sentics understanding and produces a multimedia summary for a given mood. It extracts concepts and mood tags from visual content and textual metadata of UGCs, and exploits them in supporting several significant multimedia-related services such as a musical multimedia summary. Moreover, EventSensor supports sentics-based event summarization by leveraging EventBuilder as its semantics engine component. Experimental results confirm that both EventBuilder and EventSensor outperform their baselines and efficiently summarize knowledge structures on the YFCC100M dataset.

Keywords Event analysis • Event detection • Event summarization • Flickr photos • Multimodal analysis • EventBuilder • EventSensor

3.1 Introduction

The amount of UGC (*e.g.*, UGIs and UGVs) has increased dramatically in recent years due to the ubiquitous availability of smartphones, digital cameras, and affordable network infrastructures. An interesting recent trend is that social media companies such as Flickr and YouTube, instead of producing content by themselves, create opportunities for a user to generate multimedia content. Thus, capturing multimedia content anytime and anywhere, and then instantly sharing them

R. Shah, R. Zimmermann, *Multimodal Analysis of User-Generated Multimedia Content*, Socio-Affective Computing 6, DOI 10.1007/978-3-319-61807-4_3

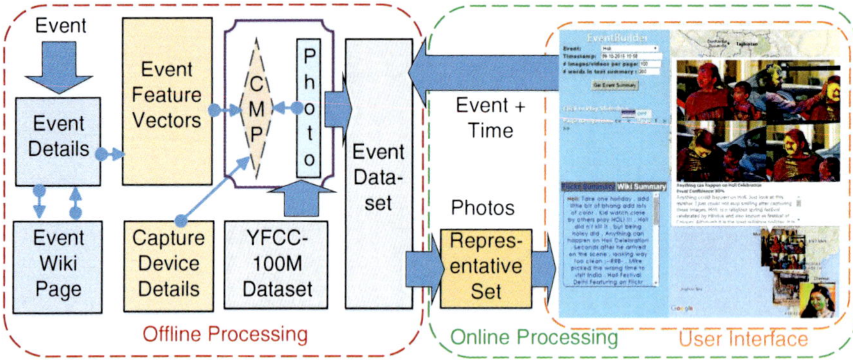

Fig. 3.1 System framework of the EventBuilder system

on social media platforms, have become a very popular activity. Since UGC belong to different interesting events (*e.g.*, festivals, games, and protests), they are now an intrinsic part of humans' daily life. For instance, on a very popular photo sharing website Instagram,[1] over 1 billion photos have been uploaded so far. Moreover, the website has more than 400 million monthly active users [11]. However, it is difficult to automatically extract knowledge structures from multimedia content due to the following reasons: (i) the difficulty in capturing the semantics and sentics of UGC, (ii) the existence of noise in textual metadata, and (iii) challenges in handling big datasets. First, aiming at the understanding of semantics and summarizing knowledge structures of multimedia content, we present the EventBuilder[2] system [182, 186]. It enables users to automatically obtain multimedia summaries for a given event from a large multimedia collection in real-time (see Fig. 3.1). This system leverages information from social media platforms such as Wikipedia and Flickr to provide useful summaries of the event. We perform extensive experiments of EventBuilder on a collection of 100 million photos and videos (the YFCC100M dataset) from Flickr and compare results with a baseline. In the baseline system, we select UGIs that contain the input event name in their metadata (*e.g.*, descriptions, titles, and tags). Experimental results confirm that the proposed algorithm in EventBuilder efficiently summarizes knowledge structures and outperforms the baseline. Next, we describe how our approach solves above mentioned problems. All notations used in this chapter are listed in Table 3.1.

Advancements in technologies have enabled mobile devices to collect a significant amount of contextual information (*e.g.*, spatial, temporal, and other sensory data) in conjunction with UGC. We argue that the multimodal analysis of UGC is very helpful in semantics and sentics understanding because often multimedia content is unstructured and difficult to access in a meaningful way from only one

[1] https://instagram.com/

[2] https://eventbuilder.geovid.org

Table 3.1 Notations used in the event understanding chapter

symbols	meanings
D_{YFCC}	The YFCC100M dataset, a collection of 100 million *UGIs* and *UGVs* from Flickr
i	A UGI in D_{YFCC}
e	An event in D_{YFCC}
N_e	Feature vector w.r.t. (with respect to) event name for the event e
T_e	Feature vector w.r.t. temporal information for e
S_e	Feature vector w.r.t. spatial information for e
K_e	Feature vector w.r.t. keywords of the event e
D	The list of 1080 camera models from Flickr which are ranked based on their sensor sizes
N_i	Feature vector w.r.t. event name for the UGI i
T_i	Feature vector w.r.t. temporal information for i
S_i	Feature vector w.r.t. spatial information for i
K_i	Feature vector w.r.t. keywords of the UGI i
$\xi\ (N_i, N_e)$	The similarity score of i with e w.r.t. event name
$\lambda\ (T_i, T_e)$	The similarity score of i with e w.r.t. temporal information
$\gamma(S_i, S_e)$	The similarity score of i with e w.r.t. spatial information
$\mu(K_i, K_e)$	The similarity score of i with e w.r.t. keywords
$\rho(D_i, D)$	The similarity score of i with e w.r.t. camera model information
D_{Event}	Event dataset for the event e
$w_{k\,k=1}^{\ m}$	Weights for m different modalities
$u(i, e)$	The relevance score of the UGI i for the event e
δ	Threshold for event detection
$\mathscr{R}$	Representative set for the event e consisting of UGIs with relevance scores $u(i, e)$ above δ in D_{Event}
$\mathscr{T}$	The set of all sentences which are extracted from descriptions of UGIs in $\mathscr{R}$ and the content of Wikipedia articles on the event e
$\mathscr{S}$	Text summary for the event e
$\lvert\mathscr{S}\rvert$	The current word count of the text summary $\mathscr{S}$
$\mathscr{L}$	The word limit for the text summary $\mathscr{S}$
c_k	A concept in the textual metadata of the UGI i
$\mathscr{K}$	The set of all *concepts* (*e.g.*, c_k) of e
y_k	The weight of a concept c_k in the textual metadata of the UGI i
$\mathscr{Y}$	The set of weights (*e.g.*, y_k) for $\mathscr{K}$
s	A sentence in $\mathscr{T}$
$\varphi(s)$	The score of the sentence s, which is the sum of weights of all concepts it covers
$\tau(i)$	The upload timestamp of the UGI i
$\omega(s)$	A binary indicator variable which indicates if the sentence s is selected in the summary or not
$\psi(i)$	A binary indicator variable which specifies if the UGI i has descriptions or not
$\phi(c, s)$	Equals to 1 if the sentence s consists concept c (*i.e.*, c is a sub-string in s), or otherwise 0
$\beta(s, i)$	Equals to 1 if the sentence s is part of the description (a list of sentences) of i, or otherwise 0

(continued)

Table 3.1 (continued)

symbols	meanings
$\bar{C}$	The set of 30,000 common and common-sense concepts, called, SenticNet-3
d	Textual description of a UGI i
C_P	A list of SenticNet-3 concepts for the UGI i
C_V	Visual concepts of the UGI i from D_{YFCC}
C_T	Textual concepts of the UGI i from its textual description
C_{FUSED}	Concepts derived from the fusion of C_V and C_T for the UGI i
E	EmoSenticNet, which maps 13,000 concepts of SenticNet-3 to mood tags such as anger, disgust, joy, sad, surprise, and fear
$\bar{E}$	EmoSenticSpace, which provides a 100D feature vector space for each concept in $\bar{C}$
M_P	A six dimensional sentics vector for the UGI i

modality [187, 188, 242, 243]. Since multimodal information augments knowledge bases by inferring semantics from the unstructured multimedia content and contextual information [180, 183, 184], we leverage it in the EventBuilder system. EventBuilder has the following three novel characteristics: (i) leveraging Wikipedia as event background knowledge to obtain additional contextual information about an input event, (ii) visualizing an interesting event in real-time with a diverse set of social media activities, and (iii) producing text summaries for the event from the description of UGIs and Wikipedia texts by solving an optimization problem.

Next, aiming at understanding sentiments and producing a sentics-based multimedia summary from a multimedia collection, we introduce the EventSensor[3] system. EventSensor leverages EventBuilder as its semantics engine to produce sentics-based event summarization. It leverages multimodal information for sentiment analysis from UGC. Specifically, it extracts concepts from the visual content and textual metadata of a UGI and exploits them to determine the sentics details of the UGI. A concept is a knowledge structure which provides important cues about sentiments. For instance, the concept *"grow movement"* indicates anger and struggle. Concepts are tags that describe multimedia content, hence, events. Thus, it would be beneficial to consider tag ranking and recommendation techniques [181, 185] in an efficient event understanding. We computed textual concepts (*e.g.*, grow movement, fight as a community, and high court injunction) from the textual metadata such as description and tags by the semantic parser provided by Poria et al. [143] (see Sect. 1.4.2 for details). Visual concepts are tags derived from the visual content of UGIs by using a convolutional network that indicates the presence of concepts such as people, buildings, food, and cars. The YFCC100M dataset provides the visual concepts of all UGIs as metadata. On this basis, we propose a novel algorithm to fuse concepts derived from the textual and visual content of a UGI. Subsequently, we exploit existing knowledge bases such as

[3]http://pilatus.d1.comp.nus.edu.sg:8080/EventSensor/

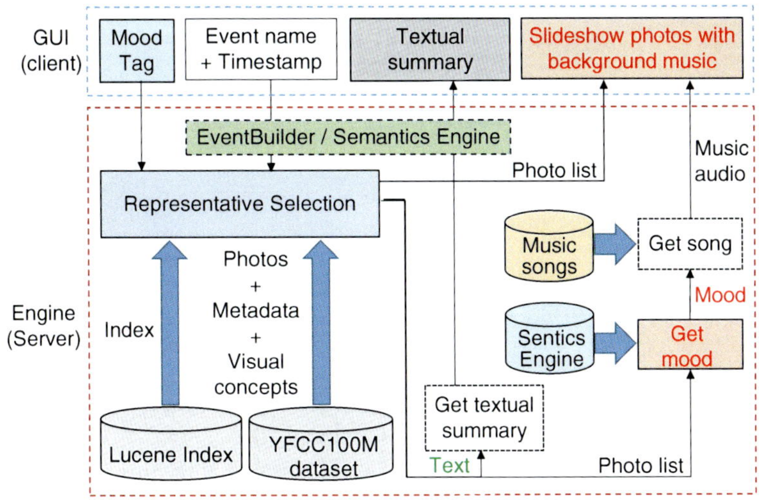

Fig. 3.2 System framework of the EventSensor system

SenticNet-3, EmoSenticNet, EmoSenticSpace, and WordNet to determine to the sentics details of the UGI (see Sects. 1.4.3 and 1.4.4). Such knowledge bases help us to build a sentics engine which is helpful in providing sentics-based services. For instance, the sentics engine is used for a mood-related soundtrack generation in our system (see Fig. 3.2). A mood-based sound that matches with emotions in UGIs is a very important aspect and contributes greatly to the appeal of a UGV (*i.e.*, a slideshow of UGIs) when it is being viewed. Thus, the UGV with a matching soundtrack has more appeal for viewing and sharing on social media websites than the normal slideshow of UGIs without an interesting sound. Therefore, people often create such music soundtrack by adding matching soundtracks to the slideshow of UGIs and share them on social media. However, adding soundtracks to UGIs is not easy due to the following reasons. Firstly, traditionally it is tedious, time-consuming, and not scalable for a user to add custom soundtracks to UGIs from a large multimedia collection. Secondly, it is difficult to extract moods of UGIs automatically. Finally, an important aspect is that a good soundtrack should match and enhance the overall moods of UGIs and meet the user's preferences. Thus, this necessitates to constructing a summarization system that enhances the experience of a multimedia summary by adding matching soundtracks to the UGIs. To this end, we present the EventSensor system that produces a musical multimedia summary (a slideshow of UGIs with matching soundtracks) based on the determined moods of UGIs.

Figure 3.3 shows the framework of our sentics engine. It provides better sentics analysis of multimedia content leveraging multimodal information. Our system exploits knowledge structures from the following knowledge bases: (i) SenticNet-3, (ii) EmoSenticNet, (iii) EmoSenticSpace, and (iv) WordNet to determine moods from UGIs. SenticNet-3 is a publicly available resource for concept-level sentiment analysis [41]. It consists of 30,000 common and commonsense concepts $\bar{C}$ such as *food*, *party*, and *accomplish goal*. Moreover, it associates each concept with five

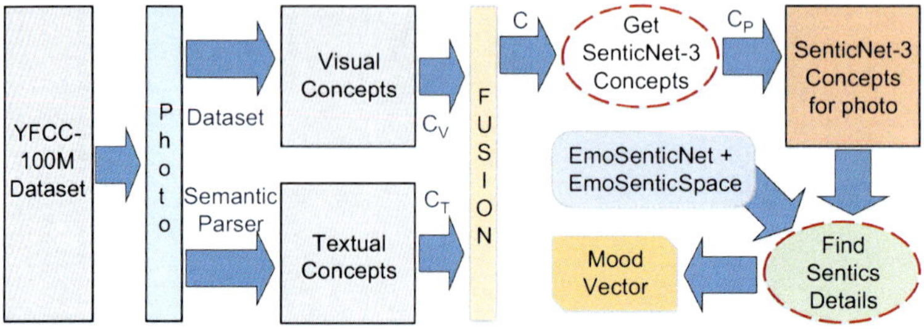

Fig. 3.3 System framework of sentics engine

other semantically related concepts in $\bar{C}$ and sentics information such as pleasantness, attention, sensitivity, aptitude, and polarity, as described in the Hourglass of Emotions model [40]. EmoSenticNet maps 13,000 concepts of $\bar{C}$ to affective labels such as anger, disgust, joy, sadness, surprise, and fear. For an effective sentics understanding, it is essential to know affective labels for the rest of the SenticNet-3 concepts [155]. Thus we leverage EmoSenticSpace which provides a 100-dimensional vector space for each concept in $\bar{C}$, to determine the missing sentics information based on neighbor voting (see Fig. 3.6). We determine 100 neighbors for each concept using the cosine similarity metric. Moreover, we use the WordNet library to leverage semantics details of different words. Additionally, we perform semantics analysis on the textual metadata of UGIs to extract knowledge structures (textual concepts) for a better understanding using the semantic parser provided by Poria et al. [143] (see Fig. 3.3). This parser deconstructs natural language text into concepts based on dependency relations between clauses (see Sect. 1.4.2 for details). To leverage such knowledge structures in determining the sentics details of a UGI, we propose an algorithm to establish an association between the determined (visual and textual) concepts and $\bar{C}$ (see Algorithm 3.2). The proposed sentics engine is very useful for providing sentics-based multimedia-related services.

We organize this chapter as follows. In Sect. 3.2, we describe the EventBuilder and EventSensor systems. Next, we present the evaluation results in Sect. 3.3. Finally, we conclude the chapter with a summary in Sect. 3.4.

3.2 System Overview

3.2.1 EventBuilder

Figure 3.1 shows the system framework of the EventBuilder system which produces a multimedia summary of an event in two steps: (i) it performs offline event detections and (ii) it then produces online event summaries. In particular, first, it performs event-wise classification and indexing of all UGIs on social media datasets such as the YFCC100M dataset (D_{YFCC}). Let ξ, λ, γ, μ, and ρ be similarity functions for a given UGI i and an event e corresponding to event name N, temporal information T, spatial information S, keywords K, and camera model D, respectively. The relevance score $u(i, e)$ of the UGI i for the event e is computed by a linear combination of similarity scores as follows in Eq. 3.1:

$$u(i,e) = w_1\,\xi(N_i,N_e) + w_2\,\lambda(T_i,T_e) + w_3\,\gamma(S_i,S_e) + w_4\,\mu(K_i,K_e) \\ + w_5\,\rho(D_i,D) \tag{3.1}$$

where $w_{k_{k=1}}^{5}$ are weights for different similarity scores such that $\Sigma_{k=1}^{5} w_k = 1$. Since an event is a thing that takes place at some locations, in some particular times, and involves some activities, we consider spatial, temporal, and other event related keywords in the calculation of event score from UGIs. Moreover, we allocate only 5% of the total score for the camera model based on the heuristic that a good camera captures a better quality UGI for the attractive visualization of the event. We set the weights as follows: $w_1 = 0{:}40$, $w_2 = 0{:}20$, $w_3 = 0{:}15$, $w_4 = 0{:}20$, and $w_5 = 0{:}05$, after initial experiments on a development set with 1000 UGIs for event detection. We construct the event dataset D_{Event} by indexing only those UGIs of D_{YFCC} whose scores $u(i, e)$ are above the threshold δ. All similarity scores, thresholds, and other scores are normalized to values [0, 1]. For instance, Fig. 3.4 shows the summary of an event, named Holi, which is a very famous festival in India. Our EventBuilder system visualizes the event summary on a Google Map in real-time since a huge number of UGIs are geo-tagged on social media websites such as Flickr. The top left portion of the EventBuilder interface enables users to set the input parameters for an event, and the right portion visualizes the multimedia summary of UGIs belong to Holi event (*i.e.*, the representative UGIs from D_{Holi}). Similar to the Google Map characteristics, EventBuilder enables a user to zoom in or zoom out to see an overview of the event geographically. Finally, the left portion shows the text summaries of the event.

Figure 3.5 shows the computation of the relevance score of the UGI[4] i for the event e, named *Olympics*, in the YFCC100M dataset. Similarity functions compute similarity scores of the UGI for the event by comparing feature vectors of the event with feature vectors of the UGI. For instance, the UGI in Fig. 3.5, has an event name (*e.g.*, *Olympics* in this case), is captured during London 2012 Olympics in the city of London (see Table 3.2), and consists of several keywords that match with keywords of the *Olympics* event (see Table 3.3). We compute the score of camera model similarity by matching the camera model which captured the UGI with the list of 1080 camera models from Flickr that are ranked based on their sensor sizes. However, we later realize that camera model D does not play any role in event detection. Thus, the similarity score $\rho(D_i; D)$ should not be included in the formula

[4]Flickr URL: https://www.flickr.com/photos/16687586@N00/8172648222/ and download URL: http://farm9.staticflickr.com/8349/8172648222`4afa16993b.jpg

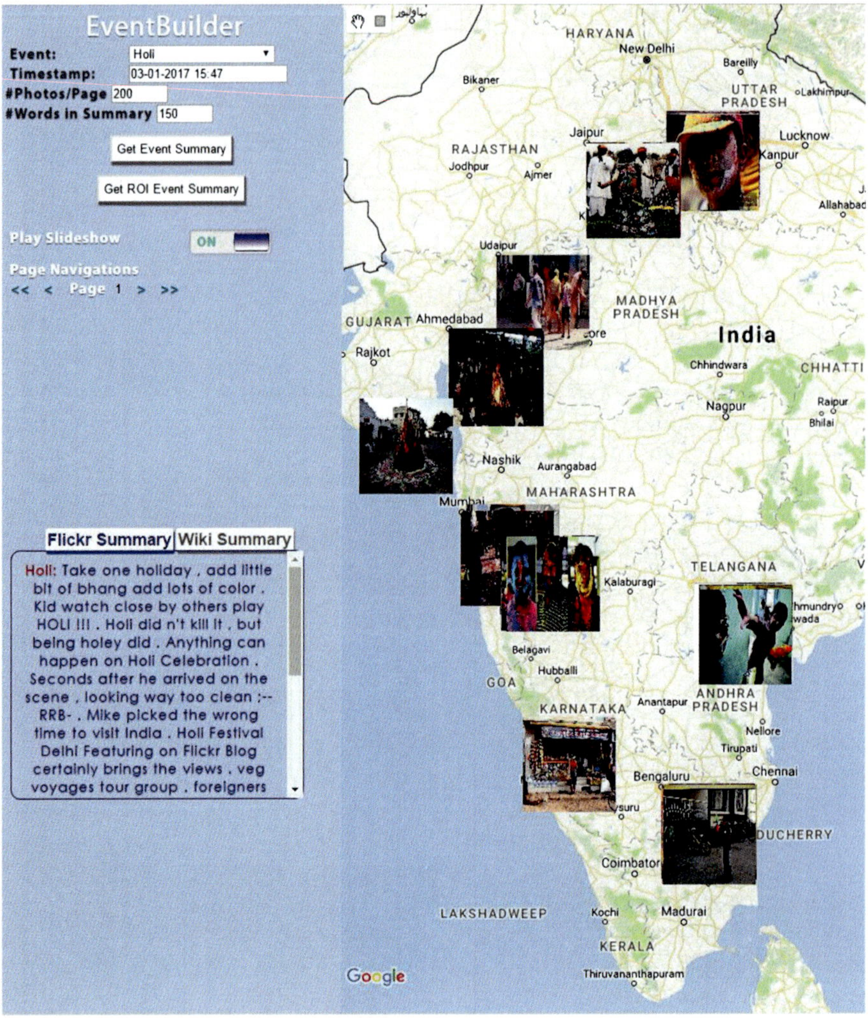

Fig. 3.4 The multimedia summary produced by EventBuilder for the Holi event. The *top left* portion shows the input parameters to the EventBuilder system and *bottom left* shows the text summaries for the event. *Right* portion shows the multimedia summary of UGIs on the Google map for the event Holi

of relevance score $u(i; e)$ of a UGI i for an event e. In our future work on the event detection, we plan to use updated formula without the similarity score $\rho(D_i; D)$.

We select a representative set of UGIs $\mathscr{R}$ that have event scores above a predefined threshold δ for visualization on the Google Maps. Since EventBuilder detects events from UGC offline rather than at search time, it is time-efficient and scales well to large repositories. Moreover, it can work well for new events by constructing feature vectors for those events and leveraging information from

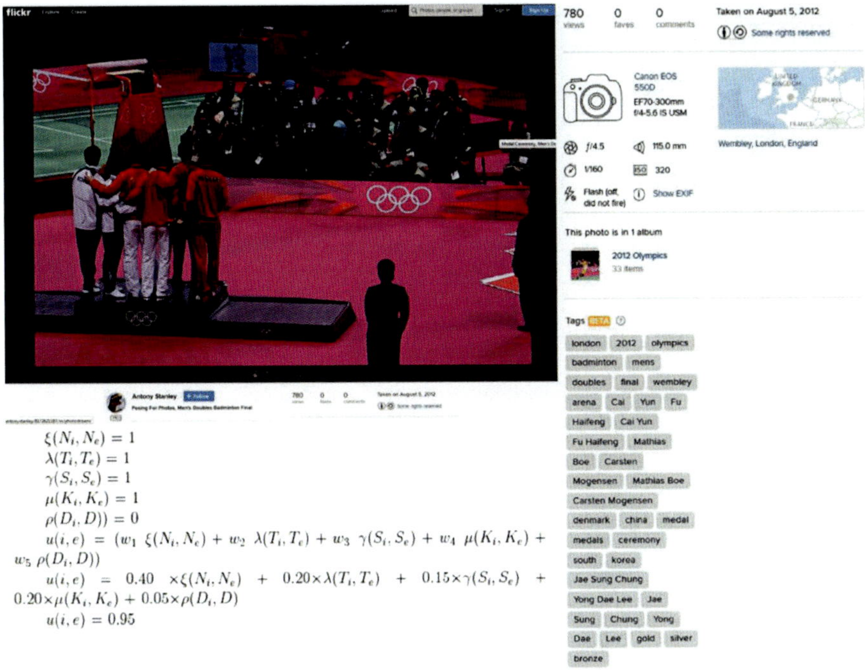

Fig. 3.5 Event score calculation for a photo in the YFCC100M dataset

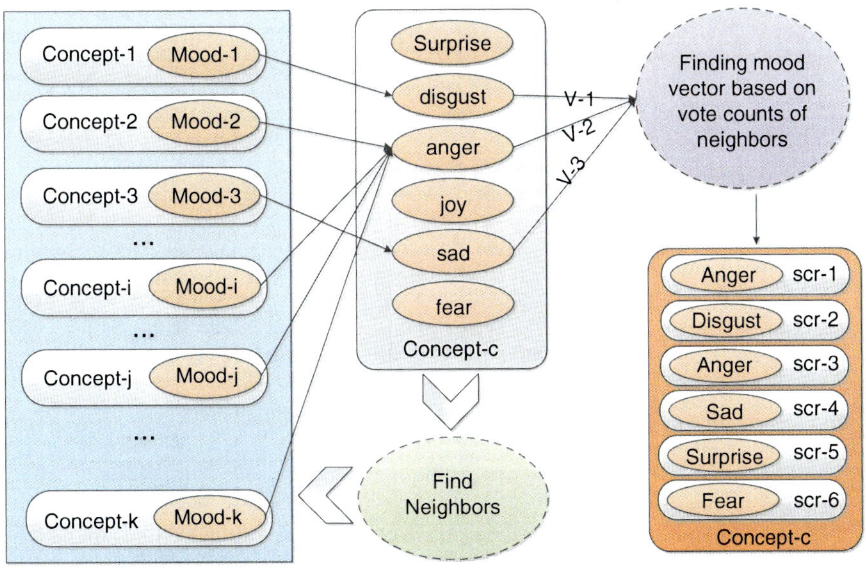

Fig. 3.6 System framework of determining mood vectors for SenticNet-3 concepts

Table 3.2 Metadata used to compute spatial and temporal feature vectors for the summer Olympics event

Venue	City	City GPS	Duration
England	London	(51.538611, −0.016389)	27-07-12 to 12-08-12
China	Beijing	(39.991667, 116.390556)	08-08-08 to 24-08-08
Greece	Athens	(38.036111, 23.787500)	13-08-04 to 29-08-04
Australia	Sydney	(−33.847222, 151.063333)	15-09-00 to 01-10-00
United States	Atlanta	(33.748995, −84.387982)	19-07-96 to 04-08-96
Spain	Barcelona	(41.385063, 2.173403)	25-07-92 to 09-08-92
South Korea	Seoul	(37.566535, 126.977969)	17-09-88 to 02-10-88
United States	Los Angeles	(34.052234, −118.243684)	28-07-84 to 12-08-84
Russia	Moscow	(55.755826, 37.617300)	19-07-80 to 03-08-80
Canada	Montreal	(45.501688, −73.567256)	17-07-76 to 01-08-76
Germany	London	(48.135125, 11.581980)	26-08-72 to 11-09-72
...	...	...	...

Table 3.3 Metadata used to compute event name and keywords feature vectors for the Olympics event

Event name	Event keywords
Olympics, Winter Olympics, Summer Olympics	Archery, Athletics, Badminton, Basketball, Beach Volleyball, Boxing, Canoe Slalom, Canoe Sprint, Cycling BMX, Cycling Mountain Bike, Cycling Road, Cycling Track, Diving, Equestrian, Equestrian, Equestrian, Dressage, Eventing, Jumping, Fencing, Football, Golf, Gymnastics Artistic, Gymnastics Rhythmic, Handball, Hockey, Judo, Modern Pentathlon, Rowing, Rugby, Sailing, Shooting, Swimming, Synchronized Swimming, Table Tennis, Taekwondo, Tennis, Trampoline, Triathlon, Volleyball, Water Polo, Weightlifting, Wrestling Freestyle, Wrestling Greco-Roman, International Olympic Committee, paralympic, teenage athletes, professional athletes, corporate sponsorship, international sports federations, Olympic rituals, Olympic program, Olympic flag, athletic festivals, competition, Olympic stadium, Olympic champion, Olympic beginnings, Olympic beginnings, Olympic association, Olympic year, international federation, exclusive sponsorship rights, Marathon medals, artistic gymnastics, Olympic sports, gold medal, silver medal, bronze medal, canadian sprinter, anti-doping, drug tests, Alpine Skiing, Biathlon, Bobsleigh, Cross Country Skiing, Curling, Figure skating, Freestyle Skiing, Ice Hockey, Luge, Nordic Combined, Short Track Speed Skating, Skeleton, Ski Jumping, Snowboard, Speed skating, *etc*

Wikipedia which we use as background knowledge for more contextual information about an event to be summarized. An event summarization system can schedule event detection algorithms for newly uploaded UGC at regular time intervals to update event datasets. After event detections, EventBuilder generates text

Table 3.4 Matrix model for event summarization

Sentences/Concepts	c_1	c_2	c_3	c_4	...	$c_{\vert\mathscr{K}\vert}$
s_1	1	1	0	1	...	1
s_2	0	1	1	0	...	0
...	...	...	...	...	...	...
$s_{\vert\mathscr{T}\vert}$	1	0	0	1	...	1

summaries from $\mathscr{R}$ for the event e. It produces two text summaries during online processing for the given event and timestamp: (i) a Flickr summary from the description of multimedia content and (ii) a Wikipedia summary from Wikipedia articles on the event. The Flickr summary is considered as a baseline for the textual summary of the event and compared with the Wikipedia summary during evaluation. We consider multimedia items that are uploaded before the given timestamp to produce event summaries in real-time. EventBuilder leverages multimodal information such as the metadata (*e.g.*, spatial- and temporal information, user tags, and descriptions) of UGIs and Wikipedia texts of the event using a feature-pivot approach in the event detection and summarization. EventBuilder produces text summaries of an event in the following two steps. First, the identification of important *concepts* (*i.e.*, important event-related information using [58]) which should be described in the event summary. Second, the composition of the text summary which covers the maximal number of important *concepts* by selecting the minimal number of sentences from available texts, within the desired summary length. Hence, text summaries of the event can be formulated as a maximum coverage problem (see Table 3.4). However, this problem can be reduced to the well-known set cover (NP-hard) problem. Thus, we solve this problem in polynomial time by approximation because NP-hard problems can only be solved by approximation algorithms [71].

Let $\mathscr{T}$ be the set of all sentences which are extracted from descriptions of UGIs in the representative set $\mathscr{R}$ and contents of Wikipedia articles on the event e. A text summary $\mathscr{S}$ for the event e is produced from the sentences in $\mathscr{T}$. Let $\mid \mathscr{S} \mid$ and $\overline{\mathscr{L}}$ be the current word count and the word limit for the summary $\mathscr{S}$, respectively. Let $\mathscr{K}$ and $\mathscr{Y}$ be the set of all *concepts* (c_k) of the event e and the set of corresponding weights (y_k), respectively. Let $\varphi(s)$ be the score of a sentence s, which is the sum of weights of all concepts it covers. Let $\tau(i)$ be the upload time of i. Let $\omega(s)$ be a binary indicator variable which indicates if s is selected in summary or not. Let $\psi(i)$ be a binary indicator variable which specifies if i has a description or not. Let $\phi(c; s)$ equals to 1 if the sentence s consists concept c (*i.e.*, c is a substring in s), or otherwise 0. Similarly, $\beta\,(s;\,i)$ equals to 1 if the sentence s is part of the description (a list of sentences) of i, or otherwise 0. The event summary $\mathscr{S}$ which cover important *concepts*, is produced by extracting some sentences from $\mathscr{T}$. With the above notations and functions, we write the problem formulation for the event summarization as follows:

$$\min \sum_{(s\in\mathscr{T})\wedge(i\in\mathscr{R})} \omega(s)\beta(s,i) \tag{3.2a}$$

$$s.t. \sum_{s\in\mathscr{T}} \omega(s)\phi(c,s) \geq 1, \forall c\in\mathscr{K} \tag{3.2b}$$

$$\varphi(s) \geq \eta,\ \ \forall s\in\mathscr{T} \tag{3.2c}$$

$$|\ \mathscr{S}\ | \leq \overline{\mathscr{L}}, \tag{3.2d}$$

The objective function in Eq. (3.2a) solves the problem of event summarization and selects the minimal number of sentences which cover the maximal number of important concepts within the desired length of a summary. Eqs. (3.2b) and (3.2c) ensure that each concept is covered by at least one sentence with a score above the threshold η. Eq. (3.2d) assures that the length constraint of the summarization is met. Moreover, while choosing the set of all sentences $\mathscr{T}$ from the representative set of UGC $\mathscr{R}$, we use the following filters: (i) the UGI i has a description (*i.e.*, $\psi(i) = 1$, $\forall\, i\in\mathscr{R}$) and (ii) the UGI i is uploaded before the given timestamp τ (*i.e.*, $\tau(i) \leq \tau$, $\forall\, i\in\mathscr{R}$).

Algorithm 3.1 Event summarization algorithm

procedure EVENTSUMMARIZATION
 INPUT: An event e and a timestamp τ
 OUTPUT: A text summary $\mathscr{S}$
 $\overline{\mathscr{K}} = []$, $|\mathscr{S}| = 0$, $\overline{\mathscr{S}} = 200$ ▷ initialization.
 $(\mathscr{K}, \mathscr{Y})$ = getEventConceptsAndWeights(e) ▷ see [58].
 D_{Event} = getEventDataset(e) ▷ pre-processed event dataset.
 $\mathscr{R}$ = getRepresentativeSet(e, D_{Event}) ▷ representative UGIs.
 $\mathscr{T}$ = getSentences(e, $\mathscr{R}$) ▷ user descriptions and Wikipedia texts.
 while $((|\mathscr{S}| \leq \overline{\mathscr{L}}) \wedge (\mathscr{K} \neq \overline{\mathscr{K}}))$ **do** ▷ $\overline{\mathscr{K}}$ is covered concepts.
  c = getUncoveredConcept($\mathscr{K}$, $\overline{\mathscr{K}}$) ▷ cover important concept c first.
  s = getSentence(c, $\mathscr{Y}$, $\overline{\mathscr{K}}$, $\mathscr{T}$) ▷ $c \in s \wedge \varphi(s)$ is max.
  updateCoveredConceptList(s, $\overline{\mathscr{K}}$) ▷ add all $c \in s$ to $\overline{\mathscr{K}}$.
  addToEventTextSummary(s, $\mathscr{S}$) ▷ add sentence s to summary.
  for each sentence $s \in \mathscr{T}$ **do** ▷ say, $s \in$ the UGI i.
   updateScore(s, $\mathscr{Y}$, $\overline{\mathscr{K}}$) ▷ $\varphi(s) = \mathrm{u}(i,e) \times \Sigma_{c\in s, c\notin\overline{\mathscr{K}}}\, y$.
  end for ▷ $\mathrm{u}(i,e)$ is the relevance score of i for e (see Equation 3.1)
 end while
end procedure

EventBuilder solves the above optimization problem for text summarization based on a greedy algorithm using event-based features which represent *concepts* (*i.e.*, important event-related information), as described by Filatova and Hatzivassiloglou [58]. *Concepts* associate actions described in texts extracted from user descriptions and Wikipedia articles through verbs or action nouns labeling the event itself. First, EventBuilder extracts important *concepts* (*e.g.*, *kid-play-holi* for an event named *Holi*) from textual metadata. Next, it solves the optimization problem by selecting the minimal number of sentences which cover the maximal number of important concepts from matrix constructed by the textual metadata and the extracted concepts. Every time, when a new sentence is added to the summary $\mathscr{S}$, we check whether it contains enough new important *concepts* to avoid redundancy. We have formulated the problem of event text summarization regarding a matrix model, as shown in Table 3.4. Sentences and important concepts are mapped onto a $\mid \mathscr{T} \mid \times \mid \mathscr{K} \mid$ matrix. An entry of this matrix is 1 if the concept (column) is present in the sentence (row). Otherwise, it is 0. We take advantage of this model matrix to avoid redundancy by globally selecting the sentences that cover the most important concepts (*i.e.*, information) present in user descriptions and Wikipedia articles. Using the matrix defined above, it is possible to formulate the event summarization problem as equivalent to extracting the minimal number of sentences which cover all the important concepts. In our approximation algorithm, we constrain the total length of the summary on the total weight of covered concepts, to handle the cost of long summaries. However, the greedy algorithm for the set cover problem is not directly applicable to event summarization, since unlike the event summarization which assigns different weights to concepts based on their importance, the set cover assumes that any combination of sets is equally good as long as they cover the same total weight of concepts. Moreover, another constraint of the event summarization task is that it aims for the summary to be the desired length instead of a fixed total number of words. Our adaptive greedy algorithm for the event summarization is motivated by the summarization algorithm presented by Filatova and Hatzivassiloglou [58].

Algorithm 3.1 presents our summarization algorithm. First, It determines all event-related important concepts and their weights, as described by Filatova and Hatzivassiloglou [58]. Next, it extracts all sentences from user descriptions of UGIs in the representative set $\mathscr{R}$ and texts in the Wikipedia article on an event e. We multiply the sum of weights of all concepts a sentence s covers with the score of the UGI (which this sentence belongs to) to compute the score of the sentence s. Since each concept has different importance, we cover important concepts first. We consider only those sentences that contain the concept with the highest weight and has not yet been covered. Among these sentences, we choose the sentence with the highest total score and add it to the final event summary. Then we add the concepts which are covered by this sentence to the list of covered concepts $\bar{\mathscr{K}}$ in the final summary. Before adding further sentences to the event summary, we re-calculate the scores of all sentences by not considering the weight of all the concepts that are already covered in the event summary. We continue adding sentences to $\mathscr{S}$ until we obtain a summary of the desired length $\overline{\mathscr{L}}$ or a summary covering all concepts. Using this text summarization algorithm, EventBuilder produces two text summaries derived from user descriptions and Wikipedia articles, respectively.

Algorithm 3.2 SenticNet-3 concepts extraction

```
procedure ConceptExtraction
    INPUT: Textual description d of a UGI i
    OUTPUT: A list of SenticNet-3 concepts C_P for the UGI i
    C_P = ∅                          ▷ initialize the set of SenticNet-3 concepts for the UGI i.
    C_V = getVisualConcepts(i)       ▷ read visual concepts of i from database.
    C_T = semanticParserAPI(d)       ▷ get textual concepts from descriptions.
    C_FUSED = conceptFusion(C_T, C_V )   ▷ see Algorithm 3.
    C̄ = concepts(SenticNet-3)        ▷ a set of all SenticNet-3 concepts.
    for each concept c ∈ C_F do      ▷ check each concept of the UGI i.
        if (c ∈ C̄) then              ▷ check if c ∈ SenticNet-3.
            addConcept(c, C_P)       ▷ c is a SenticNet-3 concept.
        else
            W = splitIntoWords(c)    ▷ split the concept c.
            for each word w ∈ W do   ▷ w is a word (concept).
                if (w ∈ C̄) then      ▷ check if w ∈ SenticNet-3.
                    addConcept(w, C_P)   ▷ add w to C_P.
                else if (WordNetSynset(w) ∈ C̄) then   ▷ using WordNet.
                    addConcept(WordNetSynSet(w), C_P)  ▷ add synset.
                end if
            end for
        end if
    end for
    return C_P                       ▷ A set of SenticNet-3 concepts for the UGI i.
end procedure
```

3.2.2 EventSensor

Figure 3.2 depicts the architecture of the EventSensor system. It consists two components: (i) a client which accepts a user's inputs such as a mood tag, an event name, and a timestamp, and (ii) a backend server which consists semantics and sentics engines. EventSensor leverages the semantics engine (EventBuilder) to obtain the representative set of UGIs $\mathscr{R}$ for a given event and timestamp. Subsequently, it uses its sentics engine to generate a mood-based event summarization. It attaches soundtracks to the slideshow of UGIs in $\mathscr{R}$. The soundtracks are selected corresponding to the most frequent mood tags of UGIs derived from the sentics engine. Moreover, the semantics engine helps in generating text summaries for the given event and timestamp. If the user selects a mood tag as an input, EventSensor retrieves $\mathscr{R}$ from a database indexed with mood tags. Next, the sentics engine produces a musical multimedia summary for the input mood tag by attaching matching soundtracks to the slideshow of UGIs in $\mathscr{R}$.

Figure 3.3 shows the system framework of the sentics engine in the EventSensor system. The sentics engine is helpful in providing significant multimedia-related services to users from multimedia content aggregated on social media. It leverages multimodal information to perform sentiments analysis which is helpful in providing such mood-related services. Specifically, we exploit concepts (knowledge structures) from the visual content and textual metadata of UGC. We extract visual concepts for each multimedia item of a dataset and determine concepts from the textual metadata of multimedia content using the semantic parser API [143]. Next, we fuse the extracted visual and textual concepts, as described in Algorithm 3.3. We propose this novel fusion algorithm based on the importance of different metadata in determining the sentics information of UGC on an evaluation set of 60 photos (see Sect. 3.3.2). Further, we use it in calculating the accuracy of sentics information for different metadata such as descriptions, tags, and titles of UGIs (see Sect. 3.3 for more details). After determining fused concepts C_{FUSED} for the multimedia content, we compute the corresponding SenticNet-3 concepts since they bridge the conceptual and affective gap and contain sentics information.

Algorithm 3.2 describes our approach to establishing an association between concepts C_{FUSED} extracted by the semantic parser and the concepts of SenticNet-3 $\bar{C}$. It checks if concepts in C_{FUSED} are present in $\bar{C}$. For each concept in C_{FUSED}, we add it to C_P if it is present in SenticNet-3. Otherwise, we split it into words W and repeat the process. We add the words (concepts) of W that are present in $\bar{C}$ to C_P, and repeat the process for the WordNet synsets of the rest of the words. For each SenticNet-3 concept in C_P of a UGI i, Algorithm 3.4 determines the corresponding mood tag by referring to the EmoSenticNet E and EmoSenticSpace $\bar{E}$ knowledge bases [155]. E maps 13,000 concepts of SenticNet-3 to mood tags such as anger, disgust, joy, sad, surprise, and fear. However, we do not know the the mood tags of the remaining 17,000 concepts in SenticNet-3 $\bar{C}$. To determine their sentics information, first, we find their neighbors using EmoSenticSpace. $\bar{E}$ provides a 100D feature vector space for each concept in $\bar{C}$. We find 100 neighbors that have mood information (*i.e.*, from E) for each concept using the cosine similarity metric and determine its six-dimensional mood vector based on a voting count, as described in Fig. 3.6. Finally, we find the mood vector M_P of the UGI i by combining the mood vectors of all concepts in C_P using an arithmetic mean. Experimental results indicate that the arithmetic mean of different mood vectors for concepts performs better than their geometric and harmonic means.

Algorithm 3.3 Fusion of concepts

procedure CONCEPTFUSION
  INPUT: Textual concepts C_T and visual concepts C_V of a UGI i
  OUTPUT: A list of fused concepts C_{FUSED} for i
  $C_{FUSED} = \emptyset$ ▷ initialize the set of fused concepts for the UGI i.
  if (hasTags(i)) **then** ▷ check if the UGI i has tags.
    C_{FUSED} = getTagConcepts(C_T) ▷ Since tags has the highest accuracy, see Figure 3.10.
  else if (hasDescription(i) $\wedge$ hasVisualConcepts(i)) **then**
    C_{FUSED} = getDescrConcepts(C_T) ▷ get concepts from descriptions.
    $C_{FUSED} = C_{FUSED} \bigcup C_V$ ▷ Since it has the second highest accuracy
  else if (hasTitle(i) $\wedge$ hasVisualConcepts(i)) **then**
    C_{FUSED} = getTitleConcepts(C_T) ▷ get concepts from descriptions.
    $C_{FUSED} = C_{FUSED} \bigcup C_V$ ▷ Since it has the third highest accuracy
  else if (hasVisualConcepts(i)) **then**
    $C_{FUSED} = C_V$ ▷ Since it has the fourth highest accuracy, see Figure 3.10.
  else if (hasDescription(i)) **then**
    C_{FUSED} = getDescrConcepts(C_T) ▷ Since it has 5th highest accuracy.
  else if (hasTitle(i)) **then** ▷ check if the UGI i has title.
    C_{FUSED} = getTitleConcepts(C_T) ▷ Since it has the lowest accuracy.
  end if
  return C_{FUSED} ▷ A set of fused concepts for the UGI i.
end procedure

Semantics and sentics information computed in earlier steps are very useful in providing different multimedia-related services to users. For instance, we provide multimedia summaries from UGIs aggregated on social media such as Flickr. Once the affective information is known, it can be used to provide different services related to affect. For instance, we can query Last.fm to retrieve songs for the determined mood tags and enable users to obtain a musical multimedia summary. To show the effectiveness of our system, we present a musical multimedia summarization by adding a matching soundtrack to the slideshow of UGIs. Since determining the sentics (mood tag) from the multimedia content is the main contribution of this chapter, we randomly select a soundtrack corresponding to the determined mood tag from a music dataset annotated with mood tags (see Sect. 3.3 for more details about the music dataset).

Algorithm 3.4 Sentics computation

procedure SENTICSEXTRACTION
  INPUT: A list of SenticNet-3 concepts C_P of a UGI i
  OUTPUT: A mood vector M_P for the UGI i
  $\overline{M_P}$ = [0,0,0,0,0,0] ▷ initialize the mood vector for the UGI i.
  for each $\overline{c} \in C_P$ **do** ▷ $\overline{c}$ is a SenticNet-3 concept of the UGI i.
    if ($\overline{c} \in E$) **then** ▷ check if $\overline{c} \in$ *EmoSenticNet* whose moods are known.
      addMood(mood($\overline{c}$), $\overline{M_P}$) ▷ add mood vector for $\overline{c}$ to M_P.
    else
      S = findNeighbor($\overline{c}$, $\overline{E}$) ▷ finding neighbors from EmoSenticSpace
      M_S = [0,0,0,0,0,0] ▷ initialize the mood vector of $\overline{c}$.
      for each $s \in S$ **do** ▷ s is a neighbor concept of $\overline{c}$.
        addMood(mood(s), M_S) ▷ add neighbor's mood vector.
      end for
      $m = \frac{M_S}{|S|}$ ▷ using arithmetic mean.
      addMood(m, $\overline{M_P}$) ▷ m is the mood vector of $\overline{c}$.
    end if
  end for
  $M_P = \frac{\overline{M_P}}{|C_P|}$ ▷ using arithmetic mean.
  return M_P ▷ M_P is a sentics vector for the UGI i.
end procedure

3.3 Evaluation

Dataset We used the YFCC100M [201] (Yahoo! Flickr Creative Commons 100 M) dataset D_{YFCC} which consists of 100 million multimedia items (approximately 99.2 million UGIs and 0.8 million UGVs) from Flickr. The reason for selecting this dataset is its volume, modalities, and metadata. For instance, each media of the dataset consists of several metadata annotations such as user tags, spatial information, and temporal information. These media are captured from the 1990's onwards and uploaded between 2004 and 2014. It includes media from top cities such as Paris, Tokyo, London, New York City, Hong Kong, and San Francisco. Moreover, all media are labeled with automatically added tags derived by using a convolutional neural network which indicates the presence of a variety of concepts, such as people, animals, objects, food, events, architecture, and scenery. There are a total 1756 visual concepts present in this dataset. For the music dataset, we used the ISMIR'04 dataset of 729 songs from the ADVISOR system [187] for generating a musical multimedia summary. This dataset is annotated with the 20 most frequent mood tags (*e.g.*, happy, sad, dreamy, and fun) of Last.fm. Based on the classification of emotional tags in earlier work [40, 187, 217], we clustered the 20 mood tags of Last.fm into the six mood categories (*i.e.*, anger, disgust, joy, sad, surprise, and fear). We used these six mood categories in this study (see Table 3.5). This music dataset consists of songs from all main music genres such as classical, electronic, jazz, metal, pop, punk, rock, and world. For the detection of seven events (*Holi, Eyjafjallajkull Eruption, Occupy Movement, Hanami, Olympic Games, Batkid,* and *Byron Bay Bluesfest*), as described in the ACM Multimedia

Table 3.5 Mapping between moods of EmoSenticNet and Last.fm

EmoSenticNet	Last.fm
Anger	Anger, Aggressive
Disgust	Intense
Joy	Happy, Fun, Gay, Playful, Sweet, Soothing, Calm, Sleepy
Sad	Sad, Melancholy, Depressing, Heavy
Surprise	Quirky, Dreamy, Sentimental
Fear	Bittersweet, Quiet

Grand Challenge 2015 for an event detection and summarization task [182], we processed all 100 million UGIs and UGVs. In pre-processing step, we compute scores of all UGIs/UGVs in the YFCC100M dataset for all seven events, as mentioned above. Table 3.6 describes the statistics of the number of UGIs/UGVs from the YFCC100M dataset for these events. The higher relevance score $u(i; e)$ of a UGI/UGV i with an event e indicates the higher likelihood that the UGI/UGV belongs to the event. For efficient and fast processing, we compute relevance scores, concepts, and mood tags of all photos and build Apache Lucene indices for them during pre-processing. Moreover, we also collected contextual information such as spatial, temporal, keywords, and other event-related metadata for these events. For instance, Tables 3.2 and 3.3 show the spatial, temporal, and keywords metadata for the Olympics event. Furthermore, we collected the information of 1080 camera models from Flickr that are ranked based on their sensor sizes. In the real-time prototype system for EventSensor, we used 113,259 UGIs which have high relevance scores for the above seven events.

Evaluators Table 3.7 shows the different user groups who participated in our evaluation. *Group-A* has totally 63 working professionals and students (the most of them are Information Technology professionals) from citizens of 11 countries such as Singapore, India, USA, Germany, and China. All *Group-A* users were given a brief introduction to events mentioned above used for the event detection. Most of the users out of 10 users in *Group-B* are different international students of National University Singapore. Since the *Group-B* users are asked to evaluate the text summaries for different events; they were not provided any prior introduction about the seven events mentioned above. Finally, the users in *Group-C* are invited to assign emotional mood tags from six categories such as anger, disgust, joy, sad, surprise, and fear to UGIs from Flickr. There are totally 20 users who are working professionals and students from different institutes and countries. Since our approach to determining sentics details of UGIs is based on leveraging multimodal information, we asked users to use all the available information such as tags, descriptions, locations, visual content, and title in finalizing their decision to assign mood tags to UGIs.

Table 3.6 The number of UGIs for various events with different scores

Event name	Scores u(i; e)	Number of photos
Holi	u(i; e) $\geq$ 0.90	1
	0.80 $\leq$ u(i; e) $\leq$ 0.90	153
	0.70 $\leq$ u(i; e) $\leq$ 0.80	388
	0.50 $\leq$ u(i; e) $\leq$ 0.70	969
	0.30 $\leq$ u(i; e) $\leq$ 0.50	6808
Eyjafjallajkull Eruption	u(i; e) $\geq$ 0.90	47
	0.80 $\leq$ u(i; e) $\leq$ 0.90	149
	0.70 $\leq$ u(i; e) $\leq$ 0.80	271
	0.50 $\leq$ u(i; e) $\leq$ 0.70	747
	0.30 $\leq$ u(i; e) $\leq$ 0.50	7136
Occupy Movement	u(i; e) $\geq$ 0.90	599
	0.80 $\leq$ u(i; e) $\leq$ 0.90	2290
	0.70 $\leq$ u(i; e) $\leq$ 0.80	4036
	0.50 $\leq$ u(i; e) $\leq$ 0.70	37,187
	0.30 $\leq$ u(i; e) $\leq$ 0.50	4,317,747
Hanami	u(i; e) $\geq$ 0.90	558
	0.80 $\leq$ u(i; e) $\leq$ 0.90	3417
	0.70 $\leq$ u(i; e) $\leq$ 0.80	3538
	0.50 $\leq$ u(i; e) $\leq$ 0.70	12,990
	0.30 $\leq$ u(i; e) $\leq$ 0.50	464,710
Olympic Games	u(i; e) $\geq$ 0.90	232
	0.80 $\leq$ u(i; e) $\leq$ 0.90	6278
	0.70 $\leq$ u(i; e) $\leq$ 0.80	10,329
	0.50 $\leq$ u(i; e) $\leq$ 0.70	23,971
	0.30 $\leq$ u(i; e) $\leq$ 0.50	233,082
Batkid	u(i; e) $\geq$ 0.90	0
	0.80 $\leq$ u(i; e) $\leq$ 0.90	17
	0.70 $\leq$ u(i; e) $\leq$ 0.80	23
	0.50 $\leq$ u(i; e) $\leq$ 0.70	7
	0.30 $\leq$ u(i; e) $\leq$ 0.50	780
Byron Bay Bluesfest	u(i; e) $\geq$ 0.90	96
	0.80 $\leq$ u(i; e) $\leq$ 0.90	56
	0.70 $\leq$ u(i; e) $\leq$ 0.80	80
	0.50 $\leq$ u(i; e) $\leq$ 0.70	1652
	0.30 $\leq$ u(i; e) $\leq$ 0.50	25,299

Table 3.7 Users/evaluators details for user studies

Group type	No. of evaluators	No. of responses	No. of Accepted responses
Group-A	63	441	364
Group-B	10	70	70
Group-C	20	120	109

3.3.1 EventBuilder

Event Detection To evaluate the proposed automatic event detection system, we performed an extensive user study on results derived from the baseline and EventBuilder. Since the most of the existing multimedia search and retrieval systems are based on keyword searching, we select a UGC that consists an event name in its metadata as the result of the baseline system. We introduced the following single and inter-annotation consistency check. We added redundancy and kept questions in a random order for a consistency check. Moreover, we added a check to reject bad responses by adding a few questions which were trivial to answer. We rejected the responses which did not fulfill the above criteria. We randomly selected four UGIs for each of the seven events listed in the Table 3.8 and repeated one UGI for evaluation consistency check. The seven events listed in the Table 3.8 are the same events that are described in the ACM Multimedia Grand Challenge 2015 for an event detection and summarization [182]. Thus, for the event detection evaluation, we give five questions each for seven events (*i.e.*, total 35 questions) to each evaluator. For each question, we showed two UGIs to an evaluator. The baseline system produced the first UGI i_1, *i.e.*, the UGI i_1 consists the name of a given event in its metadata. The EventBuilder system produced the second UGI i_2, *i.e.*, the UGI i_2 has significantly high relevance score $u(i_2; e)$ than other UGIs for a given event e. We asked evaluators from *Group-A* to select UGIs which are relevant to the event. We created a survey form using a Google Form [15] for the evaluation (see Fig. 3.7). Specifically, for given two photos from two algorithms (baseline and EventBuilder) for an event, we ask evaluators to select one of the following options: (i) *Photo A* only, *i.e.*, if only *Photo A* is relevant to the given event, (ii) *Photo B* only, *i.e.*, if only *Photo B* is relevant to the given event, (iii) Both *Photos A and B*, *i.e.*, if both *Photos A and B* are relevant to the given event, and (iv) None of the Photos, *i.e.*, if none of the photos are relevant to the given event. We received a total 63 responses from 63 users of 11 countries (*e.g.*, India, Singapore, USA, and Germany) and accepted 52 responses.

Table 3.8 Results for event text summaries of 150 words from 10 users. R1, R2, and R3 are ratings for informative, experience, and acceptance, respectively

	Baseline			EventBuilder		
Flickr event name	R1	R2	R3	R1	R2	R3
Holi	3.7	3.3	3.4	4.3	4.0	4.3
Olympic games	3.4	3.1	3.3	3.6	4.1	4.0
Eyjafjallajkull Eruption	3	2.9	3.2	4.1	4.1	4.2
Batkid	2.5	2.4	3	3.6	3.6	3.6
Occupy movement	3.6	3.1	3.5	3.8	3.9	4.1
Byron Bay Bluesfest	2.6	2.6	2.8	3.6	3.6	3.9
Hanami	3.9	3.9	4	4.1	3.9	4.1
All events	3.243	3.043	3.314	**3.871**	**3.886**	**4.029**

Batkid

Batkid is the superhero name of Miles Scott, an American child and cancer survivor. His wish was to be "Batkid", a sidekick of the eponymous comic book superhero Batman, the subject of books, radio, television, and films. Once the request went out, thousands of volunteers, city officials, businesses and supporters rallied to turn San Francisco, California into "Gotham City" - the fictional home city of Batman - on November 15, 2013 for one of the largest and most elaborate Make-A-Wish projects ever staged.
More details: https://en.wikipedia.org/wiki/Batkid

Photo A

Photo B

Q-1: Which of the above photos are relevant to the Batkid event?
*

- Photo A only
- Photo B only
- Both Photos A and B
- None of the Photos

BACK NEXT

Fig. 3.7 User interface of the survey for the evaluation of event detection

Table 3.9 Evaluation results for event detection from 52 users

Method	Precision	Recall	F measure	Cosine similarity
Baseline	0.315247	0.735577	0.441346	0.747602
EventBuilder	0.682005	0.707265	0.694405	0.832872

Since full details (both content and contextual information) of all UGIs used in the user study was known, it was easy to assign a ground truth to them. We compared the responses of users with the ground truth based on two metrics (i) precision, recall, and F-measure, and (ii) cosine similarity. These scores represent the degree of agreement among users with results produced by the baseline and EventBuilder systems. Experimental results confirm that users agree more with the results produced by EventBuilder as compared to the baseline (see Table 3.9). We use the following equations to compute precision, recall, F-measure, and cosine similarity:

$$precision = \frac{\#[G \wedge U]}{\mid U \mid}, \tag{3.3}$$

$$recall = \frac{\#[G \wedge U]}{\mid G \mid}, \tag{3.4}$$

$$F\,measure = \frac{2 \times precision \times recall}{precision + recall}, \tag{3.5}$$

$$cos\,ine\,similarity = \frac{G \cdot U}{\parallel G \parallel \parallel U \parallel} \tag{3.6}$$

where G and U are feature vectors for the ground truth and a user's response, respectively. ||U|| is the total number of questions for seven events (as listed in Table 3.8) in the user study, and ||G|| is the number of UGIs (questions) which are relevant to each event. #[G Λ U] represents how many times the user is in agreement with the ground truth. G and U are the magnitude of the feature vectors for G and U, respectively. Experimental results in Table 3.9 confirms that EventBuilder outperforms its baseline by 11.41% in event detection.

Event Summarization. To evaluate text summaries generated by the EventBuilder system, we conducted a user study (see Fig. 3.8) based on three perspectives that users should consider. First, informativeness, which indicates to what degree a user feels that the summary captures the essence of the event. Second, experience, which indicates if the user thinks the summary is helpful for understanding the event. Third, acceptance, which indicates if a user would be willing to use this event summarization function if Flickr were to incorporate it into their system. We asked ten evaluators from *Group-B* to assess the text summaries and provide scores from 1 to 5 (a higher score indicating better satisfaction). The default event summary length $\overline{\mathscr{L}}$ was 150 words during evaluation since the length of abstracts ranges from 150 to 300 words. However, the size of the summary is an input parameter to the system, and a user can change it anytime. For instance,

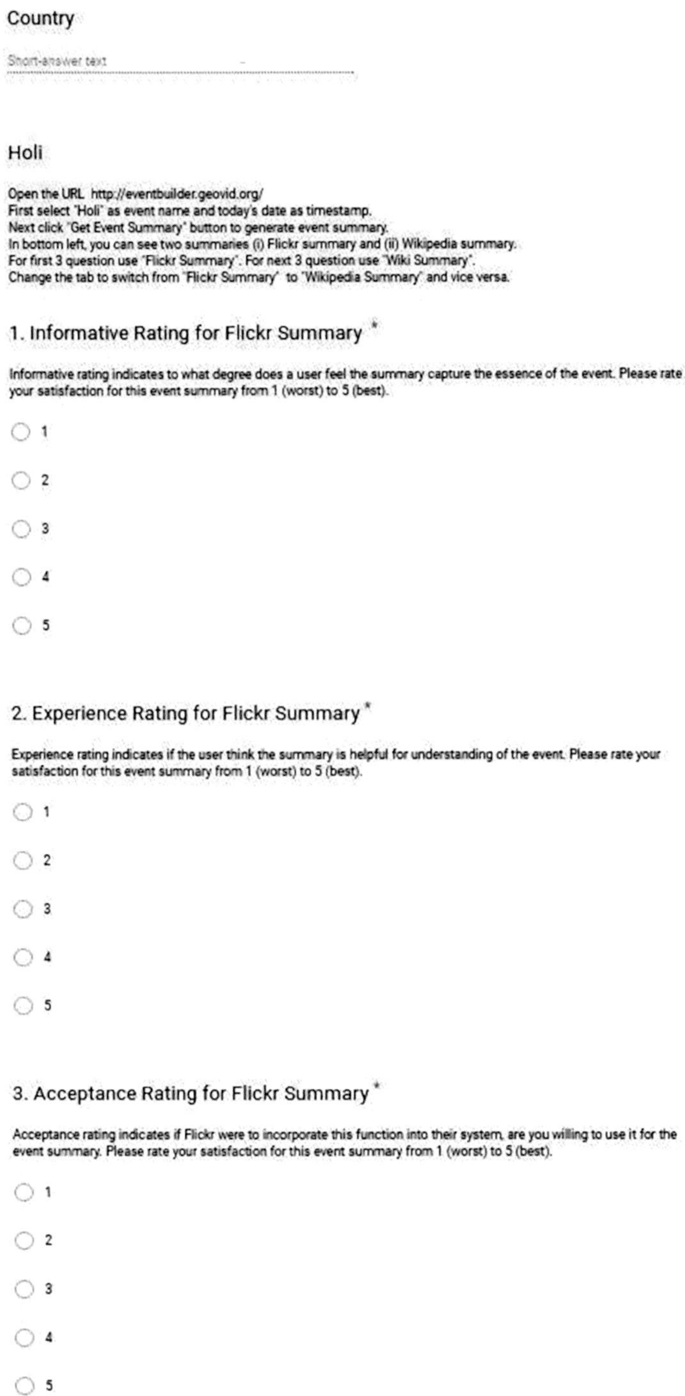

Country

Short-answer text

Holi

Open the URL http://eventbuilder.geovid.org/
First select "Holi" as event name and today's date as timestamp.
Next click "Get Event Summary" button to generate event summary.
In bottom left, you can see two summaries (i) Flickr summary and (ii) Wikipedia summary.
For first 3 question use "Flickr Summary". For next 3 question use "Wiki Summary".
Change the tab to switch from "Flickr Summary" to "Wikipedia Summary" and vice versa.

1. Informative Rating for Flickr Summary *

Informative rating indicates to what degree does a user feel the summary capture the essence of the event. Please rate your satisfaction for this event summary from 1 (worst) to 5 (best).

○ 1
○ 2
○ 3
○ 4
○ 5

2. Experience Rating for Flickr Summary *

Experience rating indicates if the user think the summary is helpful for understanding of the event. Please rate your satisfaction for this event summary from 1 (worst) to 5 (best).

○ 1
○ 2
○ 3
○ 4
○ 5

3. Acceptance Rating for Flickr Summary *

Acceptance rating indicates if Flickr were to incorporate this function into their system, are you willing to use it for the event summary. Please rate your satisfaction for this event summary from 1 (worst) to 5 (best).

○ 1
○ 2
○ 3
○ 4
○ 5

Fig. 3.8 User interface (UI) of the survey for the evaluation of Flickr text summary. Similar UI is used to evaluate the text summary produced by EventBuilder

Table 3.10 Event summaries from Wikipedia for two different summary sizes, *i.e.*, for 150 and 300 words

Size	Wikipedia Summary
150 words	The next morning is a free-for-all carnival of colours, where participants play, chase and colour each other with dry powder and coloured water, with some carrying water guns and coloured water-filled balloons for their water fight. Holi celebrations start with a Holika bonfire on the night before Holi where people gather, sing and dance. Holi is celebrated at the approach of vernal equinox, on the Phalguna Purnima Full Moon. The festival signifies the victory of good over evil, the arrival of spring, end of winter, and for many a festive day to meet others, play and laugh, forget and forgive, and repair ruptured relationships. Groups carry drums and musical instruments, go from place to place, sing and dance. People visit family, friends and foes to throw colours on each other, laugh and chit-chat, then share Holi delicacies, food and drinks
300 words	The next morning is a free-for-all carnival of colours, where participants play, chase and colour each other with dry powder and coloured water, with some carrying water guns and coloured water-filled balloons for their water fight. Holi celebrations start with a Holika bonfire on the night before Holi where people gather, sing and dance. Holi is celebrated at the approach of vernal equinox, on the Phalguna Purnima Full Moon. The festival signifies the victory of good over evil, the arrival of spring, end of winter, and for many a festive day to meet others, play and laugh, forget and forgive, and repair ruptured relationships. Groups carry drums and musical instruments, go from place to place, sing and dance. People visit family, friends and foes to throw colours on each other, laugh and chit-chat, then share Holi delicacies, food and drinks. For example, Bhang, an intoxicating ingredient made from cannabis leaves, is mixed into drinks and sweets and consumed by many. The festival date varies every year, per the Hindu calendar, and typically comes in March, sometimes February in the Gregorian Calendar. Holi is a spring festival, also known as the festival of colours or the festival of love. It is an ancient Hindu religious festival which has become popular with non-Hindus in many parts of South Asia, as well as people of other communities outside Asia. In the evening, after sobering up, people dress up, and visit friends and family

Table 3.10 shows the Wikipedia summary of the event, named Holi, for two different summary sizes, *i.e.*, for 150 and 300 words. We asked users to rate both the Flickr summary (baseline) which is derived from descriptions of UGIs and the Wikipedia summary (EventBuilder) which is derived from Wikipedia articles on events. The reason we compare the Flickr summary with the Wikipedia summary because we want to compare the information (the summary of an event) we get from what users think with the most accurate information derived from the available knowledge bases such as Wikipedia about the event. Moreover, since the evaluation of a textual summary of the event is a very subjective process, we want only to compare the textual summaries of the event derived from a strong baseline and our EventBuilder system leveraging knowledge bases such as Wikipedia. For instance, we did not consider a very simple baseline such as randomly selecting sentences till the summary length is achieved. Instead, we consider the event confidence of a UGI as well as the confidence scores of the sentences in the description of the UGI. Table 3.8 indicates that users think that the Wikipedia summary is more informative

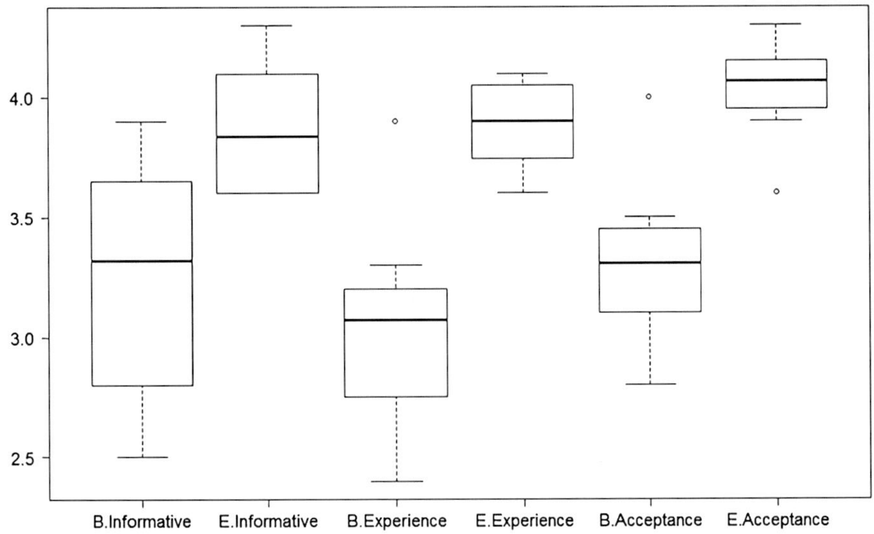

Fig. 3.9 Boxplot for the informative, experience, and acceptance ratings of text summaries, where prefix *B* and *E* in x-axis indicate baseline and EventBuilder, respectively. In y-axis, ratings range from 1 to 5, with a higher score indicate better satisfaction

than the Flickr summary (the proposed baseline) and can help them to obtain a better overview of the events. The box plot in Fig. 3.9 corresponds to the experimental results (users rating) in Table 3.8. It confirms that EventBuilder outperforms the baseline on the following three metrics: (i) informativeness, (ii) experience, and (iii) acceptance. Particularly, EventBuilder outperforms its baseline for text summaries of events by (i) 19.36% regarding informative rating, (ii) 27.70% regarding experience rating, and (ii) 21.58% regarding acceptance rating (see Table 3.8 and Fig. 3.9). Median scores of EventBuilder for three metrics mentioned above are much higher than that of the baseline. Moreover, the box plots for EventBuilder is comparatively shorter than that of the baseline. Thus, this suggests that overall users have a high level of agreement with each other for EventBuilder as compared to that of the baseline.

Despite the Wikipedia summary is more informative than the baseline, the Flickr summary is also very helpful since it gives an overview of what users think about the events. Tables 3.11 and 3.12 show text summaries produced for the Olympics event at the timestamp *2015-03-16 12:36:57* by the EventBuilder system using the descriptions of UGIs which are detected for the Olympics event, and using Wikipedia articles on the Olympics event, respectively.

Table 3.11 Event text summary derived from descriptions of UGIs for the Olympics event with 200 words as desired summary length

Event name	Timestamp	Text summary from UGIs
Olympics	2015-03-16 12:36:57	Felix Sanchez wins Olympic Gold. A day to remember, the Olympic Stadium, Tuesday 7th August 2012. One of the Magic light boxes by Tait-technologies from the opening/closing ceremony, made in Belgium. One of the cyclists participates in the men 's road time trials at the London 2012 Olympics. Two kids observe the Olympic cycle road time trial from behind the safety of the barriers. Lin Dan China celebrates winning. The Gold Medal. Mo Farah receiving his gold medal. Germany run out 3-1 winners. Details of players/scores included in some pictures. Veronica Campbell-Brown beats Carmelita Jeter in her 200 m Semi-Final. Jason Kenny beats Bernard Esterhuizen. Elisa Di Francisca, Arianna Errigo, Valentina Vezzali and Ilaria Salvatori of Italy celebrate winning the Gold Medal in the Women's Team Foil. Team USA celebrates after defeating Brazil in the Beijing Olympic quarterfinal match. Peter Charles all went clear to snatch gold. Wow, an athlete not wearing bright yellow Nike running spikes. Mauro Sarmiento Italy celebrates winning Bronze. BMX cross at the London 2012 Olympics with the velodrome in the background

Table 3.12 Event text summary derived from Wikipedia for the Olympics event with 200 words as desired summary length

Event Name	Timestamp	Text summary from Wikipedia
Olympics	2015-03-16 12:36:57	The IOC also determines the Olympic program, consisting of the sports to be contested at the Games. Their creation was inspired by the ancient Olympic Games, which were held in Olympia, Greece, from the eighth century BC to the fourth century AD. As a result, the Olympics has shifted away from pure amateurism, as envisioned by Coubertin, to allowing participation of professional athletes. The Olympic Games are held every 4 years, with the Summer and Winter Games alternating by occurring every 4 years but 2 years apart. The IOC is the governing body of the Olympic Movement, with the Olympic Charter defining its structure and authority. Baron Pierre de Coubertin founded the International Olympic Committee IOC in 1894. The modern Olympic Games French: Jeux olympiques are the leading international sporting event featuring summer and winter sports competitions in which thousands of athletes from around the world participate in a variety of competitions. This growth has created numerous challenges and controversies, including boycotts, doping, bribery, and a terrorist attack in 1972. Every 2 years the Olympics and its media exposure provide unknown athletes with the chance to attain national and sometimes international fame

3.3.2 EventSensor

To evaluate the EventSensor system, we extracted those UGC (UGIs and UGVs) of the YFCC100M dataset that contains keywords related to mood tags such as anger, disgust, joy, sad, surprise, and fear, or their synonyms. In this way, we found 1.2 million UGC. Next, we randomly selected 10 UGIs for each of the above six mood tags that have a title, description, and tags metadata. Subsequently, we randomly divided these UGIs into six sets with 10 UGIs each and assigned them to random evaluators. Similar to the EventBuilder user study, we added redundancy to provide a consistency check. We assigned these UGIs to 20 users from *Group-C*. We received an average of 17.5 responses for each UGI in the six sets. From the accepted responses, we created a six-dimensional mood vector for each UGI as ground truth and compared it with the computed mood vectors of different approaches using cosine similarity. In EventSensor, we investigated the importance of different metadata (*i.e.*, user tags, title, description, and visual concepts) in determining the affective cues from the multimedia content. Figure 3.10 with 95% confidence interval shows the accuracy (agreement with the affective information derived from crowdsourcing) of sentics analysis when different metadata and their combinations are considered in the analysis. Experimental results indicate that the feature based on user tags is salient and the most useful in determining sentics details of UGIs.

Experimental results indicate that user tags are most useful in determining sentics details of a UGI. The probable reasons that why considering user tags

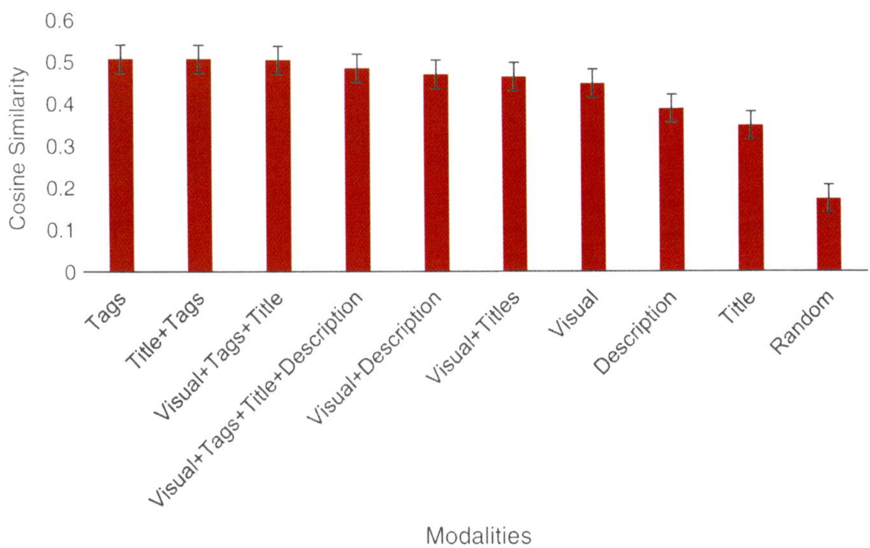

Fig. 3.10 Evaluation results for EventSensor. It shows cosine similarities between ground truth and mood vectors determined from different modalities

alone in the sentics analysis, perform better than other modalities are as follows. First, semantics understanding is easier from user tags as compared to other metadata. Second, users' tags indicate important information about the multimedia content. Third, usually users' tags are less noisy than other metadata. Since most UGIs on social media do not contain information such as user tags, description, and title, it is essential to consider a fusion technique that provides the most accurate sentics information irrespective of which metadata a UGI contains. Thus, we proposed an approach to fuse information from different modalities for an efficient sentics analysis (see Algorithm 3.3). We performed the fusion of mood vectors based on arithmetic, geometric, and harmonic means, and found that the fusion based on the arithmetic mean performs better than the other two means. In the future, we would like to leverage map matching techniques [244] and SMS/MMS based FAQ retrieval techniques [189, 190] for a better event understanding.

3.4 Summary

We presented two real-time multimedia summarization systems, called EventBuilder and EventSensor, based on the semantics and sentics understanding of UGIs, respectively. EventBuilder performs semantics analysis on multimedia content from social media such as Flickr and produces multimedia summaries for a given event. Knowledge structures derived from different modalities are beneficial in better semantics understanding from UGIs. The proposed EventBuilder system produces summaries for the event in the following two steps. First, it performs event detection from the large collection of UGIs by computing event relevance scores using our proposed method and builds event-wise indices based on their scores. Next, it generates event summaries for the given event and timestamps in real-time based on scores and descriptions of UGIs in the event dataset and facilitates efficient access to a large collection of UGIs. Since sentiment analysis for multimedia content is very useful in many multimedia-based services, we perform sentics analysis on UGIs using our proposed EventSensor system. EventSensor enables users to obtain sentics-based multimedia summaries such as the slideshow of UGIs with matching soundtracks. If a user selects a mood tag as input then soundtracks corresponding to the input mood tag are selected. If the user chooses an event as input then soundtracks corresponding to the most frequent mood tags of UGIs in the representative set are attached to the slideshow. Experimental results on the YFCC100M dataset confirm that our systems outperform their baselines. Particularly, EventBuilder outperforms its baseline by 11.41% regarding event detection, and outperform its baseline for text summaries of events by (i) 19.36% regarding informative rating, (ii) 27.70% regarding experience rating, and (ii) 21.58% regarding acceptance rating. Furthermore, our EventSensor system found that the feature based on user tags is salient among other metadata (*i.e.*, user tags, title, description, and visual concepts) in determining sentics details of UGIs. Chap. 8 describes the future work to improve these multimedia summarization systems further.

References

1. Apple Denies Steve Jobs Heart Attack Report: "It Is Not True". http://www.businessinsider.com/2008/10/apple-s-steve-jobs-rushed-to-er-after-heart-attack-says-cnn-citizen-journalist/. October 2008. Online: Last Accessed Sept 2015.
2. SVM^{hmm}: Sequence Tagging with Structural Support Vector Machines. https://www.cs.cornell.edu/people/tj/svmlight/svmhmm.html. August 2008. Online: Last Accessed May 2016.
3. NPTEL. 2009, December. http://www.nptel.ac.in. Online; Accessed Apr 2015.
4. iReport at 5: Nearly 900,000 contributors worldwide. http://www.niemanlab.org/2011/08/ireport-at-5-nearly-900000-contributors-worldwide/. August 2011. Online: Last Accessed Sept 2015.
5. Meet the million: 999,999 iReporters + you! http://www.ireport.cnn.com/blogs/ireport-blog/2012/01/23/meet-the-million-999999-ireporters-you. January 2012. Online: Last Accessed Sept 2015.
6. 5 Surprising Stats about User-generated Content. 2014, April. http://www.smartblogs.com/social-media/2014/04/11/6.-surprising-stats-about-user-generated-content/. Online: Last Accessed Sept 2015.
7. The Citizen Journalist: How Ordinary People are Taking Control of the News. 2015, June. http://www.digitaltrends.com/features/the-citizen-journalist-how-ordinary-people-are-tak ing-control-of-the-news/. Online: Last Accessed Sept 2015.
8. Wikipedia API. 2015, April. http://tinyurl.com/WikiAPI-AI. API: Last Accessed Apr 2015.
9. Apache Lucene. 2016, June. https://lucene.apache.org/core/. Java API: Last Accessed June 2016.
10. By the Numbers: 14 Interesting Flickr Stats. 2016, May. http://www.expandedramblings.com/index.php/flickr-stats/. Online: Last Accessed May 2016.
11. By the Numbers: 180+ Interesting Instagram Statistics (June 2016). 2016, June. http://www.expandedramblings.com/index.php/important-instagram-stats/. Online: Last Accessed July 2016.
12. Coursera. 2016, May. https://www.coursera.org/. Online: Last Accessed May 2016.
13. FourSquare API. 2016, June. https://developer.foursquare.com/. Last Accessed June 2016.
14. Google Cloud Vision API. 2016, December. https://cloud.google.com/vision/. Online: Last Accessed Dec 2016.
15. Google Forms. 2016, May. https://docs.google.com/forms/. Online: Last Accessed May 2016.
16. MIT Open Course Ware. 2016, May. http://www.ocw.mit.edu/. Online: Last Accessed May 2016.
17. Porter Stemmer. 2016, May. https://tartarus.org/martin/PorterStemmer/. Online: Last Accessed May 2016.
18. SenticNet. 2016, May. http://www.sentic.net/computing/. Online: Last Accessed May 2016.
19. Sentics. 2016, May. https://en.wiktionary.org/wiki/sentics. Online: Last Accessed May 2016.
20. VideoLectures.Net. 2016, May. http://www.videolectures.net/. Online: Last Accessed May, 2016.
21. YouTube Statistics. 2016, July. http://www.youtube.com/yt/press/statistics.html. Online: Last Accessed July, 2016.
22. Abba, H.A., S.N.M. Shah, N.B. Zakaria, and A.J. Pal. 2012. Deadline based performance evalu-ation of job scheduling algorithms. In *Proceedings of the IEEE International Conference on Cyber-Enabled Distributed Computing and Knowledge Discovery*, 106–110.
23. Achanta, R.S., W.-Q. Yan, and M.S. Kankanhalli. (2006). Modeling Intent for Home Video Repurposing. In *Proceedings of the IEEE MultiMedia*, (1):46–55.
24. Adcock, J., M. Cooper, A. Girgensohn, and L. Wilcox. 2005. Interactive Video Search using Multilevel Indexing. In *Proceedings of the Springer Image and Video Retrieval*, 205–214.

25. Agarwal, B., S. Poria, N. Mittal, A. Gelbukh, and A. Hussain. 2015. Concept-level Sentiment Analysis with Dependency-based Semantic Parsing: A Novel Approach. In *Proceedings of the Springer Cognitive Computation,* 1–13.
26. Aizawa, K., D. Tancharoen, S. Kawasaki, and T. Yamasaki. 2004. Efficient Retrieval of Life Log based on Context and Content. In *Proceedings of the ACM Workshop on Continuous Archival and Retrieval of Personal Experiences*, 22–31.
27. Altun, Y., I. Tsochantaridis, and T. Hofmann. 2003. Hidden Markov Support Vector Machines. In *Proceedings of the International Conference on Machine Learning*, 3–10.
28. Anderson, A., K. Ranghunathan, and A. Vogel. 2008. Tagez: Flickr Tag Recommendation. In *Proceedings of the Association for the Advancement of Artificial Intelligence.*
29. Atrey, P.K., A. El Saddik, and M.S. Kankanhalli. 2011. Effective Multimedia Surveillance using a Human-centric Approach. *Proceedings of the Springer Multimedia Tools and Applications* 51 (2): 697–721.
30. Barnard, K., P. Duygulu, D. Forsyth, N. De Freitas, D.M. Blei, and M.I. Jordan. 2003. Matching Words and Pictures. *Proceedings of the Journal of Machine Learning Research* 3: 1107–1135.
31. Basu, S., Y. Yu, V.K. Singh, and R. Zimmermann. 2016. Videopedia: Lecture Video Recommendation for Educational Blogs Using Topic Modeling. In *Proceedings of the Springer International Conference on Multimedia Modeling*, 238–250.
32. Basu, S., Y. Yu, and R. Zimmermann. 2016. Fuzzy Clustering of Lecture Videos based on Topic Modeling. In *Proceedings of the IEEE International Workshop on Content-Based Multimedia Indexing*, 1–6.
33. Basu, S., R. Zimmermann, K.L. OHalloran, S. Tan, and K. Marissa. 2015. Performance Evaluation of Students Using Multimodal Learning Systems. In *Proceedings of the Springer International Conference on Multimedia Modeling,* 135–147.
34. Beeferman, D., A. Berger, and J. Lafferty. 1999. Statistical Models for Text Segmentation. *Proceedings of the Springer Machine Learning* 34 (1–3): 177–210.
35. Bernd, J., D. Borth, C. Carrano, J. Choi, B. Elizalde, G. Friedland, L. Gottlieb, K. Ni, R. Pearce, D. Poland, et al. 2015. Kickstarting the Commons: The YFCC100M and the YLI Corpora. In *Proceedings of the ACM Workshop on Community-Organized Multimodal Mining: Opportunities for Novel Solutions*, 1–6.
36. Bhatt, C.A., and M.S. Kankanhalli. 2011. Multimedia Data Mining: State of the Art and Challenges. *Proceedings of the Multimedia Tools and Applications* 51 (1): 35–76.
37. Bhatt, C.A., A. Popescu-Belis, M. Habibi, S. Ingram, S. Masneri, F. McInnes, N. Pappas, and O. Schreer. 2013. Multi-factor Segmentation for Topic Visualization and Recommendation: the MUST-VIS System. In *Proceedings of the ACM International Conference on Multimedia*, 365–368.
38. Bhattacharjee, S., W.C. Cheng, C.-F. Chou, L. Golubchik, and S. Khuller. 2000. BISTRO: A Frame-work for Building Scalable Wide-area Upload Applications. *Proceedings of the ACM SIGMETRICS Performance Evaluation Review* 28 (2): 29–35.
39. Cambria, E., J. Fu, F. Bisio, and S. Poria. 2015. AffectiveSpace 2: Enabling Affective Intuition for Concept-level Sentiment Analysis. In *Proceedings of the AAAI Conference on Artificial Intelligence*, 508–514.
40. Cambria, E., A. Livingstone, and A. Hussain. 2012. The Hourglass of Emotions. In *Proceedings of the Springer Cognitive Behavioural Systems*, 144–157.
41. Cambria, E., D. Olsher, and D. Rajagopal. 2014. SenticNet 3: A Common and Common-sense Knowledge Base for Cognition-driven Sentiment Analysis. In *Proceedings of the AAAI Conference on Artificial Intelligence*, 1515–1521.
42. Cambria, E., S. Poria, R. Bajpai, and B. Schuller. 2016. SenticNet 4: A Semantic Resource for Sentiment Analysis based on Conceptual Primitives. In *Proceedings of the International Conference on Computational Linguistics (COLING)*, 2666–2677.

43. Cambria, E., S. Poria, F. Bisio, R. Bajpai, and I. Chaturvedi. 2015. The CLSA Model: A Novel Framework for Concept-Level Sentiment Analysis. In *Proceedings of the Springer Computational Linguistics and Intelligent Text Processing*, 3–22.
44. Cambria, E., S. Poria, A. Gelbukh, and K. Kwok. 2014. Sentic API: A Common-sense based API for Concept-level Sentiment Analysis. *CEUR Workshop Proceedings* 144: 19–24.
45. Cao, J., Z. Huang, and Y. Yang. 2015. Spatial-aware Multimodal Location Estimation for Social Images. In *Proceedings of the ACM Conference on Multimedia Conference*, 119–128.
46. Chakraborty, I., H. Cheng, and O. Javed. 2014. Entity Centric Feature Pooling for Complex Event Detection. In *Proceedings of the Workshop on HuEvent at the ACM International Conference on Multimedia*, 1–5.
47. Che, X., H. Yang, and C. Meinel. 2013. Lecture Video Segmentation by Automatically Analyzing the Synchronized Slides. In *Proceedings of the ACM International Conference on Multimedia*, 345–348.
48. Chen, B., J. Wang, Q. Huang, and T. Mei. 2012. Personalized Video Recommendation through Tripartite Graph Propagation. In *Proceedings of the ACM International Conference on Multimedia*, 1133–1136.
49. Chen, S., L. Tong, and T. He. 2011. Optimal Deadline Scheduling with Commitment. In *Proceedings of the IEEE Annual Allerton Conference on Communication, Control, and Computing*, 111–118.
50. Chen, W.-B., C. Zhang, and S. Gao. 2012. Segmentation Tree based Multiple Object Image Retrieval. In *Proceedings of the IEEE International Symposium on Multimedia*, 214–221.
51. Chen, Y., and W.J. Heng. 2003. Automatic Synchronization of Speech Transcript and Slides in Presentation. *Proceedings of the IEEE International Symposium on Circuits and Systems* 2: 568–571.
52. Cohen, J. 1960. A Coefficient of Agreement for Nominal Scales. *Proceedings of the Durham Educational and Psychological Measurement* 20 (1): 37–46.
53. Cristani, M., A. Pesarin, C. Drioli, V. Murino, A. Rodà,M. Grapulin, and N. Sebe. 2010. Toward an Automatically Generated Soundtrack from Low-level Cross-modal Correlations for Automotive Scenarios. In *Proceedings of the ACM International Conference on Multimedia*, 551–560.
54. Dang-Nguyen, D.-T., L. Piras, G. Giacinto, G. Boato, and F.G. De Natale. 2015. A Hybrid Approach for Retrieving Diverse Social Images of Landmarks. In *Proceedings of the IEEE International Conference on Multimedia and Expo (ICME)*, 1–6.
55. Fabro, M. Del, A. Sobe, and L. Böszörmenyi. 2012. Summarization of Real-life Events based on Community-contributed Content. In *Proceedings of the International Conferences on Advances in Multimedia*, 119–126.
56. Du, L., W.L. Buntine, and M. Johnson. 2013. Topic Segmentation with a Structured Topic Model. In *Proceedings of the North American Chapter of the Association for Computational Linguistics: Human Language Technologies*, 190–200.
57. Fan, Q., K. Barnard, A. Amir, A. Efrat, and M. Lin. 2006. Matching Slides to Presentation Videos using SIFT and Scene Background Matching. In *Proceedings of the ACM International Conference on Multimedia*, 239–248.
58. Filatova, E. and V. Hatzivassiloglou. 2004. Event-based Extractive Summarization. In *Proceedings of the ACL Workshop on Summarization*, 104–111.
59. Firan, C.S., M. Georgescu, W. Nejdl, and R. Paiu. 2010. Bringing Order to Your Photos: Event-driven Classification of Flickr Images based on Social Knowledge. In *Proceedings of the ACM International Conference on Information and Knowledge Management*, 189–198.
60. Gao, S., C. Zhang, and W.-B. Chen. 2012. An Improvement of Color Image Segmentation through Projective Clustering. In *Proceedings of the IEEE International Conference on Information Reuse and Integration*, 152–158.
61. Garg, N. and I. Weber. 2008. Personalized, Interactive Tag Recommendation for Flickr. In *Proceedings of the ACM Conference on Recommender Systems*, 67–74.

62. Ghias, A., J. Logan, D. Chamberlin, and B.C. Smith. 1995. Query by Humming: Musical Information Retrieval in an Audio Database. In *Proceedings of the ACM International Conference on Multimedia*, 231–236.
63. Golder, S.A., and B.A. Huberman. 2006. Usage Patterns of Collaborative Tagging Systems. *Proceedings of the Journal of Information Science* 32 (2): 198–208.
64. Gozali, J.P., M.-Y. Kan, and H. Sundaram. 2012. Hidden Markov Model for Event Photo Stream Segmentation. In *Proceedings of the IEEE International Conference on Multimedia and Expo Workshops*, 25–30.
65. Guo, Y., L. Zhang, Y. Hu, X. He, and J. Gao. 2016. Ms-celeb-1m: Challenge of recognizing one million celebrities in the real world. *Proceedings of the Society for Imaging Science and Technology Electronic Imaging* 2016 (11): 1–6.
66. Hanjalic, A., and L.-Q. Xu. 2005. Affective Video Content Representation and Modeling. *Proceedings of the IEEE Transactions on Multimedia* 7 (1): 143–154.
67. Haubold, A. and J.R. Kender. 2005. Augmented Segmentation and Visualization for Presentation Videos. In *Proceedings of the ACM International Conference on Multimedia*, 51–60.
68. Healey, J.A., and R.W. Picard. 2005. Detecting Stress during Real-world Driving Tasks using Physiological Sensors. *Proceedings of the IEEE Transactions on Intelligent Transportation Systems* 6 (2): 156–166.
69. Hefeeda, M., and C.-H. Hsu. 2010. On Burst Transmission Scheduling in Mobile TV Broadcast Networks. *Proceedings of the IEEE/ACM Transactions on Networking* 18 (2): 610–623.
70. Hevner, K. 1936. Experimental Studies of the Elements of Expression in Music. *Proceedings of the American Journal of Psychology* 48: 246–268.
71. Hochbaum, D.S. 1996. Approximating Covering and Packing Problems: Set Cover, Vertex Cover, Independent Set, and related Problems. In *Proceedings of the PWS Approximation algorithms for NP-hard problems*, 94–143.
72. Hong, R., J. Tang, H.-K. Tan, S. Yan, C. Ngo, and T.-S. Chua. 2009. Event Driven Summarization for Web Videos. In *Proceedings of the ACM SIGMM Workshop on Social Media*, 43–48.
73. P. ITU-T Recommendation. 1999. Subjective Video Quality Assessment Methods for Multimedia Applications.
74. Jiang, L., A.G. Hauptmann, and G. Xiang. 2012. Leveraging High-level and Low-level Features for Multimedia Event Detection. In *Proceedings of the ACM International Conference on Multimedia*, 449–458.
75. Joachims, T., T. Finley, and C.-N. Yu. 2009. Cutting-plane Training of Structural SVMs. *Proceedings of the Machine Learning Journal* 77 (1): 27–59.
76. Johnson, J., L. Ballan, and L. Fei-Fei. 2015. Love Thy Neighbors: Image Annotation by Exploiting Image Metadata. In *Proceedings of the IEEE International Conference on Computer Vision*, 4624–4632.
77. Johnson, J., A. Karpathy, and L. Fei-Fei. 2015. Densecap: Fully Convolutional Localization Networks for Dense Captioning. *In Proceedings of the arXiv preprint arXiv:1511.07571*.
78. Jokhio, F., A. Ashraf, S. Lafond, I. Porres, and J. Lilius. 2013. Prediction-based dynamic resource allocation for video transcoding in cloud computing. In *Proceedings of the IEEE International Conference on Parallel, Distributed and Network-Based Processing*, 254–261.
79. Kaminskas, M., I. Fernández-Tobías, F. Ricci, and I. Cantador. 2014. Knowledge-based Identification of Music Suited for Places of Interest. *Proceedings of the Springer Information Technology & Tourism* 14 (1): 73–95.
80. Kaminskas, M. and F. Ricci. 2011. Location-adapted Music Recommendation using Tags. In *Proceedings of the Springer User Modeling, Adaption and Personalization*, 183–194.
81. Kan, M.-Y. 2001. Combining Visual Layout and Lexical Cohesion Features for Text Segmentation. *In Proceedings of the Citeseer*.
82. Kan, M.-Y. 2003. *Automatic Text Summarization as Applied to Information Retrieval*. PhD thesis, Columbia University.

83. Kan, M.-Y., J.L. Klavans, and K.R. McKeown.1998. Linear Segmentation and Segment Significance. *In Proceedings of the arXiv preprint cs/9809020.*
84. Kan, M.-Y., K.R. McKeown, and J.L. Klavans. 2001. Applying Natural Language Generation to Indicative Summarization. *Proceedings of the ACL European Workshop on Natural Language Generation* 8: 1–9.
85. Kang, H.B. 2003. Affective Content Detection using HMMs. In *Proceedings of the ACM International Conference on Multimedia*, 259–262.
86. Kang, Y.-L., J.-H. Lim, M.S. Kankanhalli, C.-S. Xu, and Q. Tian. 2004. Goal Detection in Soccer Video using Audio/Visual Keywords. *Proceedings of the IEEE International Conference on Image Processing* 3: 1629–1632.
87. Kang, Y.-L., J.-H. Lim, Q. Tian, and M.S. Kankanhalli. 2003. Soccer Video Event Detection with Visual Keywords. In *Proceedings of the Joint Conference of International Conference on Information, Communications and Signal Processing, and Pacific Rim Conference on Multimedia*, 3:1796–1800.
88. Kankanhalli, M.S., and T.-S. Chua. 2000. Video Modeling using Strata-based Annotation. *Proceedings of the IEEE MultiMedia* 7 (1): 68–74.
89. Kennedy, L., M. Naaman, S. Ahern, R. Nair, and T. Rattenbury. 2007. How Flickr Helps us Make Sense of the World: Context and Content in Community-contributed Media Collections. In *Proceedings of the ACM International Conference on Multimedia*, 631–640.
90. Kennedy, L.S., S.-F. Chang, and I.V. Kozintsev. 2006. To Search or to Label?: Predicting the Performance of Search-based Automatic Image Classifiers. In *Proceedings of the ACM International Workshop on Multimedia Information Retrieval*, 249–258.
91. Kim, Y.E., E.M. Schmidt, R. Migneco, B.G. Morton, P. Richardson, J. Scott, J.A. Speck, and D. Turnbull. 2010. Music Emotion Recognition: A State of the Art Review. In *Proceedings of the International Society for Music Information Retrieval*, 255–266.
92. Klavans, J.L., K.R. McKeown, M.-Y. Kan, and S. Lee. 1998. Resources for Evaluation of Summarization Techniques. *In Proceedings of the arXiv preprint cs/9810014.*
93. Ko, Y. 2012. A Study of Term Weighting Schemes using Class Information for Text Classification. In *Proceedings of the ACM Special Interest Group on Information Retrieval*, 1029–1030.
94. Kort, B., R. Reilly, and R.W. Picard. 2001. An Affective Model of Interplay between Emotions and Learning: Reengineering Educational Pedagogy-Building a Learning Companion. *Proceedings of the IEEE International Conference on Advanced Learning Technologies* 1: 43–47.
95. Kucuktunc, O., U. Gudukbay, and O. Ulusoy. 2010. Fuzzy Color Histogram-based Video Segmentation. *Proceedings of the Computer Vision and Image Understanding* 114 (1): 125–134.
96. Kuo, F.-F., M.-F. Chiang, M.-K. Shan, and S.-Y. Lee. 2005. Emotion-based Music Recommendation by Association Discovery from Film Music. In *Proceedings of the ACM International Conference on Multimedia*, 507–510.
97. Lacy, S., T. Atwater, X. Qin, and A. Powers. 1988. Cost and Competition in the Adoption of Satellite News Gathering Technology. *Proceedings of the Taylor & Francis Journal of Media Economics* 1 (1): 51–59.
98. Lambert, P., W. De Neve, P. De Neve, I. Moerman, P. Demeester, and R. Van de Walle. 2006. Rate-distortion performance of H. 264/AVC compared to state-of-the-art video codecs. *Proceedings of the IEEE Transactions on Circuits and Systems for Video Technology* 16 (1): 134–140.
99. Laurier, C., M. Sordo, J. Serrà, and P. Herrera. 2009. Music Mood Representations from Social Tags. In *Proceedings of the International Society for Music Information Retrieval*, 381–386.
100. Li, C.T. and M.K. Shan. 2007. Emotion-based Impressionism Slideshow with Automatic Music Accompaniment. In *Proceedings of the ACM International Conference on Multimedia*, 839–842.

101. Li, J., and J.Z. Wang. 2008. Real-time Computerized Annotation of Pictures. *Proceedings of the IEEE Transactions on Pattern Analysis and Machine Intelligence* 30 (6): 985–1002.
102. Li, X., C.G. Snoek, and M. Worring. 2009. Learning Social Tag Relevance by Neighbor Voting. *Proceedings of the IEEE Transactions on Multimedia* 11 (7): 1310–1322.
103. Li, X., T. Uricchio, L. Ballan, M. Bertini, C.G. Snoek, and A.D. Bimbo. 2016. Socializing the Semantic Gap: A Comparative Survey on Image Tag Assignment, Refinement, and Retrieval. *Proceedings of the ACM Computing Surveys (CSUR)* 49 (1): 14.
104. Li, Z., Y. Huang, G. Liu, F. Wang, Z.-L. Zhang, and Y. Dai. 2012. Cloud Transcoder: Bridging the Format and Resolution Gap between Internet Videos and Mobile Devices. In *Proceedings of the ACM International Workshop on Network and Operating System Support for Digital Audio and Video*, 33–38.
105. Liang, C., Y. Guo, and Y. Liu. 2008. Is Random Scheduling Sufficient in P2P Video Streaming? In *Proceedings of the IEEE International Conference on Distributed Computing Systems*, 53–60. IEEE.
106. Lim, J.-H., Q. Tian, and P. Mulhem. 2003. Home Photo Content Modeling for Personalized Event-based Retrieval. *Proceedings of the IEEE MultiMedia* 4: 28–37.
107. Lin, M., M. Chau, J. Cao, and J.F. Nunamaker Jr. 2005. Automated Video Segmentation for Lecture Videos: A Linguistics-based Approach. *Proceedings of the IGI Global International Journal of Technology and Human Interaction* 1 (2): 27–45.
108. Liu, C.L., and J.W. Layland. 1973. Scheduling Algorithms for Multiprogramming in a Hard-real-time Environment. *Proceedings of the ACM Journal of the ACM* 20 (1): 46–61.
109. Liu, D., X.-S. Hua, L. Yang, M. Wang, and H.-J. Zhang. 2009. Tag Ranking. In *Proceedings of the ACM World Wide Web Conference*, 351–360.
110. Liu, T., C. Rosenberg, and H.A. Rowley. 2007. Clustering Billions of Images with Large Scale Nearest Neighbor Search. In *Proceedings of the IEEE Winter Conference on Applications of Computer Vision*, 28–28.
111. Liu, X. and B. Huet. 2013. Event Representation and Visualization from Social Media. In *Proceedings of the Springer Pacific-Rim Conference on Multimedia*, 740–749.
112. Liu, Y., D. Zhang, G. Lu, and W.-Y. Ma. 2007. A Survey of Content-based Image Retrieval with High-level Semantics. *Proceedings of the Elsevier Pattern Recognition* 40 (1): 262–282.
113. Livingston, S., and D.A.V. BELLE. 2005. The Effects of Satellite Technology on Newsgathering from Remote Locations. *Proceedings of the Taylor & Francis Political Communication* 22 (1): 45–62.
114. Long, R., H. Wang, Y. Chen, O. Jin, and Y. Yu. 2011. Towards Effective Event Detection, Tracking and Summarization on Microblog Data. In *Proceedings of the Springer Web-Age Information Management*, 652–663.
115. L. Lu, H. You, and H. Zhang. 2001. A New Approach to Query by Humming in Music Retrieval. In *Proceedings of the IEEE International Conference on Multimedia and Expo*, 22–25.
116. Lu, Y., H. To, A. Alfarrarjeh, S.H. Kim, Y. Yin, R. Zimmermann, and C. Shahabi. 2016. GeoUGV: User-generated Mobile Video Dataset with Fine Granularity Spatial Metadata. In *Proceedings of the ACM International Conference on Multimedia Systems*, 43.
117. Mao, J., W. Xu, Y. Yang, J. Wang, Z. Huang, and A. Yuille. 2014. Deep Captioning with Multimodal Recurrent Neural Networks (M-RNN). *In Proceedings of the arXiv preprint arXiv:1412.6632*.
118. Matusiak, K.K. 2006. Towards User-centered Indexing in Digital Image Collections. *Proceedings of the OCLC Systems & Services: International Digital Library Perspectives* 22 (4): 283–298.
119. McDuff, D., R. El Kaliouby, E. Kodra, and R. Picard. 2013. Measuring Voter's Candidate Preference Based on Affective Responses to Election Debates. In *Proceedings of the IEEE Humaine Association Conference on Affective Computing and Intelligent Interaction*, 369–374.

120. McKeown, K.R., J.L. Klavans, and M.-Y. Kan. Method and System for Topical Segmentation, Segment Significance and Segment Function, 29 2002. US Patent 6,473,730.
121. Mezaris, V., A. Scherp, R. Jain, M. Kankanhalli, H. Zhou, J. Zhang, L. Wang, and Z. Zhang. 2011. Modeling and Rrepresenting Events in Multimedia. In *Proceedings of the ACM International Conference on Multimedia*, 613–614.
122. Mezaris, V., A. Scherp, R. Jain, and M.S. Kankanhalli. 2014. Real-life Events in Multimedia: Detection, Representation, Retrieval, and Applications. *Proceedings of the Springer Multimedia Tools and Applications* 70 (1): 1–6.
123. Miller, G., and C. Fellbaum. 1998. *Wordnet: An Electronic Lexical Database*. Cambridge, MA: MIT Press.
124. Miller, G.A. 1995. WordNet: A Lexical Database for English. *Proceedings of the Communications of the ACM* 38 (11): 39–41.
125. Moxley, E., J. Kleban, J. Xu, and B. Manjunath. 2009. Not All Tags are Created Equal: Learning Flickr Tag Semantics for Global Annotation. In *Proceedings of the IEEE International Conference on Multimedia and Expo*, 1452–1455.
126. Mulhem, P., M.S. Kankanhalli, J. Yi, and H. Hassan. 2003. Pivot Vector Space Approach for Audio-Video Mixing. *Proceedings of the IEEE MultiMedia* 2: 28–40.
127. Naaman, M. 2012. Social Multimedia: Highlighting Opportunities for Search and Mining of Multimedia Data in Social Media Applications. *Proceedings of the Springer Multimedia Tools and Applications* 56 (1): 9–34.
128. Natarajan, P., P.K. Atrey, and M. Kankanhalli. 2015. Multi-Camera Coordination and Control in Surveillance Systems: A Survey. *Proceedings of the ACM Transactions on Multimedia Computing, Communications, and Applications* 11 (4): 57.
129. Nayak, M.G. 2004. *Music Synthesis for Home Videos*. PhD thesis.
130. Neo, S.-Y., J. Zhao, M.-Y. Kan, and T.-S. Chua. 2006. Video Retrieval using High Level Features: Exploiting Query Matching and Confidence-based Weighting. In *Proceedings of the Springer International Conference on Image and Video Retrieval*, 143–152.
131. Ngo, C.-W., F. Wang, and T.-C. Pong. 2003. Structuring Lecture Videos for Distance Learning Applications. In *Proceedings of the IEEE International Symposium on Multimedia Software Engineering*, 215–222.
132. Nguyen, V.-A., J. Boyd-Graber, and P. Resnik. 2012. SITS: A Hierarchical Nonparametric Model using Speaker Identity for Topic Segmentation in Multiparty Conversations. In *Proceedings of the Annual Meeting of the Association for Computational Linguistics*, 78–87.
133. Nwana, A.O. and T. Chen. 2016. Who Ordered This?: Exploiting Implicit User Tag Order Preferences for Personalized Image Tagging. *In Proceedings of the arXiv preprint arXiv:1601.06439*.
134. Papagiannopoulou, C. and V. Mezaris. 2014. Concept-based Image Clustering and Summarization of Event-related Image Collections. In *Proceedings of the Workshop on HuEvent at the ACM International Conference on Multimedia*, 23–28.
135. Park, M.H., J.H. Hong, and S.B. Cho. 2007. Location-based Recommendation System using Bayesian User's Preference Model in Mobile Devices. In *Proceedings of the Springer Ubiquitous Intelligence and Computing*, 1130–1139.
136. Petkos, G., S. Papadopoulos, V. Mezaris, R. Troncy, P. Cimiano, T. Reuter, and Y. Kompatsiaris. 2014. Social Event Detection at MediaEval: a Three-Year Retrospect of Tasks and Results. In *Proceedings of the Workshop on Social Events in Web Multimedia at ACM International Conference on Multimedia Retrieval*.
137. Pevzner, L., and M.A. Hearst. 2002. A Critique and Improvement of an Evaluation Metric for Text Segmentation. *Proceedings of the Computational Linguistics* 28 (1): 19–36.
138. Picard, R.W., and J. Klein. 2002. Computers that Recognise and Respond to User Emotion: Theoretical and Practical Implications. *Proceedings of the Interacting with Computers* 14 (2): 141–169.

139. Picard, R.W., E. Vyzas, and J. Healey. 2001. Toward machine emotional intelligence: Analysis of affective physiological state. *Proceedings of the IEEE Transactions on Pattern Analysis and Machine Intelligence* 23 (10): 1175–1191.
140. Poisson, S.D. and C.H. Schnuse. 1841. Recherches Sur La Pprobabilité Des Jugements En Mmatieré Criminelle Et En Matieré Civile. Meyer.
141. Poria, S., E. Cambria, R. Bajpai, and A. Hussain. 2017. A Review of Affective Computing: From Unimodal Analysis to Multimodal Fusion. *Proceedings of the Elsevier Information Fusion* 37: 98–125.
142. Poria, S., E. Cambria, and A. Gelbukh. 2016. Aspect Extraction for Opinion Mining with a Deep Convolutional Neural Network. *Proceedings of the Elsevier Knowledge-Based Systems* 108: 42–49.
143. Poria, S., E. Cambria, A. Gelbukh, F. Bisio, and A. Hussain. 2015. Sentiment Data Flow Analysis by Means of Dynamic Linguistic Patterns. *Proceedings of the IEEE Computational Intelligence Magazine* 10 (4): 26–36.
144. Poria, S., E. Cambria, and A.F. Gelbukh. 2015. Deep Convolutional Neural Network Textual Features and Multiple Kernel Learning for Utterance-level Multimodal Sentiment Analysis. In *Proceedings of the EMNLP*, 2539–2544.
145. Poria, S., E. Cambria, D. Hazarika, N. Mazumder, A. Zadeh, and L.-P. Morency. 2017. Context-Dependent Sentiment Analysis in User-Generated Videos. *In Proceedings of the Association for Computational Linguistics.*
146. Poria, S., E. Cambria, D. Hazarika, and P. Vij. 2016. A Deeper Look into Sarcastic Tweets using Deep Convolutional Neural Networks. In *Proceedings of the International Conference on Computational Linguistics (COLING).*
147. Poria, S., E. Cambria, N. Howard, G.-B. Huang, and A. Hussain. 2016. Fusing Audio Visual and Textual Clues for Sentiment Analysis from Multimodal Content. *Proceedings of the Elsevier Neurocomputing* 174: 50–59.
148. Poria, S., E. Cambria, N. Howard, and A. Hussain. 2015. Enhanced SenticNet with Affective Labels for Concept-based Opinion Mining: Extended Abstract. In *Proceedings of the International Joint Conference on Artificial Intelligence.*
149. Poria, S., E. Cambria, A. Hussain, and G.-B. Huang. 2015. Towards an Intelligent Framework for Multimodal Affective Data Analysis. *Proceedings of the Elsevier Neural Networks* 63: 104–116.
150. Poria, S., E. Cambria, L.-W. Ku, C. Gui, and A. Gelbukh. 2014. A Rule-based Approach to Aspect Extraction from Product Reviews. In *Proceedings of the Second Workshop on Natural Language Processing for Social Media (SocialNLP)*, 28–37.
151. Poria, S., I. Chaturvedi, E. Cambria, and F. Bisio. 2016. Sentic LDA: Improving on LDA with Semantic Similarity for Aspect-based Sentiment Analysis. In *Proceedings of the IEEE International Joint Conference on Neural Networks (IJCNN)*, 4465–4473.
152. Poria, S., I. Chaturvedi, E. Cambria, and A. Hussain. 2016. Convolutional MKL based Multimodal Emotion Recognition and Sentiment Analysis. In *Proceedings of the IEEE International Conference on Data Mining (ICDM)*, 439–448.
153. Poria, S., A. Gelbukh, B. Agarwal, E. Cambria, and N. Howard. 2014. Sentic Demo: A Hybrid Concept-level Aspect-based Sentiment Analysis Toolkit. *In Proceedings of the ESWC.*
154. Poria, S., A. Gelbukh, E. Cambria, D. Das, and S. Bandyopadhyay. 2012. Enriching SenticNet Polarity Scores Through Semi-Supervised Fuzzy Clustering. In *Proceedings of the IEEE International Conference on Data Mining Workshops (ICDMW)*, 709–716.
155. Poria, S., A. Gelbukh, E. Cambria, A. Hussain, and G.-B. Huang. 2014. EmoSenticSpace: A Novel Framework for Affective Common-sense Reasoning. *Proceedings of the Elsevier Knowledge-Based Systems* 69: 108–123.
156. Poria, S., A. Gelbukh, E. Cambria, P. Yang, A. Hussain, and T. Durrani. 2012. Merging SenticNet and WordNet-Affect Emotion Lists for Sentiment Analysis. *Proceedings of the IEEE International Conference on Signal Processing (ICSP)* 2: 1251–1255.

157. Poria, S., A. Gelbukh, A. Hussain, S. Bandyopadhyay, and N. Howard. 2013. Music Genre Classification: A Semi-Supervised Approach. In *Proceedings of the Springer Mexican Conference on Pattern Recognition*, 254–263.
158. Poria, S., N. Ofek, A. Gelbukh, A. Hussain, and L. Rokach. 2014. Dependency Tree-based Rules for Concept-level Aspect-based Sentiment Analysis. In *Proceedings of the Springer Semantic Web Evaluation Challenge*, 41–47.
159. Poria, S., H. Peng, A. Hussain, N. Howard, and E. Cambria. 2017. Ensemble Application of Convolutional Neural Networks and Multiple Kernel Learning for Multimodal Sentiment Analysis. *In Proceedings of the Elsevier Neurocomputing*.
160. Pye, D., N.J. Hollinghurst, T.J. Mills, and K.R. Wood. 1998. Audio-visual Segmentation for Content-based Retrieval. In *Proceedings of the International Conference on Spoken Language Processing*.
161. Qiao, Z., P. Zhang, C. Zhou, Y. Cao, L. Guo, and Y. Zhang. 2014. Event Recommendation in Event-based Social Networks.
162. Raad, E.J. and R. Chbeir. 2014. Foto2Events: From Photos to Event Discovery and Linking in Online Social Networks. In *Proceedings of the IEEE Big Data and Cloud Computing*, 508–515, .
163. Radsch, C.C. 2013. The Revolutions will be Blogged: Cyberactivism and the 4th Estate in Egypt. Doctoral Disseration. American University.
164. Rae, A., B. Sigurbjörnssön, and R. van Zwol. 2010. Improving Tag Recommendation using Social Networks. In *Proceedings of the Adaptivity, Personalization and Fusion of Heterogeneous Information*, 92–99.
165. Rahmani, H., B. Piccart, D. Fierens, and H. Blockeel. 2010. Three Complementary Approaches to Context Aware Movie Recommendation. In *Proceedings of the ACM Workshop on Context-Aware Movie Recommendation*, 57–60.
166. Rattenbury, T., N. Good, and M. Naaman. 2007. Towards Automatic Extraction of Event and Place Semantics from Flickr Tags. In *Proceedings of the ACM Special Interest Group on Information Retrieval*.
167. Rawat, Y. and M. S. Kankanhalli. 2016. ConTagNet: Exploiting User Context for Image Tag Recommendation. In *Proceedings of the ACM International Conference on Multimedia*, 1102–1106.
168. Repp, S., A. Groß, and C. Meinel. 2008. Browsing within Lecture Videos based on the Chain Index of Speech Transcription. *Proceedings of the IEEE Transactions on Learning Technologies* 1 (3): 145–156.
169. Repp, S. and C. Meinel. 2006. Semantic Indexing for Recorded Educational Lecture Videos. In *Proceedings of the IEEE International Conference on Pervasive Computing and Communications Workshops*, 5.
170. Repp, S., J. Waitelonis, H. Sack, and C. Meinel. 2007. Segmentation and Annotation of Audiovisual Recordings based on Automated Speech Recognition. In *Proceedings of the Springer Intelligent Data Engineering and Automated Learning*, 620–629.
171. Russell, J.A. 1980. A Circumplex Model of Affect. *Proceedings of the Journal of Personality and Social Psychology* 39: 1161–1178.
172. Sahidullah, M., and G. Saha. 2012. Design, Analysis and Experimental Evaluation of Block based Transformation in MFCC Computation for Speaker Recognition. *Proceedings of the Speech Communication* 54: 543–565.
173. J. Salamon, J. Serra, and E. Gomez´. Tonal Representations for Music Retrieval: From Version Identification to Query-by-Humming. *In* Proceedings of the Springer International Journal of Multimedia Information Retrieval, 2(1):45–58, 2013.
174. Schedl, M. and D. Schnitzer. 2014. Location-Aware Music Artist Recommendation. In *Proceedings of the Springer MultiMedia Modeling*, 205–213.
175. M. Schedl and F. Zhou. 2016. Fusing Web and Audio Predictors to Localize the Origin of Music Pieces for Geospatial Retrieval. In *Proceedings of the Springer European Conference on Information Retrieval*, 322–334.

176. Scherp, A., and V. Mezaris. 2014. Survey on Modeling and Indexing Events in Multimedia. *Proceedings of the Springer Multimedia Tools and Applications* 70 (1): 7–23.
177. Scherp, A., V. Mezaris, B. Ionescu, and F. De Natale. 2014. HuEvent '14: Workshop on Human-Centered Event Understanding from Multimedia. In *Proceedings of the ACM International Conference on Multimedia*, 1253–1254, .
178. Schmitz, P. 2006. Inducing Ontology from Flickr Tags. In *Proceedings of the Collaborative Web Tagging Workshop at ACM World Wide Web Conference*, volume 50.
179. Schuller, B., C. Hage, D. Schuller, and G. Rigoll. 2010. Mister DJ, Cheer Me Up!: Musical and Textual Features for Automatic Mood Classification. *Proceedings of the Journal of New Music Research* 39 (1): 13–34.
180. Shah, R.R., M. Hefeeda, R. Zimmermann, K. Harras, C.-H. Hsu, and Y. Yu. 2016. NEWSMAN: Uploading Videos over Adaptive Middleboxes to News Servers In Weak Network Infrastructures. In *Proceedings of the Springer International Conference on Multimedia Modeling*, 100–113.
181. Shah, R.R., A. Samanta, D. Gupta, Y. Yu, S. Tang, and R. Zimmermann. 2016. PROMPT: Personalized User Tag Recommendation for Social Media Photos Leveraging Multimodal Information. In *Proceedings of the ACM International Conference on Multimedia*, 486–492.
182. Shah, R.R., A.D. Shaikh, Y. Yu, W. Geng, R. Zimmermann, and G. Wu. 2015. EventBuilder: Real-time Multimedia Event Summarization by Visualizing Social Media. In *Proceedings of the ACM International Conference on Multimedia*, 185–188.
183. Shah, R.R., Y. Yu, A.D. Shaikh, S. Tang, and R. Zimmermann. 2014. ATLAS: Automatic Temporal Segmentation and Annotation of Lecture Videos Based on Modelling Transition Time. In *Proceedings of the ACM International Conference on Multimedia*, 209–212.
184. Shah, R.R., Y. Yu, A.D. Shaikh, and R. Zimmermann. 2015. TRACE: A Linguistic-based Approach for Automatic Lecture Video Segmentation Leveraging Wikipedia Texts. In *Proceedings of the IEEE International Symposium on Multimedia*, 217–220.
185. Shah, R.R., Y. Yu, S. Tang, S. Satoh, A. Verma, and R. Zimmermann. 2016. Concept-Level Multimodal Ranking of Flickr Photo Tags via Recall Based Weighting. In *Proceedings of the MMCommon's Workshop at ACM International Conference on Multimedia*, 19–26.
186. Shah, R.R., Y. Yu, A. Verma, S. Tang, A.D. Shaikh, and R. Zimmermann. 2016. Leveraging Multimodal Information for Event Summarization and Concept-level Sentiment Analysis. In *Proceedings of the Elsevier Knowledge-Based Systems*, 102–109.
187. Shah, R.R., Y. Yu, and R. Zimmermann. 2014. ADVISOR: Personalized Video Soundtrack Recommendation by Late Fusion with Heuristic Rankings. In *Proceedings of the ACM International Conference on Multimedia*, 607–616.
188. Shah, R.R., Y. Yu, and R. Zimmermann. 2014. User Preference-Aware Music Video Generation Based on Modeling Scene Moods. In *Proceedings of the ACM International Conference on Multimedia Systems*, 156–159.
189. Shaikh, A.D., M. Jain, M. Rawat, R.R. Shah, and M. Kumar. 2013. Improving Accuracy of SMS Based FAQ Retrieval System. In *Proceedings of the Springer Multilingual Information Access in South Asian Languages*, 142–156.
190. Shaikh, A.D., R.R. Shah, and R. Shaikh. 2013. SMS based FAQ Retrieval for Hindi, English and Malayalam. In *Proceedings of the ACM Forum on Information Retrieval Evaluation*, 9.
191. Shamma, D.A., R. Shaw, P.L. Shafton, and Y. Liu. 2007. Watch What I Watch: Using Community Activity to Understand Content. In *Proceedings of the ACM International Workshop on Multimedia Information Retrieval*, 275–284.
192. Shaw, B., J. Shea, S. Sinha, and A. Hogue. 2013. Learning to Rank for Spatiotemporal Search. In *Proceedings of the ACM International Conference on Web Search and Data Mining*, 717–726.
193. Sigurbjörnsson, B. and R. Van Zwol. 2008. Flickr Tag Recommendation based on Collective Knowledge. In *Proceedings of the ACM World Wide Web Conference*, 327–336.

194. Snoek, C.G., M. Worring, and A.W.Smeulders. 2005. Early versus Late Fusion in Semantic Video Analysis. In *Proceedings of the ACM International Conference on Multimedia*, 399–402.
195. Snoek, C.G., M. Worring, J.C. Van Gemert, J.-M. Geusebroek, and A.W. Smeulders. 2006. The Challenge Problem for Automated Detection of 101 Semantic Concepts in Multimedia. In *Proceedings of the ACM International Conference on Multimedia*, 421–430.
196. Soleymani, M., J.J.M. Kierkels, G. Chanel, and T. Pun. 2009. A Bayesian Framework for Video Affective Representation. In *Proceedings of the IEEE International Conference on Affective Computing and Intelligent Interaction and Workshops*, 1–7.
197. Stober, S., and A. . Nürnberger. 2013. Adaptive Music Retrieval–a State of the Art. *Proceedings of the Springer Multimedia Tools and Applications* 65 (3): 467–494.
198. Stoyanov, V., N. Gilbert, C. Cardie, and E. Riloff. 2009. Conundrums in Noun Phrase Coreference Resolution: Making Sense of the State-of-the-art. In *Proceedings of the ACL International Joint Conference on Natural Language Processing of the Asian Federation of Natural Language Processing*, 656–664.
199. Stupar, A. and S. Michel. 2011. Picasso: Automated Soundtrack Suggestion for Multi-modal Data. In *Proceedings of the ACM Conference on Information and Knowledge Management*, 2589–2592.
200. Thayer, R.E. 1989. *The Biopsychology of Mood and Arousal*. New York: Oxford University Press.
201. Thomee, B., B. Elizalde, D.A. Shamma, K. Ni, G. Friedland, D. Poland, D. Borth, and L.-J. Li. 2016. YFCC100M: The New Data in Multimedia Research. *Proceedings of the Communications of the ACM* 59 (2): 64–73.
202. Tirumala, A., F. Qin, J. Dugan, J. Ferguson, and K. Gibbs. 2005. Iperf: The TCP/UDP Bandwidth Measurement Tool. http://dast.nlanr.net/Projects/Iperf/
203. Torralba, A., R. Fergus, and W.T. Freeman. 2008. 80 Million Tiny Images: A Large Data set for Nonparametric Object and Scene Recognition. *Proceedings of the IEEE Transactions on Pattern Analysis and Machine Intelligence* 30 (11): 1958–1970.
204. Toutanova, K., D. Klein, C.D. Manning, and Y. Singer. 2003. Feature-Rich Part-of-Speech Tagging with a Cyclic Dependency Network. In *Proceedings of the North American Chapter of the Association for Computational Linguistics on Human Language Technology*, 173–180.
205. Toutanova, K. and C.D. Manning. 2000. Enriching the Knowledge Sources used in a Maximum Entropy Part-of-Speech Tagger. In *Proceedings of the Annual Meeting of the Association for Computational Linguistics*, 63–70.
206. Utiyama, M. and H. Isahara. 2001. A Statistical Model for Domain-Independent Text Segmentation. In *Proceedings of the Annual Meeting on Association for Computational Linguistics*, 499–506.
207. Vishal, K., C. Jawahar, and V. Chari. 2015. Accurate Localization by Fusing Images and GPS Signals. In *Proceedings of the IEEE Computer Vision and Pattern Recognition Workshops*, 17–24.
208. Wang, C., F. Jing, L. Zhang, and H.-J. Zhang. 2008. Scalable Search-based Image Annotation. *Proceedings of the Springer Multimedia Systems* 14 (4): 205–220.
209. Wang, H.L., and L.F. Cheong. 2006. Affective Understanding in Film. *Proceedings of the IEEE Transactions on Circuits and Systems for Video Technology* 16 (6): 689–704.
210. Wang, J., J. Zhou, H. Xu, T. Mei, X.-S. Hua, and S. Li. 2014. Image Tag Refinement by Regularized Latent Dirichlet Allocation. *Proceedings of the Elsevier Computer Vision and Image Understanding* 124: 61–70.
211. Wang, P., H. Wang, M. Liu, and W. Wang. 2010. An Algorithmic Approach to Event Summarization. In *Proceedings of the ACM Special Interest Group on Management of Data*, 183–194.
212. Wang, X., Y. Jia, R. Chen, and B. Zhou. 2015. Ranking User Tags in Micro-Blogging Website. In *Proceedings of the IEEE ICISCE*, 400–403.

213. Wang, X., L. Tang, H. Gao, and H. Liu. 2010. Discovering Overlapping Groups in Social Media. In *Proceedings of the IEEE International Conference on Data Mining*, 569–578.
214. Wang, Y. and M.S. Kankanhalli. 2015. Tweeting Cameras for Event Detection. In *Proceedings of the IW3C2 International Conference on World Wide Web*, 1231–1241.
215. Webster, A.A., C.T. Jones, M.H. Pinson, S.D. Voran, and S. Wolf. 1993. Objective Video Quality Assessment System based on Human Perception. In *Proceedings of the IS&T/SPIE's Symposium on Electronic Imaging: Science and Technology*, 15–26. International Society for Optics and Photonics.
216. Wei, C.Y., N. Dimitrova, and S.-F. Chang. 2004. Color-mood Analysis of Films based on Syntactic and Psychological Models. In *Proceedings of the IEEE International Conference on Multimedia and Expo*, 831–834.
217. Whissel, C. 1989. The Dictionary of Affect in Language. In *Emotion: Theory, Research and Experience. Vol. 4. The Measurement of Emotions*, ed. R. Plutchik and H. Kellerman, 113–131. New York: Academic.
218. Wu, L., L. Yang, N. Yu, and X.-S. Hua. 2009. Learning to Tag. In *Proceedings of the ACM World Wide Web Conference*, 361–370.
219. Xiao, J., W. Zhou, X. Li, M. Wang, and Q. Tian. 2012. Image Tag Re-ranking by Coupled Probability Transition. In *Proceedings of the ACM International Conference on Multimedia*, 849–852.
220. Xie, D., B. Qian, Y. Peng, and T. Chen. 2009. A Model of Job Scheduling with Deadline for Video-on-Demand System. In *Proceedings of the IEEE International Conference on Web Information Systems and Mining*, 661–668.
221. Xu, M., L.-Y. Duan, C. Xu, M. Kankanhalli, and Q. Tian. 2003. Event Detection in Basketball Video using Multiple Modalities. *Proceedings of the IEEE Joint Conference of the Fourth International Conference on Information, Communications and Signal Processing, and Fourth Pacific Rim Conference on Multimedia* 3: 1526–1530.
222. Xu, M., N.C. Maddage, C. Xu, M. Kankanhalli, and Q. Tian. 2003. Creating Audio Keywords for Event Detection in Soccer Video. In *Proceedings of the IEEE International Conference on Multimedia and Expo*, 2:II–281.
223. Yamamoto, N., J. Ogata, and Y. Ariki. 2003. Topic Segmentation and Retrieval System for Lecture Videos based on Spontaneous Speech Recognition. In *Proceedings of the INTERSPEECH*, 961–964.
224. Yang, H., M. Siebert, P. Luhne, H. Sack, and C. Meinel. 2011. Automatic Lecture Video Indexing using Video OCR Technology. In *Proceedings of the IEEE International Symposium on Multimedia*, 111–116.
225. Yang, Y.H., Y.C. Lin, Y.F. Su, and H.H. Chen. 2008. A Regression Approach to Music Emotion Recognition. *Proceedings of the IEEE Transactions on Audio, Speech, and Language Processing* 16 (2): 448–457.
226. Ye, G., D. Liu, I.-H. Jhuo, and S.-F. Chang. 2012. Robust Late Fusion with Rank Minimization. In *Proceedings of the IEEE Conference on Computer Vision and Pattern Recognition*, 3021–3028.
227. Ye, Q., Q. Huang, W. Gao, and D. Zhao. 2005. Fast and Robust Text Detection in Images and Video Frames. *Proceedings of the Elsevier Image and Vision Computing* 23 (6): 565–576.
228. Yin, Y., Z. Shen, L. Zhang, and R. Zimmermann. 2015. Spatial-temporal Tag Mining for Automatic Geospatial Video Annotation. *Proceedings of the ACM Transactions on Multimedia Computing, Communications, and Applications* 11 (2): 29.
229. Yoon, S. and V. Pavlovic. 2014. Sentiment Flow for Video Interestingness Prediction. In *Proceedings of the Workshop on HuEvent at the ACM International Conference on Multimedia*, 29–34.
230. Yu, Y., K. Joe, V. Oria, F. Moerchen, J.S. Downie, and L. Chen. 2009. Multi-version Music Search using Acoustic Feature Union and Exact Soft Mapping. *Proceedings of the World Scientific International Journal of Semantic Computing* 3 (02): 209–234.

231. Yu, Y., Z. Shen, and R. Zimmermann. 2012. Automatic Music Soundtrack Generation for Out-door Videos from Contextual Sensor Information. In *Proceedings of the ACM International Conference on Multimedia*, 1377–1378.
232. Zaharieva, M., M. Zeppelzauer, and C. Breiteneder. 2013. Automated Social Event Detection in Large Photo Collections. In *Proceedings of the ACM International Conference on Multimedia Retrieval*, 167–174.
233. Zhang, J., X. Liu, L. Zhuo, and C. Wang. 2015. Social Images Tag Ranking based on Visual Words in Compressed Domain. *Proceedings of the Elsevier Neurocomputing* 153: 278–285.
234. Zhang, J., S. Wang, and Q. Huang. 2015. Location-Based Parallel Tag Completion for Geo-tagged Social Image Retrieval. In *Proceedings of the ACM International Conference on Multimedia Retrieval*, 355–362.
235. Zhang, M., J. Wong, W. Tavanapong, J. Oh, and P. de Groen. 2004. Media Uploading Systems with Hard Deadlines. In *Proceedings of the Citeseer International Conference on Internet and Multimedia Systems and Applications*, 305–310.
236. Zhang, M., J. Wong, W. Tavanapong, J. Oh, and P. de Groen. 2008. Deadline-constrained Media Uploading Systems. *Proceedings of the Springer Multimedia Tools and Applications* 38(1): 51–74.
237. Zhang, W., J. Lin, X. Chen, Q. Huang, and Y. Liu. 2006. Video Shot Detection using Hidden Markov Models with Complementary Features. *Proceedings of the IEEE International Conference on Innovative Computing, Information and Control* 3: 593–596.
238. Zheng, L., V. Noroozi, and P.S. Yu. 2017. Joint Deep Modeling of Users and Items using Reviews for Recommendation. In *Proceedings of the ACM International Conference on Web Search and Data Mining*, 425–434.
239. Zhou, X.S. and T.S. Huang. 2000. CBIR: from Low-level Features to High-level Semantics. In *Proceedings of the International Society for Optics and Photonics Electronic Imaging*, 426–431.
240. Zhuang, J. and S.C. Hoi. 2011. A Two-view Learning Approach for Image Tag Ranking. In *Proceedings of the ACM International Conference on Web Search and Data Mining*, 625–634.
241. Zimmermann, R. and Y. Yu. 2013. Social Interactions over Geographic-aware Multimedia Systems. In *Proceedings of the ACM International Conference on Multimedia*, 1115–1116.
242. Shah, R.R. 2016. Multimodal-based Multimedia Analysis, Retrieval, and Services in Support of Social Media Applications. *In Proceedings of the ACM International Conference on Multimedia*, 1425–1429.
243. Shah, R.R. 2016. Multimodal Analysis of User-Generated Content in Support of Social Media Applications. In *Proceedings of the ACM International Conference in Multimedia Retrieval*, 423–426.
244. Yin, Y., R.R. Shah, and R. Zimmermann. 2016. A General Feature-based Map Matching Framework with Trajectory Simplification. In *Proceedings of the ACM SIGSPATIAL International Workshop on GeoStreaming*, 7.

Chapter 4
Tag Recommendation and Ranking

Abstract Social media platforms such as Flickr allow users to annotate photos with descriptive keywords, called, tags with the goal of making multimedia content easily understandable, searchable, and discoverable. However, due to the manual, ambiguous, and personalized nature of user tagging, many tags of a photo are in a random order and even irrelevant to the visual content. Moreover, manual annotation is very time-consuming and cumbersome for most users. Thus, it is difficult to search and retrieve relevant photos. To this end, we compute relevance scores to predict and rank tags of photos. Specifically, first we present a tag recommendation system, called, PROMPT, that recommends personalized tags for a given photo leveraging personal and social contexts. Specifically, first, we determine a group of users who have similar tagging behavior as the user of the photo, which is very useful in recommending personalized tags. Next, we find candidate tags from visual content, textual metadata, and tags of neighboring photos, and recommends five most suitable tags. We initialize scores of the candidate tags using asymmetric tag co-occurrence probabilities and normalized scores of tags after neighbor voting, and later perform random walk to promote the tags that have many close neighbors and weaken isolated tags. Finally, we recommend top five user tags to the given photo. Next, we present a tag ranking system, called, CRAFT, based on voting from photo neighbors derived from multimodal information. Specifically, we determine photo neighbors leveraging geo, visual, and semantics concepts derived from spatial information, visual content, and textual metadata, respectively. We leverage high-level features instead traditional low-level features to compute tag relevance. Experimental results on the YFCC100M dataset confirm that PROMPT and CRAFT systems outperform their baselines.

Keywords Tag relevance • Tag recommendation • Tag ranking • Flickr photos • Multimodal analysis • PROMPT

R. Shah, R. Zimmermann, *Multimodal Analysis of User-Generated Multimedia Content*, Socio-Affective Computing 6, DOI 10.1007/978-3-319-61807-4_4

4.1 Introduction

The amount of online UGIs has increased dramatically in recent years due to the ubiquitous availability of smartphones, digital cameras, and affordable network infrastructures. For instance, over 10 billion UGIs have been uploaded so far in a famous photo sharing website Flickr which has over 112 million users, and an average of 1 million photos are uploaded daily [10]. Such UGIs belong to different interesting activities (*e.g.*, festivals, games, and protests), and are described with descriptive keywords, called tags. Similar to our work on multimedia summarization in Chapter 3, we consider the YFCC100M dataset in this study. UGIs in the YFCC100M dataset are annotated with approximately 520 million user tags (*i.e.*, around five user tags per UGI). Such tags are treated as concepts (*e.g.*, playing soccer) which describe the objective aspects of UGIs (*e.g.*, visual content and activities), and suitable for real-world tag-related applications. Thus, such rich tags (concepts) as metadata are very helpful in the analysis, search, and retrieval of UGIs on social media platforms. They are beneficial in providing several significant multimedia-related applications such as landmark recognition [89], tag recommendation [193], automatic photo tagging [203, 208], personalized information delivery [191], and tag-based photo search and group recommendation [102]. However, the manual annotation of tags is very time-consuming and cumbersome for most users. Furthermore, predicted tags for the UGI are not necessarily relevant to users' interests. Moreover, often annotated tags of a UGI are in a random order and even irrelevant to the visual content. Thus, the original tag list of a UGI may not give any information about the relevance with the UGI [109] because often user tags are ambiguous, misspelled, and incomplete. However, Mao et al. [117] presented a multimodal Recurrent Neural Network (m-RNN) model for generating image captions, which indicates that multimodal information of UGIs are very useful in tag recommendation since image captioning seems to have subsumed image tagging. Therefore, for an efficient tag-based multimedia search and retrieval, it necessitates for automatic tag recommendation and ranking systems. To this end, we present a tag recommendation system, called, PROMPT, and a tag ranking system, called CRAFT, based on leveraging multimodal information . The reason why we leverage multimodal information is that it is very useful in addressing many multimedia analytics problems [242, 243] such as event understanding [182, 186], lecture video segmentation [183, 184], news videos uploading [180], music video generation [187, 188], and SMS/MMS based FAQ retrieval [189, 190]. All notations used in this chapter are listed in Table 4.1.

4.1.1 Tag Recommendation

PROMPT stands for a personalized user tag recommendation for social media photos leveraging personal and social contexts from multimodal information. It leverages knowledge structures from multiple modalities such as the visual content and textual metadata of a given UGI to predict user tags. Johnson et al. [76]

candidate tags iteratively leveraging exemplar and concurrent similarities. Next, we recommend the top five user tags when the random walk converge. Finally, we investigate the effect of fusion by combining candidate tags derived from different approaches and next perform a random walk to recommend the top five user tags to the given photo. Experimental results on a test set of 46,700 Flickr UGIs (see Sect. 4.3 for details) confirm that our proposed approaches perform well and comparable to state-of-the-arts regarding precision, recall, and accuracy scores. Steps of predicting tags for a UGI is summarized as follows.

- First, we determine the nearest group of users from the train set with 259,149 unique users, having the similar tagging behavior as the user of a given UGI.
- Next, we compute the candidate set of UGIs and tags for the given UGI from the selected cluster.
- Finally, we compute relevance scores for tags in the candidate set using our proposed approaches and predict the top five tags with the highest relevance scores for the given UGI.

4.1.2 Tag Ranking

Our tag ranking system, called, CRAFT, stands for concept-level multimodal ranking of Flickr photo tags via recall based weighting. An earlier study [90] indicates that only 50% of the user tags are related to UGIs. Say, Fig. 4.1 depicts that all user tags of a seed UGI[1] (with the title, *"A Train crossing Forth Bridge, Scotland, United Kingdom"*) are either irrelevant or weakly relevant. Moreover, relevant visual tags such as *bridge* and *outdoor* appear later in the tag list. Furthermore, another relevant tag in this example is *train*, but it is missing in both user and visual tags. Additionally, often tags are overly personalized [63, 118], this affects the ordering of tags. Thus, it necessitates to leveraging knowledge structures from more modalities for an effective social tag ranking.

The presence of contextual information in conjunction with multimedia content is very helpful in several significant tag-related applications since real-world UGIs are complex and extracting all semantics from only one modality (say, visual content) is very difficult. It happens because suitable concepts may exhibit in different representations (say, textual metadata and location information). Since multimodal information augments knowledge bases by inferring semantics from unstructured multimedia content and contextual information, we leverage the multimodal information in computing tag relevance for UGIs. Similar to earlier work [102], we compute tag relevance based on neighbor voting neighbor voting (NV).

[1]https://www.flickr.com/photos/94132145@N04/11953270116/

Fig. 4.1 The original tag list for an exemplary UGI from Flickr. Tags in normal and italic fonts are user tags and automatically generated visual tags from visual content, respectively

Since the research focus in content-based image retrieval (CBIR) systems has shifted from leveraging low-level visual features to high-level semantics [112, 239], high-level features are now widely used in different multimedia-related applications such as event detection [74]. We determine neighbors of UGIs using three novel high-level features instead of using low-level visual features exploited in state-of-the-arts [102, 109, 233]. The proposed high-level features are constructed from concepts derived from spatial information, visual content, and textual metadata using the bag-of-words model (see Sect. 4.2.2 for details). Next, we determine improved tag ranking of a UGI by accumulating votes from its semantically similar neighbors derived from different modalities. Furthermore, we also investigate the effect of early and late fusion of knowledge structures derived from different modalities. Specifically, in the early fusion, we fuse neighbors of different modalities and perform voting on tags of a given UGI. However, in the late fusion, we perform a linear combination of tag voting from neighbors derived using different high-level features (modalities) with weights computed from recall scores of modalities. The recall score of a modality indicates the percentage of original tags covered by the modality. Experimental results on a collection of 203,840 Flickr UGIs (see Sect. 4.3 for details) confirm that our proposed new features and their late fusion based on recall weights significantly improve the tag ranking of UGIs and outperform state-of-the-arts regarding the normalized discounted cumulative gain (NDCG) score. Our contributions are summarized as follows:

- We demonstrate that high-level concepts are very useful in the tag ranking of a UGI. Even a simple neighbor voting scheme to compute tag relevance outperforms state-of-the-arts if high-level features are used instead of low-level features to determine neighbors of the UGI.

- Our experiments confirm that high-level concepts derived from different modalities such as geo, visual, and textual information complement each other in the computation of tag relevance for UGIs.
- We propose a novel late fusion technique to combine confidence scores of different modalities by employing recall-based weights.

The chapter is organized as follows. In Sect. 4.2, we describe the PROMPT and CRAFT systems. The evaluation results are presented in Sect. 4.3. Finally, we conclude the chapter with a summary in Sect. 4.4.

4.2 System Overview

4.2.1 Tag Recommendation

Figure 4.2 shows the system framework of the PROMPT system. We compute user behavior vectors for all users based on their past annotated UGIs using the bag-of-the-words model on a set of 1540 user tags U_{Tags} used in this study. We exploit user behavior vectors to perform the grouping of users in the train set and compute asymmetric tag co-occurrence scores among all 1540 user tags for each cluster during pre-processing. Moreover, the cluster center of a group is determine by averaging user behavior vectors of all users in that group. Similarly, we compute photo description vectors for UGIs using the bag-of-the-words model on U_{Tags}. However, we do not consider user tags of UGIs to construct their photo description vectors. Instead, we leverage tags derived from the title, description, and visual tags which belong to U_{Tags}. Photo description vectors are used to determine semantically similar neighbors for UGIs based on the cosine similarity metric. During online processing to predict user tags for a test UGI, we first compute its user behavior vector, and subsequently a closest matching user group from the train set. We refer to the set of UGIs and tags in the selected user group as the candidate set and further use them to predict tags of the test UGI. We exploit the following three techniques to compute tag relevance, and subsequently predict top five user tags.

Asymmetric Co-occurrence Based Relevance Scores As described in the literature [218], tag relevance is standardized into mainly asymmetric and symmetric structures. Symmetric tag co-occurrence tends to measure how similar two tags are, *i.e.*, high symmetric tag co-occurrence score between two tags indicates that they most likely to occur together. However, asymmetric tag co-occurrence suggests relative tag co-occurrence, *i.e.*, $p(t_j|t_{\bar{j}})$ is interpreted as the probability of a UGI being annotated with t_j given that it is already annotated with $t_{\bar{j}}$. Thus, asymmetric tag co-occurrence scores are beneficial in introducing diversity to tag prediction. The asymmetric tag co-occurrence score between tags $t_{\bar{j}}$ and t_j is defined as follows:

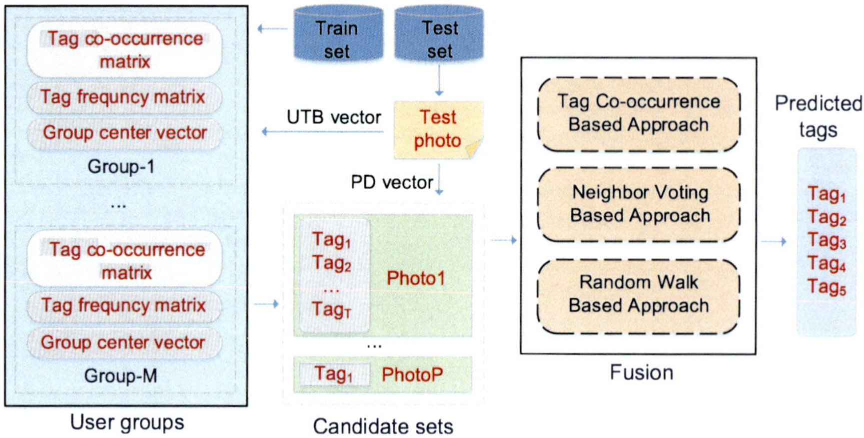

Fig. 4.2 Architecture of the PROMPT system

$$p_{\bar{j}j} = p\left(t_{t_j|\bar{j}}\right) = \frac{|t_{\bar{j}} \cap t_j|}{|t_{\bar{j}}|} \tag{4.1}$$

where $|t_{\bar{j}}|$ and $|t_{\bar{j}} \cap t_j|$ represent the number of times the tag $t_{\bar{j}}$ appears alone and with tag t_j, respectively.

Figure 4.3 describes the system framework to predict tags based on asymmetric co-occurrence scores. We first determine seed tags from the textual metadata and visual tags of a given UGI. Seed tags are the tags appeared in the title and visual tags of the UGI, which belong to the set of 1540 user tags used in this study. We add seed tags and their five most co-occurred non-seed tags to the candidate set of the UGI. For all visual tags of the UGI, their confidence scores $\sigma_{\bar{j}}$ are also given as part of the YFCC100M dataset. Initially, we set confidence scores of seed tags from the title as 1.0, and compute relevance scores r of non-seed tags in the candidate set as follows:

$$r_{\bar{j}j} = p_{\bar{j}j} \times \sigma_{\bar{j}} \tag{4.2}$$

where, $\sigma_{\bar{j}}$ is the confidence score of seed tag $t_{\bar{j}}$. This formula to compute the relevance score of tag t_j for a given tag $t_{\bar{j}}$ is justifiable because it assigns the high relevance score when the confidence of the seed tag $t_{\bar{j}}$ is high. We compute the relevance score of a seed tag by averaging the asymmetric co-occurrence scores of its five most likely co-occurred tags. In this way, we compute relevance scores of all tags in the candidate set. Next, we aggregate all tags and merge scores of common tags. Finally, we predict top five tags with the highest relevance scores from the candidate set to the UGI.

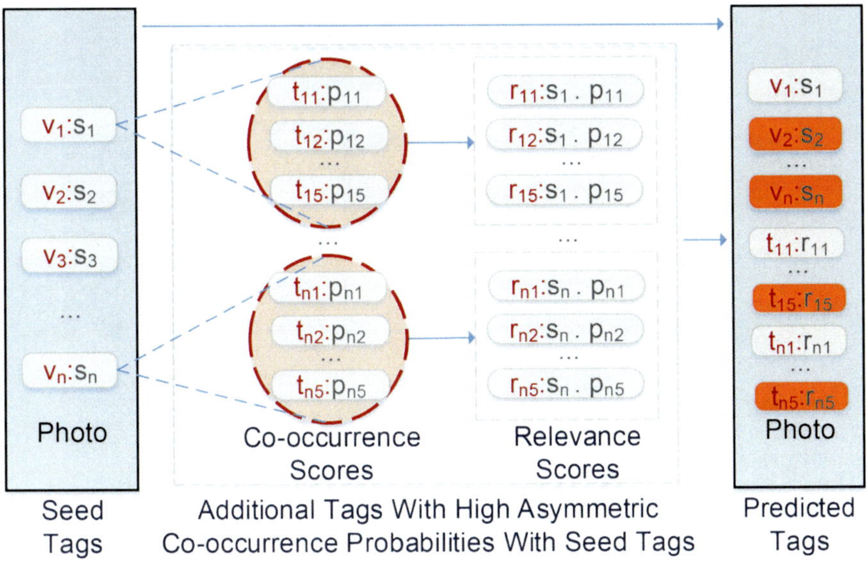

Fig. 4.3 System framework of the tag prediction system based on asymmetric co-occurrence scores

Neighbor Voting Based Relevance Scores Earlier work [102, 185] on computing tag relevance for photos confirm that a neighbor voting based approach is very useful in determining tag ranking. Leveraging personal and social contexts, we apply this approach for tag recommendation. Relevance scores of tags for a UGI[2] is computed in the following two steps (see Fig. 4.4). Firstly, k nearest neighbors of the UGI are obtained from the user group of similar tagging behaviors. Next, the relevance score of tag t for the UGI is obtained as follows:

$$z(t) = vote(t) \;\; prior(t, k) \tag{4.3}$$

where $z(t)$; is the tag t's final relevance score, $vote(t)$ represents the number of votes tag t gets from the k nearest neighbors of the UGI. $prior(t; k)$ indicates the prior frequency of the tag t and is defined as follow:

$$prior(t, k) = k \frac{M_t}{|D_{TagRecomTrain}|} \tag{4.4}$$

where $M_{t.}$ is the number of UGIs tagged with t, and $|D_{TagRecomTrain}|$ is the size of the train set.

Random Walk Based Relevance Scores Another very popular technique for tag ranking is based on a random walk. Liu et al. [109] estimates initial relevance

[2]https://www.flickr.com/photos/bt-photos/15428978066/

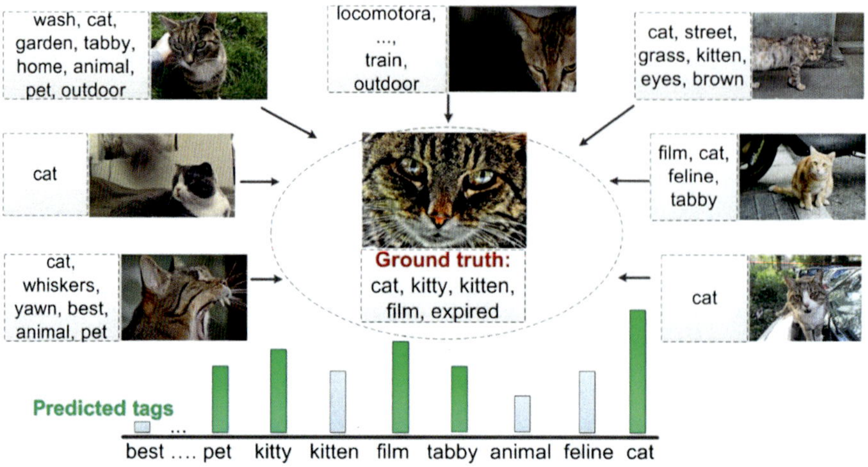

Fig. 4.4 System framework of the tag recommendation system based on neighbor voting scores

scores for tags based on probability density estimation, and then perform a random walk over a tag similarity graph to refine the relevance scores. We leverage the multimodal information of UGIs and apply this tag ranking approach for tag recommendation (see Fig. 4.5). Specifically, first, we determine candidate tags leveraging multimodal information such as the textual metadata (*e.g.*, title and description) and the visual content (*e.g.*, visual tags). We estimate the initial relevance scores of candidate tags adopting a probabilistic approach on co-occurrence of tags. We also use the normalized scores of tags derived from neighbor voting. Next, we refine relevance scores of tags by implementing a random walk process over a tag graph which is constructed by combining an exemplar-based approach and a concurrence-based approach to estimate the relationship among tags. The exemplar similarity φ_e is defined as follows:

$$\varphi_e = \exp\left(-\frac{1}{k*k}\sum_{x\in\Gamma_{t_i},\, y\in\Gamma_{t_j}}\frac{\|x-y\|^2}{\sigma^2}\right) \tag{4.5}$$

where Γ_t denotes the representative UGI collection of tag t and k is the number of nearest neighbors. Moreover, s is the radius parameter for the classical Kernel Density Estimation (KDE) [109]. Next, the concurrence similarity φ_c between tag t_i and tag t_j is defined as follows:

$$\varphi_c = \exp\left(-d\left(t_i, t_j\right)\right) \tag{4.6}$$

where the distance $d(t_i,t_j)$ between two tags t_i and t_j is defined as follows.

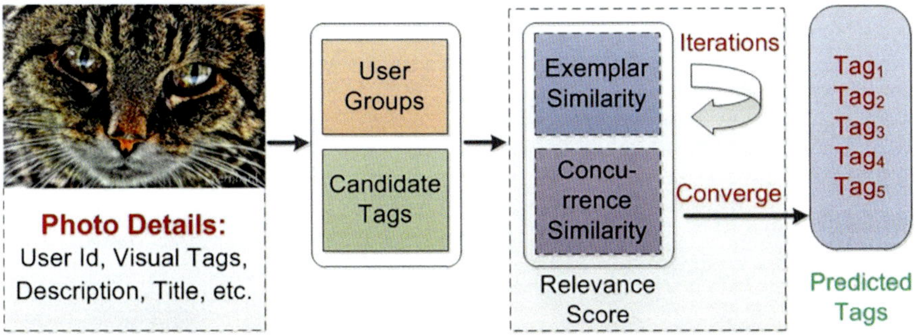

Fig. 4.5 Architecture of the tag prediction system based on random walk

$$d(t_i, t_j) = \frac{\max(\log f(t_i), \log f(t_j)) - \log f(t_i, t_j))}{logG - \min(\log f(t_i), \log f(t_j))} \tag{4.7}$$

where $f(t_i)$,$f(t_j)$, and $f(t_i,t_j)$ are the numbers of photos containing tags t_i, t_j, and both t_i and t_j, respectively, in the training dataset. Moreover, G is the number of photos in the training dataset. Finally, the exemplar similarity φ_e and concurrence similarity φ_c are combined as follows:

$$\Phi_{ij} = \lambda \cdot \varphi_e + (1 - \lambda) \cdot \varphi_c \tag{4.8}$$

where λ belongs to [0,1]. We set it to 0.5 in our study.

We use $u_k(i)$ to denote the relevance score of node i at iteration k in a tag graph with n nodes. Thus, relevance scores of all nodes in the graph at iteration k form a column vector u_k $[u_k(i)]_{n\times1}$. An element $q_{i\ j}$ of this n×n transition matrix indicates the probability of the transition from node i to node j and it is computed as follows:

$$q_{ij} = \frac{\Phi_{ij}}{\Sigma_k \Phi_{ik}} \tag{4.9}$$

The random walk process promotes tags that have many close neighbors and weakens isolated tags. This process is formulated as follows.

$$u_k(j) = a \sum_i u_{k-1}(i) q_{ij} + (1 - \alpha) w_j \tag{4.10}$$

where w_j is the initial score of a tag t_j and α is a weight parameter between (0, 1).

Fusion of Relevance Scores The final recommended tags for a given UGI is determined by fusing different approaches mentioned above. We combine candidate tags determined by asymmetric tag co-occurrence and neighbor voting schemes. Next, we initialize scores of the fused candidate tags with their normalized scores from [0,1]. Further, we perform a random walk on a tag graph which has

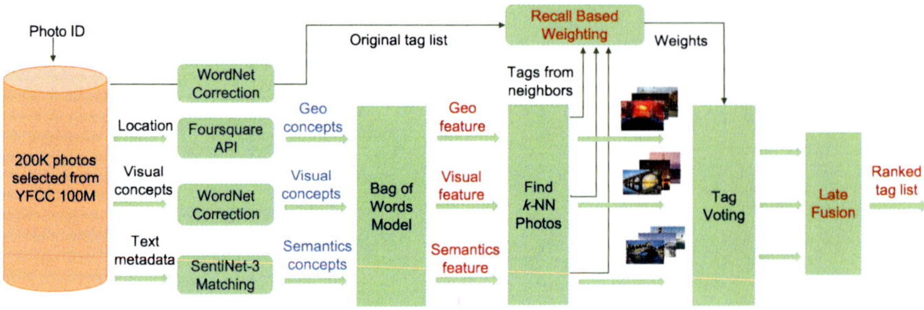

Fig. 4.6 The system framework of computing tag ranking for UGIs

the fused candidate tags as its nodes. This tag graph is constructed by combining exemplar and concurrence similarities and useful in estimating the relationship among the tags. In this way, the random walk refines relevance scores of the fused candidate tags iteratively. Finally, our PROMPT system recommends the top five tags with the highest relevance scores to the UGI, when the random walk converges.

4.2.2 *Tag Ranking*

Figure 4.6 shows the system framework of our tag ranking system. We propose three novel high-level features based on concepts derived from the following three modalities: (i) spatial information, (ii) visual content, and (iii) textual metadata. We leverage the concepts in constructing the high-level feature vectors using the bag-of-words model, and subsequently use the feature vectors in finding k nearest neighbors of UGIs. Next, we accumulate votes on tags from such neighbors and perform their fusion to compute tag relevance. We consider both early and late fusion techniques to combine confidence scores of knowledge structures derived from different modalities.

Features and Neighbors Computation A concept is a knowledge structure which is helpful in the understanding of objective aspects of multimedia content. Table 4.2 shows the ten most frequent geo, visual, and semantics concepts with their frequency in our experimental dataset of 203,840 UGIs (see Experimental dataset in Sect. 4.3 for details) that are captured by unique users. Each UGI in the experimental dataset has location information, textual description, visual tags, and at least five user tags. Leveraging high-level feature vectors that are computed from concepts mentioned above using the bag-of-words model, we determine neighbors of UGIs using the cosine similarity defined as follows:

Table 4.2 Top ten geo, visual, and semantics concepts used in the experimental set of 203,840 UGIs

Geo concepts	Count	Visual concepts	Count	Semantics concepts	Count
Home (private)	49,638	Outdoor	128,613	Photo	12,147
Cafe	41,657	Nature	65,235	Watch	10,531
Hotel	34,516	Indoor	47,298	New	10,479
Office	29,533	Architecture	46,392	Catch	10,188
Restaurant	29,156	Landscape	43,913	Regard	9744
Bar	34,505	Water	30,767	Consider	9686
Park	23,542	Vehicle	29,662	Reckon	9776
Pizza Place	19,269	People	26,333	Scene	9656
Building	17,399	Building	25,506	Take	8348
Pub	16,531	Sport	25,465	Make	8020

$$cosine\ similarity = \frac{A.B}{\|A\|\|B\|} \tag{4.11}$$

where A and B are feature vectors for two UGIs. $\|A\|$ and $\|B\|$ are the magnitudes of feature vectors for A and B, respectively.

Geo Features and Neighbors Since UGIs captured by modern devices are enriched with several contextual information such as GPS location, this work assumes that the spatial information of UGIs is known. Moreover, significant earlier work [45, 207] exist which estimate the location of a UGI, if it is not known. Thus, we select UGIs with GPS information in our tag ranking experiments. Earlier work [192] investigated the problem of mapping a noisy estimate of a user's current location to a semantically meaningful point of interest (location categories) such as a home, park, or restaurant. They suggested combining a variety of signals about a user's current context to explicitly model both places and users. Thus, in our future work, we plan to combine the user's contextual information and objects in UGIs to map the location of UGIs to geo categories accurately. The GPS location of a UGI is mapped to geo concepts (categories) using the Foursquare API [13] (see Sect. 1.4.1 for details). This API also provides distances of geo concepts such as beach, temple, and hotel on the queried GPS location, which describe the typical objects near the scene in the UGI. We treat each geo concept as a word and exploit the bag-of-words model [93] on a set of 1194 different geo concepts in this study. Next, we use the cosine similarity metric defined in Eq. 4.11 to find k nearest neighbors of UGIs in the evaluation set of 500 randomly selected UGIs from the experimental dataset of 203,840 UGIs (see Sect. 4.3 for details).

Visual Features and Neighbors For each UGI in the YFCC100M dataset, a variable number of visual concepts are provided (see Dataset in Sect. 4.3 for details). There are total 1756 visual concepts present in the collection of 100 million UGIs and UGVs. Thus, each UGI can be represented leveraging such visual concepts by the bag-of-words model. We construct a 1732-dimensional feature vector corresponding to 1732 visual concepts present in the experimental dataset. Finally, we use the cosine similarity metric to find k nearest neighbors for all UGIs in the evaluation set using Eq. 4.11.

Semantics Features and Neighbors Users annotate UGIs using textual metadata such as title, description, and tags. We only consider sentences and words which are written in English. We extract semantics concepts from the textual metadata using the semantic parser provided by Poria et al. [143] (see Sect. 1.4.2 for details). For instance, semantics concepts such as a *beach*, *corn island*, *sunset from beach*, *big corn island*, and *beach on corn island* are computed from the following sentence: *Sunset from the beach on Big Corn Island*. We also consider knowledge base related to sentiments such as SenticNet-3. SenticNet-3 is a publicly available resource for concept-level sentiment analysis [41] and consists of 30,000 common and common-sense concepts such as *food*, *party*, and *accomplish goal*. Since such common and common-sense concepts are often used in the tagging of UGIs, we leverage the SenticNet-3 knowledge base to construct a unified vector space. Earlier work [186] presented an algorithm to associate the determined semantics concepts to SenticNet-3 concepts. There are totally 13,727 SenticNet-3 concepts present in the experimental dataset. With the bag-of-words model, we construct 13,727-dimensional feature vectors for all UGIs and compute their k nearest neighbors using the cosine similarity metric.

Tag Ranking by Neighbor Voting Figure 4.7 shows neighbors derived leveraging different modalities, and their tag voting for a seed UGI (see Fig. 4.1). During the preprocessing step, all meaningless and misspelled tags of the seed UGI are removed. We consider only those words (tags) which are defined in the WordNet dictionary. If tags have more than one words, then we only keep tags whose all words are a valid WordNet word. Thus, our algorithm outputs a ranked list of tags satisfying above criteria for UGIs.

Tag Relevance Score Computation Based on Neighbor Voting The tag relevance score of a seed UGI i is computed in the following two steps. Firstly, k nearest neighbors of the seed UGI are obtained from the experimental dataset based on leveraging different modalities, as described above in Sect. 4.2.2. Next, the relevance score of the seed UGI's tag t is obtained as follows:

$$z(t) = vote(t)prior(t, k) \tag{4.12}$$

where $z(t)$; is the tag t's final relevance score, $vote(t)$ represents the number of votes tag t gets from the k nearest neighbors. $Prior(t; k)$ indicates the prior frequency of t and is defined as follow:

$$prior(t, k) = k\frac{M_t}{|D_{TagRanking}|} \tag{4.13}$$

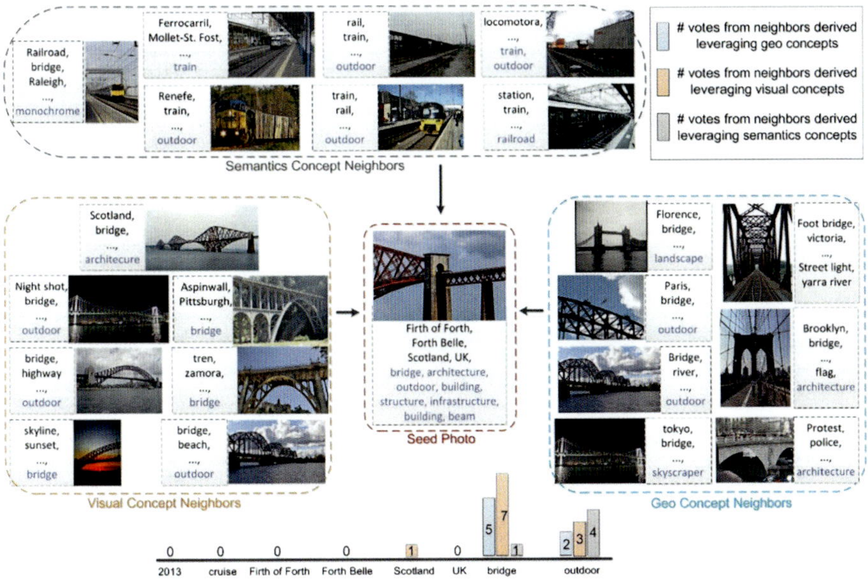

Fig. 4.7 The system framework of neighbor voting scheme for tag ranking based on geo, visual, and semantics concepts derived from different modalities

where $M_{t.}$ is the number of UGIs tagged with t, and $/D_{TagRanking}/$ is the size of the evaluation dataset for tag ranking task. For fast processing, we perform the Lucene [9] Lucene indexing of tags and UGIs. Finally, we rank tags $t_1;t_2, \ldots, t_n$ of the seed UGI based on their relevance score as follows:

$$\text{rank}(z(t_1, p), z(t_2), p), \ldots, z(t_n, p)). \tag{4.14}$$

Thus, UGIs' tag ranking based on geo, visual, and semantics concepts is accomplished. We refer to these tag ranking systems based on neighbor voting (NV) as NVGC, NVVC, and NVSC corresponding to geo, visual, and semantics concepts, respectively. However, only one modality is not enough to compute tag relevance scores because different tags are covered by diverse modalities. For instance, a geo-tagged UGI that depicts a cat in an apartment is described by tags that include several objects and concepts such as *cat, apartment, relaxing, happy*, and *home*. It is a difficult problem to rank tags of such UGI based on only one modality. Knowledge structures derived from different modalities describe different tags of the UGI. For instance, say, a cat is described by visual content, (*i.e.*, visual concepts), apartment and home are described by spatial information (*i.e.*, geo concepts), and relaxing is described by the textual metadata (*i.e.*, semantics concepts). The final score of a tag is determined by fusing the tag's scores for different modalities (say, spatial, visual, and textual content).

Let O be the original tag set for a UGI after removing non-relevant and misspelled tags. Let $\bar{G}, \bar{V}$ and $\bar{S}$ be sets of tags computed from neighbors of the

Table 4.3 The coverage of tags from spatial, visual, and textual modalities

	Tag set	Avg recall
Geo concepts	$\bar{G} \cap O$	0.419
	$(\bar{G} \setminus \bar{V}) \cap O$	0.087
	$(\bar{G} \setminus \bar{S}) \cap O$	0.139
	$(\bar{G} \setminus (\bar{V} \cup \bar{S}) \cap O$	0.031
Visual concepts	$\bar{V} \cap O$	0.659
	$(\bar{V} \setminus \bar{G}) \cap O$	0.405
	$(\bar{V} \setminus \bar{S} \cap O)$	0.592
	$(\bar{V} \setminus (\bar{G} \cup \bar{S})) \cap O$	0.183
Semantics concepts	$\bar{S} \cap O$	0.572
	$(\bar{S} \setminus \bar{V}) \cap O$	0.282
	$(\bar{S} \setminus \bar{G}) \cap O$	0.346
	$(\bar{S} \setminus \bar{G} \cup \bar{V}) \cap O$	0.107

UGI derived using geo, visual, and semantics concepts, respectively. Table 4.3 confirms that different features complement each other in tag coverage, which is helpful in computing tag relevance. For instance, 10.7% of original tags are covered by only semantics concepts, and not by other two remaining modalities (*i.e.*, geographical information, and visual content). Similarly, 18.3% of original tags are covered by only visual concepts, and not by remaining two modalities. Subsequently, 3.1% of original tags are covered by only geo concepts. Tag coverage by geo concepts is much less than that of visual and semantics concepts, probably because the location of the UGI is the location of the camera/mobile but not the location of objects in the UGI. Thus, in our future work, we plan to leverage the field-of-view (FoV) model [166, 228] to accurately determine tags based on the location of the user and objects in UGIs. Moreover, geo concepts are very useful in providing useful contextual information of a UGI and its user (photographer). Table 4.3 describes the following statistics the for three modalities mentioned above (*i.e.*, geo, visual, and textual information). First, the fraction of original tags O which are covered by only one modality (say, only $\bar{G}, \bar{V}$, or $\bar{S}$). Second, the fraction of O which are covered by only one modality (say, $\bar{G}$) but not by the one of other two modalities (*i.e.*, $\bar{V}$ or $\bar{S}$). Third, the fraction of O which are covered by only one modality (say, $\bar{G}$) but not by both of other two modalities (*i.e.*, $\bar{V}$ and $\bar{S}$). Thus, different modalities complement each other, and it is necessary to fuse them to further improve tag relevance for UGIs.

Final tag Relevance Score Computation Using Early and Late Fusion Techniques Our fusion techniques for tag ranking leverage knowledge structures derived from different modalities based on neighbor voting (NV). We refer to them as NVGVC, NVGSC, NVVSC, and NVGVSC corresponding to the fusion of geo and visual concepts, geo and semantics concepts, visual and semantics concepts, and geo, visual, and semantics concepts, respectively. During early fusion (EF), we fuse UGI neighbors derived from different modalities for a given seed UGI and pick

k nearest neighboring UGIs based on cosine similarity for voting. We use the following two approaches for late fusion. First, accumulate vote counts by using equal weights for different modalities (LFE). Second, accumulate vote counts from neighbors of different modalities with weights decided by recall score (LFR), *i.e.*, the proportion of the seed UGI's original tag covered by different modalities. Next, the relevance score of the seed UGI's tag is obtained based on late fusion as follows:

$$\bar{z}(t) = \sum_{j=1}^{\bar{m}} w_j \left(vote_j(t) - prtior(t, k) \right) \tag{4.15}$$

where $\bar{m}$ is the number of modalities, w_j are weight for different modalities such that $\sum_{j=1}^{m} w_j = 1$ and $vote_j(t)$ is the vote count from neighbors derived from the jth modality for the tag t of the UGI i.

4.3 Evaluation

4.3.1 Tag Recommendation

Dataset Similar to our work on event understanding, we used the YFCC100M dataset [201] D_{YFCC} from Flickr. It consists of 100 million multimedia items (approximately 99.2 million UGIs and 0.8 million UGVs) from Flickr. The reason for selecting this dataset is its volume, modalities, and metadata. For instance, each media of the dataset consists of several metadata annotations such as user tags, spatial information, and temporal information. These media are captured from the 1990s onwards and uploaded between 2004 and 2014. It includes media from top cities such as Paris, Tokyo, London, New York City, Hong Kong, and San Francisco. Moreover, all media are labeled with automatically added tags derived by using a convolutional neural network which indicates the presence of a variety of concepts, such as people, animals, objects, food, events, architecture, and scenery. There are a total 1756 visual concepts present in this dataset. For tag prediction, the whole dataset has been split into ten parts based the last digit prior to the @-symbol in their Flickr user identifier (NSID). Such split ensures that no user occurs in multiple partitions, thus avoiding dependencies between the different splits. Split 0 is used as the test set and the remaining nine splits as the training set.
Despite the number of objects that deep neural networks (*e.g.*, Google Cloud Vision API [14]) can identify is rapidly increasing, often the objective aspects of UGIs can not be described by the identified objects alone. Thus, we need to predict tags for UGIs from the tags that users often use to describe similar UGIs. For tag prediction in this study, a specific subset of the most frequent 1540 user tags U_{Tags} from D_{YFCC} are considered since predicting the correct tags from a virtually endless pool of possible user tags is extremely challenging. Tags in U_{Tags} fulfill the following

criteria. First, they are valid English dictionary words. Second, such tags do not refer to persons, dates, times or places. Third, they appear frequently with UGIs in the train and test sets. Finally, fourth, they were different tenses/plurals of the same word. The train set contains all UGIs from the YFCC100M that have at least one tag that appeared in U_{Tags}, and do not belong to the split 0. There are approximately 28 million UGIs present in the train set $D_{TagRecomTrain}$. The test set $D_{TagRecomTest}$ contains 46,700 UGIs from the split 0 such that each UGI has at least five tags from the list of 1540 tags. There are totally 259,149 and 7083 unique users in the train and test sets for this subtask, respectively.

Results Recommended tags for a given photo in the test set are evaluated based on the following three metrics. First, Precision@*K*, *i.e.*, proportion of the top *K* predicted tags that appear in user tags of the photo. Second, Recall@*K*, *i.e.*, proportion of the user tags that appear in the top *K* predicted tags. Finally, third, Accuracy@*K*, *i.e.*, 1 if at least one of the top *K* predicted tags is present in the user tags, 0 otherwise. PROMPT is tested for the following values of *K*: 1, 3, and 5. We implemented two baselines and proposed a few approaches to recommend personalized user tags for social media photos. In *Baseline1*, we predict the top five most frequent tags from the training set of 28 million photos to a test photo. Further, in *Baseline2*, we predict five visual tags with the highest confidence scores (already provided with the YFCC100M dataset) to a test photo. Since state-of-the-arts for tag prediction [28, 193] mostly recommend tags for photos based on input seed tags. In our PROMPT system, first, we construct a list of candidate tags using asymmetric co-occurrence, neighbor voting, probability density estimation techniques. Next, we compute tag relevance for photos through co-occurrence, neighbor voting, random walk based approaches. We further investigate the fusion of these approaches for tag recommendation.

Figures 4.8, 4.9, and 4.10 depicts scores@K for accuracy, precision, and recall, respectively, for different baselines and approaches. For all metrics, *Baseline1* (*i.e.*, recommending the five most frequent user tags) performs worst, and the combination of all three approaches (*i.e.*, co-occurrence, neighbor voting, and random walk based tag recommendation) outperforms rest. Moreover, the performance of *Baseline2* (*i.e.*, recommending the five most confident visual tags) is second from last since it only considers the visual content of a photo for tag recommendation. Intuitively, accuracy@K and recall@K increase for all approaches when we the number of recommended tags increases from 1 to 5. Moreover, precison@K decreases for all approaches when we increase the number of recommend tags. Our PROMPT system recommends user tags with 76% accuracy, 26% precision, and 20% recall for five predicted tags on the test set with 46,700 photos from Flickr. Thus, there is an improvement of 11.34%, 17.84%, and 17.5% regarding accuracy, precision, and recall evaluation metrics, respectively, in the performance of the PROMPT system as compared to the best performing state-of-the-art for tag recommendation (*i.e.*, an approach based on a random walk).

Table 4.4 depicts accuracy, precision, and recall scores when a combination of co-occurrence, voting, and a random walk is used for tag prediction. Type-1

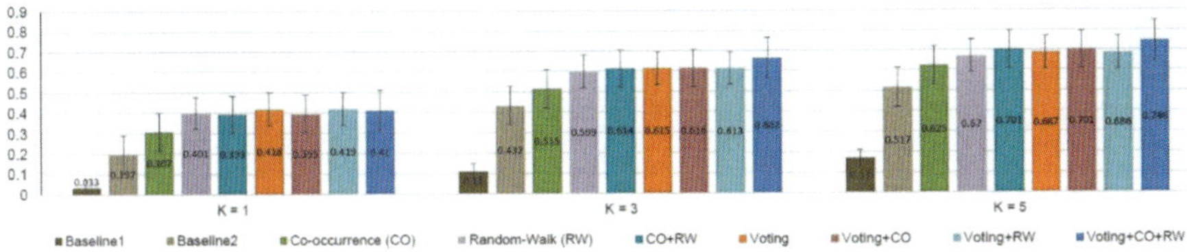

Fig. 4.8 Accuracy@K, *i.e.*, user tag prediction accuracy for K predicted tags for different approaches

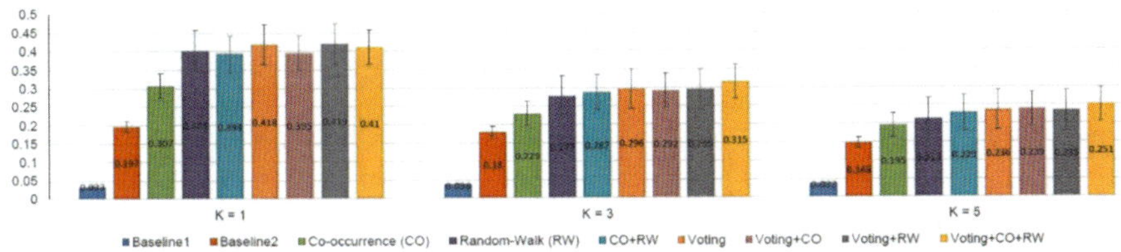

Fig. 4.9 Precision@K, *i.e.*, the precision of tag recommendation for K recommended tags for different approaches

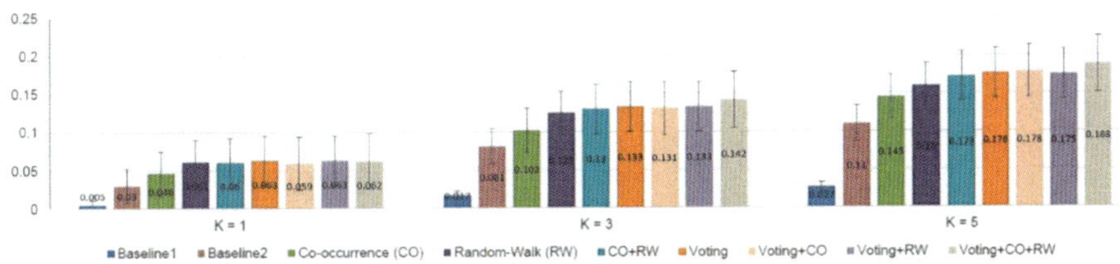

Fig. 4.10 Recall@K, *i.e.*, recall scores for K predicted tags for different approaches

Table 4.4 Results for the top K predicted tags

	Comparison type	K = 1	K = 3	K = 5
Accuracy@K	Type-1	0.410	0.662	0.746
	Type-2	0.422	0.678	0.763
Precision@K	Type-1	0.410	0.315	0.251
	Type-2	0.422	0.326	0.262
Recall@K	Type-1	0.062	0.142	0.188
	Type-2	0.064	0.147	0.197

considers a comparison as a hit if a predicted tag matches ground truth tags and Type-2 considers a comparison as a hit if either a predicted tag or its synonyms match ground truth tags. Intuitively, accuracy, precision, and recall scores are slightly improved when the Type-2 comparison is made. Results are consistent with all baselines and approaches which we used in our study for tag prediction. All results reported in Figs. 4.8, 4.9, and 4.10 correspond to the Type-1 match. Finally, Fig. 4.11 shows the ground truth user tags and the tags recommended by our system for five sample photos in the test set.

Fig. 4.11 Examples of tag prediction

4.3.2 Tag Ranking

Experimental Dataset $D_{TagRanking}$ Advancement in technologies enables users to capture several contextual information such as location, description, and tags in conjunction with a UGI. Such contextual information is very useful in the semantics understanding of the UGI. Table 4.3 indicates that concepts derived from geo, visual, and textual information are helpful in social media applications such as tag ranking and recommendation. Moreover, Li et al. [102] demonstrated that learning social tag relevance by neighbor voting from UGIs of unique users performs better than UGIs of same users. Furthermore, during pre-processing, we found that the average number of tags per UGI is five and approximately 51 million UGIs have location information. Thus in our experiment for the tag ranking problem, we only considered UGIs that have at least five user tags, description metadata, location information, and visual tags, and captured by unique users. We selected totally 203,840 such UGIs from the YFCC100M dataset which fulfilled the criteria mentioned above. We refer to this dataset as the experimental dataset $D_{TagRanking}$. It consists of 1732 visual concepts out of total 1756 visual concepts present in the YFCC100M dataset with approximately same proportion. Thus, our experimental dataset is a representative sample of 100 million media records and has 96,297 unique user tags. Moreover, we downloaded the ACC, RGB, and Tamura low-level visual features of UGIs in the experimental dataset from the Multimedia Commons website [35].

Annotators Table 4.5 shows the two group of annotators who participated in our evaluation. *Group-A* has totally four annotators and *Group-B* has totally six annotators (different from *Group-A*). Most of the annotators are students from different countries such as India, Germany, Chile, Japan, China, Taiwan, and Portugal. Annotators in *Group-A* are asked to select the five most suitable tags from the list of user tags and visual concepts. We did not tell users which are the user tags and which are the visual concepts in the list of tags for UGVs to avoid any biasedness in the annotation. We provided five tags (consist of both user tags and visual concept) selected by *Group-A* annotators to *Group-B* annotators and asked them to assign relevance scores to these tags from 1 (irrelevant) to 5 (most relevant). Intentionally, we select different annotators in *Group-A* and *Group-B* to avoid any biasedness in the annotation.

Table 4.5 Annotators details for tag ranking task

Group type	No. of evaluators	No. of responses	No. of accepted responses
Group-A	4	500	500
Group-B	6	1000	1000

Evaluation Dataset $D_{TagRankingEval}$ For the evaluation, we have randomly selected 500 UGIs from the experimental dataset $D_{TagRanking}$. We refer to this dataset as the evaluation dataset $D_{TagRankingEval}$. Table 4.5 shows two different groups of annotators from different countries such as India, Germany, Chile, Japan, China, Taiwan, and Portugal. All annotators are either students or working professionals. Our tag annotation experiment consists of two steps. Since the average number of tags for a UGI in the YFCC100M dataset is five, during the first step we assigned UGIs in the evaluation dataset to four annotators from *Group-A* and asked them to select five most relevant tags for a UGI from its user and visual tags. Experimental results indicate that the annotators have selected approximately 2.36 user tags and approximately 2.64 visual tags for each UGI. Average confidence scores of selected visual tags are above 80%. This result indicates that visual tags with scores above 80% are very useful in describing the UGI. Thus, during the second step of annotation we created tag lists of UGIs with all user tags (after removing misspelled words and preserving the original tag order) and visual tags with a confidence score above 80%. Similar to Flickr, we appended visual tags after user tags in the tag list. Next, we assigned UGIs in the evaluation dataset to six annotators from *Group-B* asking them to assign a relevance score to each tag in the tag list. We assigned each UGI to two annotators from *Group-B* to compute the inter-annotation agreement. Each tag is assigned one of the following relevance scores: most relevant (score 5), relevant (score 4), partially relevant (score 3), weakly relevant (score 2), and irrelevant (score 1). We treat this order and relevance score as a *gold standard* in our experiments. Since we assign each UGI to two annotators, we computed the Cohen Kappa's coefficient k [52] to evaluate the annotation consistency. The computed k in our experiment is 0.512 which is considered as moderate to good annotation [179]. We computed the Cohen Kappa coefficient using the following standard formula:

$$\kappa = \frac{p_o - p_e}{1 - p_e} \tag{4.16}$$

where p_o is the relative observed agreement among annotators, and p_e is the hypothetical probability of chance agreement, using the observed data to calculate the probabilities of each observer randomly say each category.

Evaluation Metric To evaluate our tag ranking system, we computed the normalized discounted cumulative gain (NDCG) for the ranked tag list of a given UGI. For the ranked tag list $t_1, t_2, \ldots, t_n$, NDCG is computed by the following formula:

$$NDCG_n = \frac{DCG_n}{IDCG_n} = \lambda_n \sum_{i=1}^{n} \frac{2^{l(k)} - 1}{\log(1 + k)} \tag{4.17}$$

where DCG_n is the Discounted Cumulative Gain and computed by the following formula:

$$DCG_n = \sum_{i=1}^{n} \frac{2^{l(k)} - 1}{\log(1 + k)} \tag{4.18}$$

where $l(k)$ is the relevance level of the kth tag and I_n ($i.e.$, $\frac{1}{IDCG_n}$) is a normalization constant so that the optimal NDCG is 1. That is, $IDCG_n$ is the maximum possible (ideal) DCG for a given set of tags and relevances. For instance, say, a UGI has five tags, t_1, t_2, t_3, t_4, and t_5 with relevance scores 1, 2, 3, 4, and 5, respectively. Thus, $IDCG_n$ is computed for the tag sequence t_5, t_4, t_3, t_2, and t_1 since it provides the highest relevance scores sequence 5, 4, 3, 2, and 1, respectively. Further, say, if our algorithm produces the sequence t_5, t_3, t_2, t_4, and t_1 as the ranked tag list, then DCG_n is computed for the following relevance scores sequence: 5, 3, 2, 4, 1. Thus, DCG_n will always be less than or equal to $IDCG_n$, and $NDCG\ _n$ will always be between zero and one (boundaries included). We compute the average of NDCG scores of all UGIs in the evaluation dataset as the system performance for different approaches.

Results Our experiments consist of two steps. In the first step, we computed tag relevance based on voting from UGI neighbors derived using three proposed high-level features. Moreover, we compare the performance of our systems (NVGC, NVVC, and NVSC) with a baseline and two state-of-the-art techniques. We consider the original list of tags for a UGI, *i.e.*, the order in which the user annotated the UGI, as a baseline for the evaluation of our tag ranking approach. For state-of-the-arts, we use the following techniques: (i) computing tag relevance based on voting from 50 neighbors derived using low-level features such as RGB moment, texture, and correlogram (NVLV) [102], and (ii) computing tag relevance based on a probabilistic random walk approach (PRW) [109]. In the second step, we investigate early and late fusion techniques (NVGVC, NVGSC, NVVSC, and NVGVSC) to compute the tag relevance leveraging our proposed high-level features, as described above in Sect. 4.2.2. Figure 4.12 confirms that late fusion based on the recall of different modality (LFR) outperforms early fusion (EF) and late fusion with equal weight (LFE). Experimental results in Fig. 4.13 confirm that our proposed high-level features and their fusion are very helpful in improving the tag relevance compared with the baseline and state-of-the-arts. The NDCG score of tags ranked by our CRAFT system is 0.886264, *i.e.*, there is an improvement of 22.24% in the NDCG score for the original order of tags (the baseline). Moreover, there is an improvement of 5.23% and 9.28% in the tag ranking performance (in terms of NDCG scores) of the CRAFT system than the following two most popular state-of-the-arts, respectively. First, a probabilistic random walk approach (PRW) [109]. Second, a neighbor voting approach (NVLV) [102]. Furthermore, our proposed recall based late fusion technique results in 9.23% improvement regarding the NDCG score than the early fusion technique.

Results in Figs. 4.12 and 4.13 correspond to 50 neighbors of UGIs. Experimental results confirm that our results are consistent with a different number of neighbors such as 50, 100, 200, 300, and 500 (see Fig. 4.14). Figure 4.15 shows the original

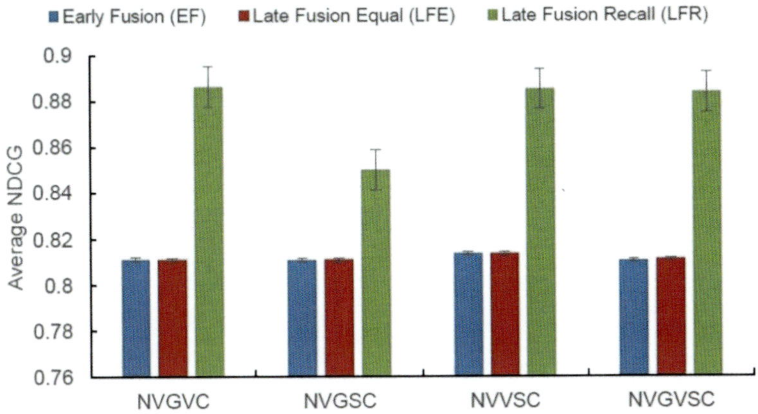

Fig. 4.12 Performance of fusion techniques

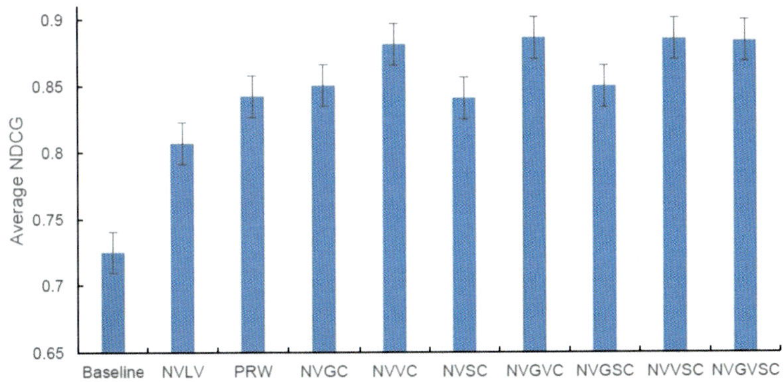

Fig. 4.13 Baseline is the original list of ranked tags (*i.e.*, the order in which a user annotated tags for a UGI). NVLV and PRW are state-of-the-art techniques based on neighbor voting and probabilistic random walk approach leveraging low-level visual features of the UGI. Other approaches are based on neighbor voting leveraging our proposed high-level features of the UGI and their fusion

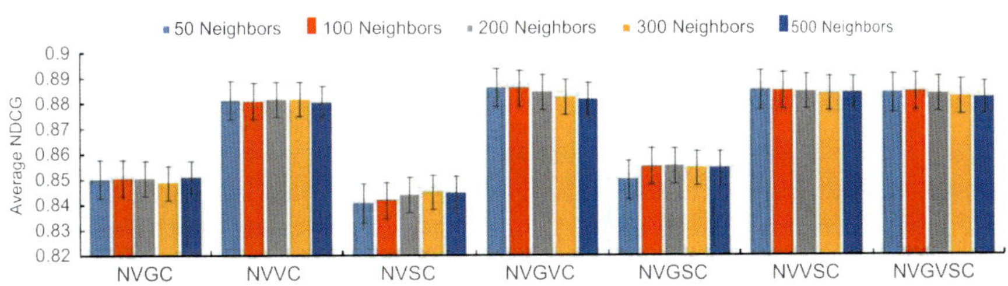

Fig. 4.14 Tag ranking performance for the different number of UGI neighbors and several approaches

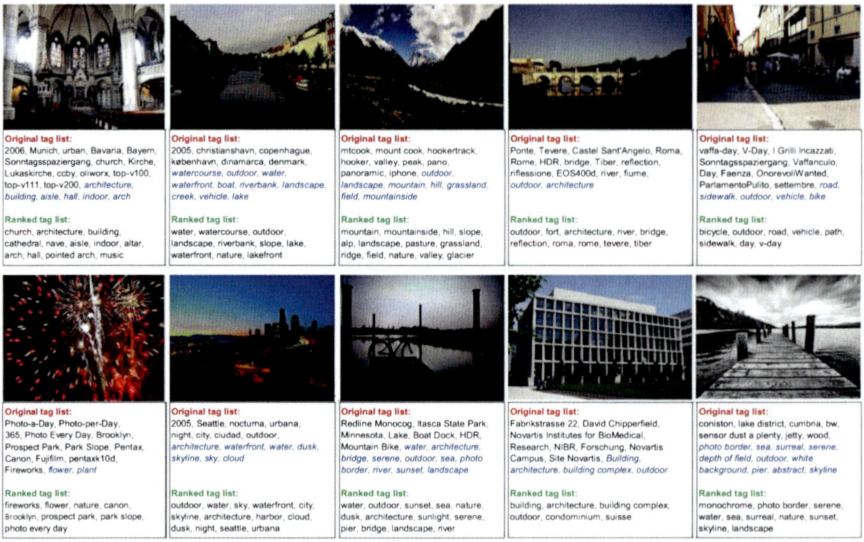

Fig. 4.15 The list of ten UGIs with original and ranked tags

and ranked tag lists of ten exemplary UGIs in the evaluation dataset. Tags in normal and italic fonts are user tags and automatically generated visual tags from visual content, respectively. Moreover, Fig. 4.15 suggests that the user tags are not sufficient to describe the objective aspects of UGIs. Visual tags are also very important in describing the visual content of UGIs. Thus, our techniques leverage both user and visual tags in tag ranking. Similar to much earlier work in tag ranking, we do not rank tags which are a noun. For instance, we ignore tags such as *Lukaskirche*, *Copenhague*, and *mount cook* in Fig. 4.15 during tag ranking. Our tag ranking method should work well for a large collection of UGIs as well since neighbors can be computed accurately and efficiently using the computed concepts and created clusters [110]. In the future, we would like to leverage map matching techniques [244] to further improve tag recommendation and ranking accuracies.

4.4 Summary

We proposed automatic tag recommendation and ranking systems. The proposed tag recommendation system, called, PROMPT first determines a group of users who have similar tagging behavior as the user of a given UGI. Next, we construct lists of candidate tags for different approaches based on co-occurrence and neighbor voting. Further, we compute relevance scores of candidate tags. Next, we perform a random walk process on a tag graph with candidate tags as its nodes. Relevance scores of candidate tags are used as initial scores for nodes and updated in every iteration based on exemplar and concurrent tag similarities. The random walk process iterates until it converges. Finally, we recommend the top five tags with the highest scores when the random walk process terminates. Experimental results confirm that our proposed approaches outperform baselines in personalized user tag recommendation. Particularly, PROMPT outperforms the best performing state-of-the-art for tag recommendation (*i.e.*, an approach based on random walk by 11.34%, 17.84%, and 17.5% regarding accuracy, precision, and recall. These approaches could be further enhanced to improve accuracy, precision, and recall in the future (see Chap. 8 for details).

The proposed tag ranking system, called, CRAFT leverage three novel high-level features based on concepts derived from different modalities. Since concepts are very useful in understanding UGIs, we first leverage them in finding semantically similar neighbors. Subsequently, we compute tag relevance based on neighbor voting using late fusion technique with weights determined by the recall of modality. Experimental results confirm that our proposed features are very useful and complement each other in determining tag relevance for UGIs. Particularly, there is an improvement of 22.24% in the NDCG score for the original order of tags (the baseline). Moreover, there is an improvement of 5.23% and 9.28% in the tag ranking performance (in terms of NDCG scores) of the CRAFT system than the following two most popular state-of-the-arts, respectively: (i) a probabilistic random walk approach (PRW) [109] and (ii) a neighbor voting approach (NVLV) [102]. Furthermore, there is an improvement of 9.23% NDCG score in the proposed recall-based late fusion technique than the early fusion technique. In our future work, we plan to investigate the fusion of knowledge structures from more modalities and employ deep neural network techniques to further improve tag relevance accuracy for UGIs (see Chapter 8 for details).

References

1. Apple Denies Steve Jobs Heart Attack Report: "It Is Not True". http://www.businessinsider.com/2008/10/apple-s-steve-jobs-rushed-to-er-after-heart-attack-says-cnn-citizen-journalist/. October 2008. Online: Last Accessed Sept 2015.
2. SVM^{hmm}: Sequence Tagging with Structural Support Vector Machines. https://www.cs.cornell.edu/people/tj/svm light/svm hmm.html. August 2008. Online: Last Accessed May 2016.
3. NPTEL. 2009, December. http://www.nptel.ac.in. Online; Accessed Apr 2015.
4. iReport at 5: Nearly 900,000 contributors worldwide. http://www.niemanlab.org/2011/08/ireport-at-5-nearly-900000-contributors-worldwide/. August 2011. Online: Last Accessed Sept 2015.
5. Meet the million: 999,999 iReporters + you! http://www.ireport.cnn.com/blogs/ireport-blog/2012/01/23/meet-the-million-999999-ireporters-you. January 2012. Online: Last Accessed Sept 2015.
6. 5 Surprising Stats about User-generated Content. 2014, April. http://www.smartblogs.com/social-media/2014/04/11/6.-surprising-stats-about-user-generated-content/. Online: Last Accessed Sept 2015.
7. The Citizen Journalist: How Ordinary People are Taking Control of the News. 2015, June. http://www.digitaltrends.com/features/the-citizen-journalist-how-ordinary-people-are-taking-control-of-the-news/. Online: Last Accessed Sept 2015.

8. Wikipedia API. 2015, April. http://tinyurl.com/WikiAPI-AI. API: Last Accessed Apr 2015.
9. Apache Lucene. 2016, June. https://lucene.apache.org/core/. Java API: Last Accessed June 2016.
10. By the Numbers: 14 Interesting Flickr Stats. 2016, May. http://www.expandedramblings.com/index.php/flickr-stats/. Online: Last Accessed May 2016.
11. By the Numbers: 180+ Interesting Instagram Statistics (June 2016). 2016, June. http://www.expandedramblings.com/index.php/important-instagram-stats/. Online: Last Accessed July 2016.
12. Coursera. 2016, May. https://www.coursera.org/. Online: Last Accessed May 2016.
13. FourSquare API. 2016, June. https://developer.foursquare.com/. Last Accessed June 2016.
14. Google Cloud Vision API. 2016, December. https://cloud.google.com/vision/. Online: Last Accessed Dec 2016.
15. Google Forms. 2016, May. https://docs.google.com/forms/. Online: Last Accessed May 2016.
16. MIT Open Course Ware. 2016, May. http://www.ocw.mit.edu/. Online: Last Accessed May 2016.
17. Porter Stemmer. 2016, May. https://tartarus.org/martin/PorterStemmer/. Online: Last Accessed May 2016.
18. SenticNet. 2016, May. http://www.sentic.net/computing/. Online: Last Accessed May 2016.
19. Sentics. 2016, May. https://en.wiktionary.org/wiki/sentics. Online: Last Accessed May 2016.
20. VideoLectures.Net. 2016, May. http://www.videolectures.net/. Online: Last Accessed May, 2016.
21. YouTube Statistics. 2016, July. http://www.youtube.com/yt/press/statistics.html. Online: Last Accessed July, 2016.
22. Abba, H.A., S.N.M. Shah, N.B. Zakaria, and A.J. Pal. 2012. Deadline based performance evalu-ation of job scheduling algorithms. In *Proceedings of the IEEE International Conference on Cyber-Enabled Distributed Computing and Knowledge Discovery*, 106–110.
23. Achanta, R.S., W.-Q. Yan, and M.S. Kankanhalli. (2006). Modeling Intent for Home Video Repurposing. In *Proceedings of the IEEE MultiMedia*, (1):46–55.
24. Adcock, J., M. Cooper, A. Girgensohn, and L. Wilcox. 2005. Interactive Video Search using Multilevel Indexing. In *Proceedings of the Springer Image and Video Retrieval*, 205–214.
25. Agarwal, B., S. Poria, N. Mittal, A. Gelbukh, and A. Hussain. 2015. Concept-level Sentiment Analysis with Dependency-based Semantic Parsing: A Novel Approach. In *Proceedings of the Springer Cognitive Computation*, 1–13.
26. Aizawa, K., D. Tancharoen, S. Kawasaki, and T. Yamasaki. 2004. Efficient Retrieval of Life Log based on Context and Content. In *Proceedings of the ACM Workshop on Continuous Archival and Retrieval of Personal Experiences*, 22–31.
27. Altun, Y., I. Tsochantaridis, and T. Hofmann. 2003. Hidden Markov Support Vector Machines. In *Proceedings of the International Conference on Machine Learning*, 3–10.
28. Anderson, A., K. Ranghunathan, and A. Vogel. 2008. Tagez: Flickr Tag Recommendation. In *Proceedings of the Association for the Advancement of Artificial Intelligence*.
29. Atrey, P.K., A. El Saddik, and M.S. Kankanhalli. 2011. Effective Multimedia Surveillance using a Human-centric Approach. *Proceedings of the Springer Multimedia Tools and Applications* 51(2): 697–721.
30. Barnard, K., P. Duygulu, D. Forsyth, N. De Freitas, D.M. Blei, and M.I. Jordan. 2003. Matching Words and Pictures. *Proceedings of the Journal of Machine Learning Research* 3: 1107–1135.
31. Basu, S., Y. Yu, V.K. Singh, and R. Zimmermann. 2016. Videopedia: Lecture Video Recommendation for Educational Blogs Using Topic Modeling. In *Proceedings of the Springer International Conference on Multimedia Modeling*, 238–250.
32. Basu, S., Y. Yu, and R. Zimmermann. 2016. Fuzzy Clustering of Lecture Videos based on Topic Modeling. In *Proceedings of the IEEE International Workshop on Content-Based Multimedia Indexing*, 1–6.

33. Basu, S., R. Zimmermann, K.L. OHalloran, S. Tan, and K. Marissa. 2015. Performance Evaluation of Students Using Multimodal Learning Systems. In *Proceedings of the Springer International Conference on Multimedia Modeling,* 135–147.
34. Beeferman, D., A. Berger, and J. Lafferty. 1999. Statistical Models for Text Segmentation. *Proceedings of the Springer Machine Learning* 34(1–3): 177–210.
35. Bernd, J., D. Borth, C. Carrano, J. Choi, B. Elizalde, G. Friedland, L. Gottlieb, K. Ni, R. Pearce, D. Poland, et al. 2015. Kickstarting the Commons: The YFCC100M and the YLI Corpora. In *Proceedings of the ACM Workshop on Community-Organized Multimodal Mining: Opportunities for Novel Solutions*, 1–6.
36. Bhatt, C.A., and M.S. Kankanhalli. 2011. Multimedia Data Mining: State of the Art and Challenges. *Proceedings of the Multimedia Tools and Applications* 51(1): 35–76.
37. Bhatt, C.A., A. Popescu-Belis, M. Habibi, S. Ingram, S. Masneri, F. McInnes, N. Pappas, and O. Schreer. 2013. Multi-factor Segmentation for Topic Visualization and Recommendation: the MUST-VIS System. In *Proceedings of the ACM International Conference on Multimedia*, 365–368.
38. Bhattacharjee, S., W.C. Cheng, C.-F. Chou, L. Golubchik, and S. Khuller. 2000. BISTRO: A Frame-work for Building Scalable Wide-area Upload Applications. *Proceedings of the ACM SIGMETRICS Performance Evaluation Review* 28(2): 29–35.
39. Cambria, E., J. Fu, F. Bisio, and S. Poria. 2015. AffectiveSpace 2: Enabling Affective Intuition for Concept-level Sentiment Analysis. In *Proceedings of the AAAI Conference on Artificial Intelligence*, 508–514.
40. Cambria, E., A. Livingstone, and A. Hussain. 2012. The Hourglass of Emotions. In *Proceedings of the Springer Cognitive Behavioural Systems*, 144–157.
41. Cambria, E., D. Olsher, and D. Rajagopal. 2014. SenticNet 3: A Common and Common-sense Knowledge Base for Cognition-driven Sentiment Analysis. In *Proceedings of the AAAI Conference on Artificial Intelligence*, 1515–1521.
42. Cambria, E., S. Poria, R. Bajpai, and B. Schuller. 2016. SenticNet 4: A Semantic Resource for Sentiment Analysis based on Conceptual Primitives. In *Proceedings of the International Conference on Computational Linguistics (COLING)*, 2666–2677.
43. Cambria, E., S. Poria, F. Bisio, R. Bajpai, and I. Chaturvedi. 2015. The CLSA Model: A Novel Framework for Concept-Level Sentiment Analysis. In *Proceedings of the Springer Computational Linguistics and Intelligent Text Processing*, 3–22.
44. Cambria, E., S. Poria, A. Gelbukh, and K. Kwok. 2014. Sentic API: A Common-sense based API for Concept-level Sentiment Analysis. *CEUR Workshop Proceedings* 144: 19–24.
45. Cao, J., Z. Huang, and Y. Yang. 2015. Spatial-aware Multimodal Location Estimation for Social Images. In *Proceedings of the ACM Conference on Multimedia Conference*, 119–128.
46. Chakraborty, I., H. Cheng, and O. Javed. 2014. Entity Centric Feature Pooling for Complex Event Detection. In *Proceedings of the Workshop on HuEvent at the ACM International Conference on Multimedia,* 1–5.
47. Che, X., H. Yang, and C. Meinel. 2013. Lecture Video Segmentation by Automatically Analyzing the Synchronized Slides. In *Proceedings of the ACM International Conference on Multimedia*, 345–348.
48. Chen, B., J. Wang, Q. Huang, and T. Mei. 2012. Personalized Video Recommendation through Tripartite Graph Propagation. In *Proceedings of the ACM International Conference on Multimedia*, 1133–1136.
49. Chen, S., L. Tong, and T. He. 2011. Optimal Deadline Scheduling with Commitment. In *Proceedings of the IEEE Annual Allerton Conference on Communication, Control, and Computing*, 111–118.
50. Chen, W.-B., C. Zhang, and S. Gao. 2012. Segmentation Tree based Multiple Object Image Retrieval. In *Proceedings of the IEEE International Symposium on Multimedia*, 214–221.
51. Chen, Y., and W.J. Heng. 2003. Automatic Synchronization of Speech Transcript and Slides in Presentation. *Proceedings of the IEEE International Symposium on Circuits and Systems* 2: 568–571.

52. Cohen, J. 1960. A Coefficient of Agreement for Nominal Scales. *Proceedings of the Durham Educational and Psychological Measurement* 20(1): 37–46.
53. Cristani, M., A. Pesarin, C. Drioli, V. Murino, A. Rodà,M. Grapulin, and N. Sebe. 2010. Toward an Automatically Generated Soundtrack from Low-level Cross-modal Correlations for Automotive Scenarios. In *Proceedings of the ACM International Conference on Multimedia*, 551–560.
54. Dang-Nguyen, D.-T., L. Piras, G. Giacinto, G. Boato, and F.G. De Natale. 2015. A Hybrid Approach for Retrieving Diverse Social Images of Landmarks. In *Proceedings of the IEEE International Conference on Multimedia and Expo (ICME)*, 1–6.
55. Fabro, M. Del, A. Sobe, and L. Böszörmenyi. 2012. Summarization of Real-life Events based on Community-contributed Content. In *Proceedings of the International Conferences on Advances in Multimedia*, 119–126.
56. Du, L., W.L. Buntine, and M. Johnson. 2013. Topic Segmentation with a Structured Topic Model. In *Proceedings of the North American Chapter of the Association for Computational Linguistics: Human Language Technologies*, 190–200.
57. Fan, Q., K. Barnard, A. Amir, A. Efrat, and M. Lin. 2006. Matching Slides to Presentation Videos using SIFT and Scene Background Matching. In *Proceedings of the ACM International Conference on Multimedia*, 239–248.
58. Filatova, E. and V. Hatzivassiloglou. 2004. Event-based Extractive Summarization. In *Proceedings of the ACL Workshop on Summarization*, 104–111.
59. Firan, C.S., M. Georgescu, W. Nejdl, and R. Paiu. 2010. Bringing Order to Your Photos: Event-driven Classification of Flickr Images based on Social Knowledge. In *Proceedings of the ACM International Conference on Information and Knowledge Management*, 189–198.
60. Gao, S., C. Zhang, and W.-B. Chen. 2012. An Improvement of Color Image Segmentation through Projective Clustering. In *Proceedings of the IEEE International Conference on Information Reuse and Integration*, 152–158.
61. Garg, N. and I. Weber. 2008. Personalized, Interactive Tag Recommendation for Flickr. In *Proceedings of the ACM Conference on Recommender Systems*, 67–74.
62. Ghias, A., J. Logan, D. Chamberlin, and B.C. Smith. 1995. Query by Humming: Musical Information Retrieval in an Audio Database. In *Proceedings of the ACM International Conference on Multimedia*, 231–236.
63. Golder, S.A., and B.A. Huberman. 2006. Usage Patterns of Collaborative Tagging Systems. *Proceedings of the Journal of Information Science* 32(2): 198–208.
64. Gozali, J.P., M.-Y. Kan, and H. Sundaram. 2012. Hidden Markov Model for Event Photo Stream Segmentation. In *Proceedings of the IEEE International Conference on Multimedia and Expo Workshops*, 25–30.
65. Guo, Y., L. Zhang, Y. Hu, X. He, and J. Gao. 2016. Ms-celeb-1m: Challenge of recognizing one million celebrities in the real world. *Proceedings of the Society for Imaging Science and Technology Electronic Imaging* 2016(11): 1–6.
66. Hanjalic, A., and L.-Q. Xu. 2005. Affective Video Content Representation and Modeling. *Proceedings of the IEEE Transactions on Multimedia* 7(1): 143–154.
67. Haubold, A. and J.R. Kender. 2005. Augmented Segmentation and Visualization for Presentation Videos. In *Proceedings of the ACM International Conference on Multimedia*, 51–60.
68. Healey, J.A., and R.W. Picard. 2005. Detecting Stress during Real-world Driving Tasks using Physiological Sensors. *Proceedings of the IEEE Transactions on Intelligent Transportation Systems* 6(2): 156–166.
69. Hefeeda, M., and C.-H. Hsu. 2010. On Burst Transmission Scheduling in Mobile TV Broadcast Networks. *Proceedings of the IEEE/ACM Transactions on Networking* 18(2): 610–623.
70. Hevner, K. 1936. Experimental Studies of the Elements of Expression in Music. *Proceedings of the American Journal of Psychology* 48: 246–268.

71. Hochbaum, D.S. 1996. Approximating Covering and Packing Problems: Set Cover, Vertex Cover, Independent Set, and related Problems. In *Proceedings of the PWS Approximation algorithms for NP-hard problems*, 94–143.
72. Hong, R., J. Tang, H.-K. Tan, S. Yan, C. Ngo, and T.-S. Chua. 2009. Event Driven Summarization for Web Videos. In *Proceedings of the ACM SIGMM Workshop on Social Media*, 43–48.
73. P. ITU-T Recommendation. 1999. Subjective Video Quality Assessment Methods for Multimedia Applications.
74. Jiang, L., A.G. Hauptmann, and G. Xiang. 2012. Leveraging High-level and Low-level Features for Multimedia Event Detection. In *Proceedings of the ACM International Conference on Multimedia*, 449–458.
75. Joachims, T., T. Finley, and C.-N. Yu. 2009. Cutting-plane Training of Structural SVMs. *Proceedings of the Machine Learning Journal* 77(1): 27–59.
76. Johnson, J., L. Ballan, and L. Fei-Fei. 2015. Love Thy Neighbors: Image Annotation by Exploiting Image Metadata. In *Proceedings of the IEEE International Conference on Computer Vision*, 4624–4632.
77. Johnson, J., A. Karpathy, and L. Fei-Fei. 2015. Densecap: Fully Convolutional Localization Networks for Dense Captioning. *In Proceedings of the arXiv preprint arXiv:1511.07571*.
78. Jokhio, F., A. Ashraf, S. Lafond, I. Porres, and J. Lilius. 2013. Prediction-based dynamic resource allocation for video transcoding in cloud computing. In *Proceedings of the IEEE International Conference on Parallel, Distributed and Network-Based Processing*, 254–261.
79. Kaminskas, M., I. Fernández-Tobías, F. Ricci, and I. Cantador. 2014. Knowledge-based Identification of Music Suited for Places of Interest. *Proceedings of the Springer Information Technology & Tourism* 14(1): 73–95.
80. Kaminskas, M. and F. Ricci. 2011. Location-adapted Music Recommendation using Tags. In *Proceedings of the Springer User Modeling, Adaption and Personalization*, 183–194.
81. Kan, M.-Y. 2001. Combining Visual Layout and Lexical Cohesion Features for Text Segmentation. *In Proceedings of the Citeseer.*
82. Kan, M.-Y. 2003. *Automatic Text Summarization as Applied to Information Retrieval*. PhD thesis, Columbia University.
83. Kan, M.-Y., J.L. Klavans, and K.R. McKeown.1998. Linear Segmentation and Segment Significance. *In Proceedings of the arXiv preprint cs/9809020*.
84. Kan, M.-Y., K.R. McKeown, and J.L. Klavans. 2001. Applying Natural Language Generation to Indicative Summarization. *Proceedings of the ACL European Workshop on Natural Language Generation* 8: 1–9.
85. Kang, H.B. 2003. Affective Content Detection using HMMs. In *Proceedings of the ACM International Conference on Multimedia*, 259–262.
86. Kang, Y.-L., J.-H. Lim, M.S. Kankanhalli, C.-S. Xu, and Q. Tian. 2004. Goal Detection in Soccer Video using Audio/Visual Keywords. *Proceedings of the IEEE International Conference on Image Processing* 3: 1629–1632.
87. Kang, Y.-L., J.-H. Lim, Q. Tian, and M.S. Kankanhalli. 2003. Soccer Video Event Detection with Visual Keywords. In *Proceedings of the Joint Conference of International Conference on Information, Communications and Signal Processing, and Pacific Rim Conference on Multimedia*, 3:1796–1800.
88. Kankanhalli, M.S., and T.-S. Chua. 2000. Video Modeling using Strata-based Annotation. *Proceedings of the IEEE MultiMedia* 7(1): 68–74.
89. Kennedy, L., M. Naaman, S. Ahern, R. Nair, and T. Rattenbury. 2007. How Flickr Helps us Make Sense of the World: Context and Content in Community-contributed Media Collections. In *Proceedings of the ACM International Conference on Multimedia*, 631–640.
90. Kennedy, L.S., S.-F. Chang, and I.V. Kozintsev. 2006. To Search or to Label?: Predicting the Performance of Search-based Automatic Image Classifiers. In *Proceedings of the ACM International Workshop on Multimedia Information Retrieval*, 249–258.

91. Kim, Y.E., E.M. Schmidt, R. Migneco, B.G. Morton, P. Richardson, J. Scott, J.A. Speck, and D. Turnbull. 2010. Music Emotion Recognition: A State of the Art Review. In *Proceedings of the International Society for Music Information Retrieval*, 255–266.
92. Klavans, J.L., K.R. McKeown, M.-Y. Kan, and S. Lee. 1998. Resources for Evaluation of Summarization Techniques. *In Proceedings of the arXiv preprint cs/9810014*.
93. Ko, Y. 2012. A Study of Term Weighting Schemes using Class Information for Text Classification. In *Proceedings of the ACM Special Interest Group on Information Retrieval*, 1029–1030.
94. Kort, B., R. Reilly, and R.W. Picard. 2001. An Affective Model of Interplay between Emotions and Learning: Reengineering Educational Pedagogy-Building a Learning Companion. *Proceedings of the IEEE International Conference on Advanced Learning Technologies* 1: 43–47.
95. Kucuktunc, O., U. Gudukbay, and O. Ulusoy. 2010. Fuzzy Color Histogram-based Video Segmentation. *Proceedings of the Computer Vision and Image Understanding* 114(1): 125–134.
96. Kuo, F.-F., M.-F. Chiang, M.-K. Shan, and S.-Y. Lee. 2005. Emotion-based Music Recommendation by Association Discovery from Film Music. In *Proceedings of the ACM International Conference on Multimedia*, 507–510.
97. Lacy, S., T. Atwater, X. Qin, and A. Powers. 1988. Cost and Competition in the Adoption of Satellite News Gathering Technology. *Proceedings of the Taylor & Francis Journal of Media Economics* 1(1): 51–59.
98. Lambert, P., W. De Neve, P. De Neve, I. Moerman, P. Demeester, and R. Van de Walle. 2006. Rate-distortion performance of H. 264/AVC compared to state-of-the-art video codecs. *Proceedings of the IEEE Transactions on Circuits and Systems for Video Technology* 16(1): 134–140.
99. Laurier, C., M. Sordo, J. Serrà, and P. Herrera. 2009. Music Mood Representations from Social Tags. In *Proceedings of the International Society for Music Information Retrieval*, 381–386.
100. Li, C.T. and M.K. Shan. 2007. Emotion-based Impressionism Slideshow with Automatic Music Accompaniment. In *Proceedings of the ACM International Conference on Multimedia*, 839–842.
101. Li, J., and J.Z. Wang. 2008. Real-time Computerized Annotation of Pictures. *Proceedings of the IEEE Transactions on Pattern Analysis and Machine Intelligence* 30(6): 985–1002.
102. Li, X., C.G. Snoek, and M. Worring. 2009. Learning Social Tag Relevance by Neighbor Voting. *Proceedings of the IEEE Transactions on Multimedia* 11(7): 1310–1322.
103. Li, X., T. Uricchio, L. Ballan, M. Bertini, C.G. Snoek, and A.D. Bimbo. 2016. Socializing the Semantic Gap: A Comparative Survey on Image Tag Assignment, Refinement, and Retrieval. *Proceedings of the ACM Computing Surveys (CSUR)* 49(1): 14.
104. Li, Z., Y. Huang, G. Liu, F. Wang, Z.-L. Zhang, and Y. Dai. 2012. Cloud Transcoder: Bridging the Format and Resolution Gap between Internet Videos and Mobile Devices. In *Proceedings of the ACM International Workshop on Network and Operating System Support for Digital Audio and Video*, 33–38.
105. Liang, C., Y. Guo, and Y. Liu. 2008. Is Random Scheduling Sufficient in P2P Video Streaming? In *Proceedings of the IEEE International Conference on Distributed Computing Systems*, 53–60. IEEE.
106. Lim, J.-H., Q. Tian, and P. Mulhem. 2003. Home Photo Content Modeling for Personalized Event-based Retrieval. *Proceedings of the IEEE MultiMedia* 4: 28–37.
107. Lin, M., M. Chau, J. Cao, and J.F. Nunamaker Jr. 2005. Automated Video Segmentation for Lecture Videos: A Linguistics-based Approach. *Proceedings of the IGI Global International Journal of Technology and Human Interaction* 1(2): 27–45.
108. Liu, C.L., and J.W. Layland. 1973. Scheduling Algorithms for Multiprogramming in a Hard-real-time Environment. *Proceedings of the ACM Journal of the ACM* 20(1): 46–61.

109. Liu, D., X.-S. Hua, L. Yang, M. Wang, and H.-J. Zhang. 2009. Tag Ranking. In *Proceedings of the ACM World Wide Web Conference*, 351–360.
110. Liu, T., C. Rosenberg, and H.A. Rowley. 2007. Clustering Billions of Images with Large Scale Nearest Neighbor Search. In *Proceedings of the IEEE Winter Conference on Applications of Computer Vision*, 28–28.
111. Liu, X. and B. Huet. 2013. Event Representation and Visualization from Social Media. In *Proceedings of the Springer Pacific-Rim Conference on Multimedia*, 740–749.
112. Liu, Y., D. Zhang, G. Lu, and W.-Y. Ma. 2007. A Survey of Content-based Image Retrieval with High-level Semantics. *Proceedings of the Elsevier Pattern Recognition* 40(1): 262–282.
113. Livingston, S., and D.A.V. BELLE. 2005. The Effects of Satellite Technology on Newsgathering from Remote Locations. *Proceedings of the Taylor & Francis Political Communication* 22(1): 45–62.
114. Long, R., H. Wang, Y. Chen, O. Jin, and Y. Yu. 2011. Towards Effective Event Detection, Tracking and Summarization on Microblog Data. In *Proceedings of the Springer Web-Age Information Management*, 652–663.
115. L. Lu, H. You, and H. Zhang. 2001. A New Approach to Query by Humming in Music Retrieval. In *Proceedings of the IEEE International Conference on Multimedia and Expo*, 22–25.
116. Lu, Y., H. To, A. Alfarrarjeh, S.H. Kim, Y. Yin, R. Zimmermann, and C. Shahabi. 2016. GeoUGV: User-generated Mobile Video Dataset with Fine Granularity Spatial Metadata. In *Proceedings of the ACM International Conference on Multimedia Systems*, 43.
117. Mao, J., W. Xu, Y. Yang, J. Wang, Z. Huang, and A. Yuille. 2014. Deep Captioning with Multimodal Recurrent Neural Networks (M-RNN). *In Proceedings of the arXiv preprint arXiv:1412.6632*.
118. Matusiak, K.K. 2006. Towards User-centered Indexing in Digital Image Collections. *Proceedings of the OCLC Systems & Services: International Digital Library Perspectives* 22(4): 283–298.
119. McDuff, D., R. El Kaliouby, E. Kodra, and R. Picard. 2013. Measuring Voter's Candidate Preference Based on Affective Responses to Election Debates. In *Proceedings of the IEEE Humaine Association Conference on Affective Computing and Intelligent Interaction*, 369–374.
120. McKeown, K.R., J.L. Klavans, and M.-Y. Kan. Method and System for Topical Segmentation, Segment Significance and Segment Function, 29 2002. US Patent 6,473,730.
121. Mezaris, V., A. Scherp, R. Jain, M. Kankanhalli, H. Zhou, J. Zhang, L. Wang, and Z. Zhang. 2011. Modeling and Rrepresenting Events in Multimedia. In *Proceedings of the ACM International Conference on Multimedia*, 613–614.
122. Mezaris, V., A. Scherp, R. Jain, and M.S. Kankanhalli. 2014. Real-life Events in Multimedia: Detection, Representation, Retrieval, and Applications. *Proceedings of the Springer Multimedia Tools and Applications* 70(1): 1–6.
123. Miller, G., and C. Fellbaum. 1998. *Wordnet: An Electronic Lexical Database*. Cambridge, MA: MIT Press.
124. Miller, G.A. 1995. WordNet: A Lexical Database for English. *Proceedings of the Communications of the ACM* 38(11): 39–41.
125. Moxley, E., J. Kleban, J. Xu, and B. Manjunath. 2009. Not All Tags are Created Equal: Learning Flickr Tag Semantics for Global Annotation. In *Proceedings of the IEEE International Conference on Multimedia and Expo*, 1452–1455.
126. Mulhem, P., M.S. Kankanhalli, J. Yi, and H. Hassan. 2003. Pivot Vector Space Approach for Audio-Video Mixing. *Proceedings of the IEEE MultiMedia* 2: 28–40.
127. Naaman, M. 2012. Social Multimedia: Highlighting Opportunities for Search and Mining of Multimedia Data in Social Media Applications. *Proceedings of the Springer Multimedia Tools and Applications* 56(1): 9–34.

128. Natarajan, P., P.K. Atrey, and M. Kankanhalli. 2015. Multi-Camera Coordination and Control in Surveillance Systems: A Survey. *Proceedings of the ACM Transactions on Multimedia Computing, Communications, and Applications* 11(4): 57.
129. Nayak, M.G. 2004. *Music Synthesis for Home Videos*. PhD thesis.
130. Neo, S.-Y., J. Zhao, M.-Y. Kan, and T.-S. Chua. 2006. Video Retrieval using High Level Features: Exploiting Query Matching and Confidence-based Weighting. In *Proceedings of the Springer International Conference on Image and Video Retrieval*, 143–152.
131. Ngo, C.-W., F. Wang, and T.-C. Pong. 2003. Structuring Lecture Videos for Distance Learning Applications. In *Proceedings of the IEEE International Symposium on Multimedia Software Engineering*, 215–222.
132. Nguyen, V.-A., J. Boyd-Graber, and P. Resnik. 2012. SITS: A Hierarchical Nonparametric Model using Speaker Identity for Topic Segmentation in Multiparty Conversations. In *Proceedings of the Annual Meeting of the Association for Computational Linguistics*, 78–87.
133. Nwana, A.O. and T. Chen. 2016. Who Ordered This?: Exploiting Implicit User Tag Order Preferences for Personalized Image Tagging. *In Proceedings of the arXiv preprint arXiv:1601.06439*.
134. Papagiannopoulou, C. and V. Mezaris. 2014. Concept-based Image Clustering and Summarization of Event-related Image Collections. In *Proceedings of the Workshop on HuEvent at the ACM International Conference on Multimedia*, 23–28.
135. Park, M.H., J.H. Hong, and S.B. Cho. 2007. Location-based Recommendation System using Bayesian User's Preference Model in Mobile Devices. In *Proceedings of the Springer Ubiquitous Intelligence and Computing*, 1130–1139.
136. Petkos, G., S. Papadopoulos, V. Mezaris, R. Troncy, P. Cimiano, T. Reuter, and Y. Kompatsiaris. 2014. Social Event Detection at MediaEval: a Three-Year Retrospect of Tasks and Results. In *Proceedings of the Workshop on Social Events in Web Multimedia at ACM International Conference on Multimedia Retrieval*.
137. Pevzner, L., and M.A. Hearst. 2002. A Critique and Improvement of an Evaluation Metric for Text Segmentation. *Proceedings of the Computational Linguistics* 28(1): 19–36.
138. Picard, R.W., and J. Klein. 2002. Computers that Recognise and Respond to User Emotion: Theoretical and Practical Implications. *Proceedings of the Interacting with Computers* 14(2): 141–169.
139. Picard, R.W., E. Vyzas, and J. Healey. 2001. Toward machine emotional intelligence: Analysis of affective physiological state. *Proceedings of the IEEE Transactions on Pattern Analysis and Machine Intelligence* 23(10): 1175–1191.
140. Poisson, S.D. and C.H. Schnuse. 1841. Recherches Sur La Pprobabilité Des Jugements En Mmatieré Criminelle Et En Matieré Civile. Meyer.
141. Poria, S., E. Cambria, R. Bajpai, and A. Hussain. 2017. A Review of Affective Computing: From Unimodal Analysis to Multimodal Fusion. *Proceedings of the Elsevier Information Fusion* 37: 98–125.
142. Poria, S., E. Cambria, and A. Gelbukh. 2016. Aspect Extraction for Opinion Mining with a Deep Convolutional Neural Network. *Proceedings of the Elsevier Knowledge-Based Systems* 108: 42–49.
143. Poria, S., E. Cambria, A. Gelbukh, F. Bisio, and A. Hussain. 2015. Sentiment Data Flow Analysis by Means of Dynamic Linguistic Patterns. *Proceedings of the IEEE Computational Intelligence Magazine* 10(4): 26–36.
144. Poria, S., E. Cambria, and A.F. Gelbukh. 2015. Deep Convolutional Neural Network Textual Features and Multiple Kernel Learning for Utterance-level Multimodal Sentiment Analysis. In *Proceedings of the EMNLP*, 2539–2544.
145. Poria, S., E. Cambria, D. Hazarika, N. Mazumder, A. Zadeh, and L.-P. Morency. 2017. Context-Dependent Sentiment Analysis in User-Generated Videos. *In Proceedings of the Association for Computational Linguistics*.

146. Poria, S., E. Cambria, D. Hazarika, and P. Vij. 2016. A Deeper Look into Sarcastic Tweets using Deep Convolutional Neural Networks. In *Proceedings of the International Conference on Computational Linguistics (COLING)*.
147. Poria, S., E. Cambria, N. Howard, G.-B. Huang, and A. Hussain. 2016. Fusing Audio Visual and Textual Clues for Sentiment Analysis from Multimodal Content. *Proceedings of the Elsevier Neurocomputing* 174: 50–59.
148. Poria, S., E. Cambria, N. Howard, and A. Hussain. 2015. Enhanced SenticNet with Affective Labels for Concept-based Opinion Mining: Extended Abstract. In *Proceedings of the International Joint Conference on Artificial Intelligence*.
149. Poria, S., E. Cambria, A. Hussain, and G.-B. Huang. 2015. Towards an Intelligent Framework for Multimodal Affective Data Analysis. *Proceedings of the Elsevier Neural Networks* 63: 104–116.
150. Poria, S., E. Cambria, L.-W. Ku, C. Gui, and A. Gelbukh. 2014. A Rule-based Approach to Aspect Extraction from Product Reviews. In *Proceedings of the Second Workshop on Natural Language Processing for Social Media (SocialNLP)*, 28–37.
151. Poria, S., I. Chaturvedi, E. Cambria, and F. Bisio. 2016. Sentic LDA: Improving on LDA with Semantic Similarity for Aspect-based Sentiment Analysis. In *Proceedings of the IEEE International Joint Conference on Neural Networks (IJCNN)*, 4465–4473.
152. Poria, S., I. Chaturvedi, E. Cambria, and A. Hussain. 2016. Convolutional MKL based Multimodal Emotion Recognition and Sentiment Analysis. In *Proceedings of the IEEE International Conference on Data Mining (ICDM)*, 439–448.
153. Poria, S., A. Gelbukh, B. Agarwal, E. Cambria, and N. Howard. 2014. Sentic Demo: A Hybrid Concept-level Aspect-based Sentiment Analysis Toolkit. *In Proceedings of the ESWC*.
154. Poria, S., A. Gelbukh, E. Cambria, D. Das, and S. Bandyopadhyay. 2012. Enriching SenticNet Polarity Scores Through Semi-Supervised Fuzzy Clustering. In *Proceedings of the IEEE International Conference on Data Mining Workshops (ICDMW)*, 709–716.
155. Poria, S., A. Gelbukh, E. Cambria, A. Hussain, and G.-B. Huang. 2014. EmoSenticSpace: A Novel Framework for Affective Common-sense Reasoning. *Proceedings of the Elsevier Knowledge-Based Systems* 69: 108–123.
156. Poria, S., A. Gelbukh, E. Cambria, P. Yang, A. Hussain, and T. Durrani. 2012. Merging SenticNet and WordNet-Affect Emotion Lists for Sentiment Analysis. *Proceedings of the IEEE International Conference on Signal Processing (ICSP)* 2: 1251–1255.
157. Poria, S., A. Gelbukh, A. Hussain, S. Bandyopadhyay, and N. Howard. 2013. Music Genre Classification: A Semi-Supervised Approach. In *Proceedings of the Springer Mexican Conference on Pattern Recognition*, 254–263.
158. Poria, S., N. Ofek, A. Gelbukh, A. Hussain, and L. Rokach. 2014. Dependency Tree-based Rules for Concept-level Aspect-based Sentiment Analysis. In *Proceedings of the Springer Semantic Web Evaluation Challenge*, 41–47.
159. Poria, S., H. Peng, A. Hussain, N. Howard, and E. Cambria. 2017. Ensemble Application of Convolutional Neural Networks and Multiple Kernel Learning for Multimodal Sentiment Analysis. *In Proceedings of the Elsevier Neurocomputing*.
160. Pye, D., N.J. Hollinghurst, T.J. Mills, and K.R. Wood. 1998. Audio-visual Segmentation for Content-based Retrieval. In *Proceedings of the International Conference on Spoken Language Processing*.
161. Qiao, Z., P. Zhang, C. Zhou, Y. Cao, L. Guo, and Y. Zhang. 2014. Event Recommendation in Event-based Social Networks.
162. Raad, E.J. and R. Chbeir. 2014. Foto2Events: From Photos to Event Discovery and Linking in Online Social Networks. In *Proceedings of the IEEE Big Data and Cloud Computing*, 508–515, .
163. Radsch, C.C. 2013. The Revolutions will be Blogged: Cyberactivism and the 4th Estate in Egypt. Doctoral Disseration. American University.

164. Rae, A., B. Sigurbjörnssön, and R. van Zwol. 2010. Improving Tag Recommendation using Social Networks. In *Proceedings of the Adaptivity, Personalization and Fusion of Heterogeneous Information*, 92–99.
165. Rahmani, H., B. Piccart, D. Fierens, and H. Blockeel. 2010. Three Complementary Approaches to Context Aware Movie Recommendation. In *Proceedings of the ACM Workshop on Context-Aware Movie Recommendation*, 57–60.
166. Rattenbury, T., N. Good, and M. Naaman. 2007. Towards Automatic Extraction of Event and Place Semantics from Flickr Tags. In *Proceedings of the ACM Special Interest Group on Information Retrieval.*
167. Rawat, Y. and M. S. Kankanhalli. 2016. ConTagNet: Exploiting User Context for Image Tag Recommendation. In *Proceedings of the ACM International Conference on Multimedia*, 1102–1106.
168. Repp, S., A. Groß, and C. Meinel. 2008. Browsing within Lecture Videos based on the Chain Index of Speech Transcription. *Proceedings of the IEEE Transactions on Learning Technologies* 1(3): 145–156.
169. Repp, S. and C. Meinel. 2006. Semantic Indexing for Recorded Educational Lecture Videos. In *Proceedings of the IEEE International Conference on Pervasive Computing and Communications Workshops*, 5.
170. Repp, S., J. Waitelonis, H. Sack, and C. Meinel. 2007. Segmentation and Annotation of Audiovisual Recordings based on Automated Speech Recognition. In *Proceedings of the Springer Intelligent Data Engineering and Automated Learning*, 620–629.
171. Russell, J.A. 1980. A Circumplex Model of Affect. *Proceedings of the Journal of Personality and Social Psychology* 39: 1161–1178.
172. Sahidullah, M., and G. Saha. 2012. Design, Analysis and Experimental Evaluation of Block based Transformation in MFCC Computation for Speaker Recognition. *Proceedings of the Speech Communication* 54: 543–565.
173. J. Salamon, J. Serra, and E. Gomez´. Tonal Representations for Music Retrieval: From Version Identification to Query-by-Humming. *In* Proceedings of the Springer International Journal of Multimedia Information Retrieval, 2(1):45–58, 2013.
174. Schedl, M. and D. Schnitzer. 2014. Location-Aware Music Artist Recommendation. In *Proceedings of the Springer MultiMedia Modeling*, 205–213.
175. M. Schedl and F. Zhou. 2016. Fusing Web and Audio Predictors to Localize the Origin of Music Pieces for Geospatial Retrieval. In *Proceedings of the Springer European Conference on Information Retrieval*, 322–334.
176. Scherp, A., and V. Mezaris. 2014. Survey on Modeling and Indexing Events in Multimedia. *Proceedings of the Springer Multimedia Tools and Applications* 70(1): 7–23.
177. Scherp, A., V. Mezaris, B. Ionescu, and F. De Natale. 2014. HuEvent '14: Workshop on Human-Centered Event Understanding from Multimedia. In *Proceedings of the ACM International Conference on Multimedia*, 1253–1254, .
178. Schmitz, P. 2006. Inducing Ontology from Flickr Tags. In *Proceedings of the Collaborative Web Tagging Workshop at ACM World Wide Web Conference*, volume 50.
179. Schuller, B., C. Hage, D. Schuller, and G. Rigoll. 2010. Mister DJ, Cheer Me Up!: Musical and Textual Features for Automatic Mood Classification. *Proceedings of the Journal of New Music Research* 39(1): 13–34.
180. Shah, R.R., M. Hefeeda, R. Zimmermann, K. Harras, C.-H. Hsu, and Y. Yu. 2016. NEWSMAN: Uploading Videos over Adaptive Middleboxes to News Servers In Weak Network Infrastructures. In *Proceedings of the Springer International Conference on Multimedia Modeling*, 100–113.
181. Shah, R.R., A. Samanta, D. Gupta, Y. Yu, S. Tang, and R. Zimmermann. 2016. PROMPT: Personalized User Tag Recommendation for Social Media Photos Leveraging Multimodal Information. In *Proceedings of the ACM International Conference on Multimedia*, 486–492.

182. Shah, R.R., A.D. Shaikh, Y. Yu, W. Geng, R. Zimmermann, and G. Wu. 2015. EventBuilder: Real-time Multimedia Event Summarization by Visualizing Social Media. In *Proceedings of the ACM International Conference on Multimedia*, 185–188.
183. Shah, R.R., Y. Yu, A.D. Shaikh, S. Tang, and R. Zimmermann. 2014. ATLAS: Automatic Temporal Segmentation and Annotation of Lecture Videos Based on Modelling Transition Time. In *Proceedings of the ACM International Conference on Multimedia*, 209–212.
184. Shah, R.R., Y. Yu, A.D. Shaikh, and R. Zimmermann. 2015. TRACE: A Linguistic-based Approach for Automatic Lecture Video Segmentation Leveraging Wikipedia Texts. In *Proceedings of the IEEE International Symposium on Multimedia*, 217–220.
185. Shah, R.R., Y. Yu, S. Tang, S. Satoh, A. Verma, and R. Zimmermann. 2016. Concept-Level Multimodal Ranking of Flickr Photo Tags via Recall Based Weighting. In *Proceedings of the MMCommon's Workshop at ACM International Conference on Multimedia*, 19–26.
186. Shah, R.R., Y. Yu, A. Verma, S. Tang, A.D. Shaikh, and R. Zimmermann. 2016. Leveraging Multimodal Information for Event Summarization and Concept-level Sentiment Analysis. In *Proceedings of the Elsevier Knowledge-Based Systems*, 102–109.
187. Shah, R.R., Y. Yu, and R. Zimmermann. 2014. ADVISOR: Personalized Video Soundtrack Recommendation by Late Fusion with Heuristic Rankings. In *Proceedings of the ACM International Conference on Multimedia*, 607–616.
188. Shah, R.R., Y. Yu, and R. Zimmermann. 2014. User Preference-Aware Music Video Generation Based on Modeling Scene Moods. In *Proceedings of the ACM International Conference on Multimedia Systems*, 156–159.
189. Shaikh, A.D., M. Jain, M. Rawat, R.R. Shah, and M. Kumar. 2013. Improving Accuracy of SMS Based FAQ Retrieval System. In *Proceedings of the Springer Multilingual Information Access in South Asian Languages*, 142–156.
190. Shaikh, A.D., R.R. Shah, and R. Shaikh. 2013. SMS based FAQ Retrieval for Hindi, English and Malayalam. In *Proceedings of the ACM Forum on Information Retrieval Evaluation,* 9.
191. Shamma, D.A., R. Shaw, P.L. Shafton, and Y. Liu. 2007. Watch What I Watch: Using Community Activity to Understand Content. In *Proceedings of the ACM International Workshop on Multimedia Information Retrieval*, 275–284.
192. Shaw, B., J. Shea, S. Sinha, and A. Hogue. 2013. Learning to Rank for Spatiotemporal Search. In *Proceedings of the ACM International Conference on Web Search and Data Mining*, 717–726.
193. Sigurbjörnsson, B. and R. Van Zwol. 2008. Flickr Tag Recommendation based on Collective Knowledge. In *Proceedings of the ACM World Wide Web Conference*, 327–336.
194. Snoek, C.G., M. Worring, and A.W.Smeulders. 2005. Early versus Late Fusion in Semantic Video Analysis. In *Proceedings of the ACM International Conference on Multimedia*, 399–402.
195. Snoek, C.G., M. Worring, J.C. Van Gemert, J.-M. Geusebroek, and A.W. Smeulders. 2006. The Challenge Problem for Automated Detection of 101 Semantic Concepts in Multimedia. In *Proceedings of the ACM International Conference on Multimedia*, 421–430.
196. Soleymani, M., J.J.M. Kierkels, G. Chanel, and T. Pun. 2009. A Bayesian Framework for Video Affective Representation. In *Proceedings of the IEEE International Conference on Affective Computing and Intelligent Interaction and Workshops*, 1–7.
197. Stober, S., and A. . Nürnberger. 2013. Adaptive Music Retrieval–a State of the Art. *Proceedings of the Springer Multimedia Tools and Applications* 65(3): 467–494.
198. Stoyanov, V., N. Gilbert, C. Cardie, and E. Riloff. 2009. Conundrums in Noun Phrase Coreference Resolution: Making Sense of the State-of-the-art. In *Proceedings of the ACL International Joint Conference on Natural Language Processing of the Asian Federation of Natural Language Processing*, 656–664.
199. Stupar, A. and S. Michel. 2011. Picasso: Automated Soundtrack Suggestion for Multi-modal Data. In *Proceedings of the ACM Conference on Information and Knowledge Management*, 2589–2592.

200. Thayer, R.E. 1989. *The Biopsychology of Mood and Arousal*. New York: Oxford University Press.
201. Thomee, B., B. Elizalde, D.A. Shamma, K. Ni, G. Friedland, D. Poland, D. Borth, and L.-J. Li. 2016. YFCC100M: The New Data in Multimedia Research. *Proceedings of the Communications of the ACM* 59(2): 64–73.
202. Tirumala, A., F. Qin, J. Dugan, J. Ferguson, and K. Gibbs. 2005. Iperf: The TCP/UDP Bandwidth Measurement Tool. http://dast.nlanr.net/Projects/Iperf/
203. Torralba, A., R. Fergus, and W.T. Freeman. 2008. 80 Million Tiny Images: A Large Data set for Nonparametric Object and Scene Recognition. *Proceedings of the IEEE Transactions on Pattern Analysis and Machine Intelligence* 30(11): 1958–1970.
204. Toutanova, K., D. Klein, C.D. Manning, and Y. Singer. 2003. Feature-Rich Part-of-Speech Tagging with a Cyclic Dependency Network. In *Proceedings of the North American Chapter of the Association for Computational Linguistics on Human Language Technology*, 173–180.
205. Toutanova, K. and C.D. Manning. 2000. Enriching the Knowledge Sources used in a Maximum Entropy Part-of-Speech Tagger. In *Proceedings of the Annual Meeting of the Association for Computational Linguistics*, 63–70.
206. Utiyama, M. and H. Isahara. 2001. A Statistical Model for Domain-Independent Text Segmentation. In *Proceedings of the Annual Meeting on Association for Computational Linguistics*, 499–506.
207. Vishal, K., C. Jawahar, and V. Chari. 2015. Accurate Localization by Fusing Images and GPS Signals. In *Proceedings of the IEEE Computer Vision and Pattern Recognition Workshops*, 17–24.
208. Wang, C., F. Jing, L. Zhang, and H.-J. Zhang. 2008. Scalable Search-based Image Annotation. *Proceedings of the Springer Multimedia Systems* 14(4): 205–220.
209. Wang, H.L., and L.F. Cheong. 2006. Affective Understanding in Film. *Proceedings of the IEEE Transactions on Circuits and Systems for Video Technology* 16(6): 689–704.
210. Wang, J., J. Zhou, H. Xu, T. Mei, X.-S. Hua, and S. Li. 2014. Image Tag Refinement by Regularized Latent Dirichlet Allocation. *Proceedings of the Elsevier Computer Vision and Image Understanding* 124: 61–70.
211. Wang, P., H. Wang, M. Liu, and W. Wang. 2010. An Algorithmic Approach to Event Summarization. In *Proceedings of the ACM Special Interest Group on Management of Data*, 183–194.
212. Wang, X., Y. Jia, R. Chen, and B. Zhou. 2015. Ranking User Tags in Micro-Blogging Website. In *Proceedings of the IEEE ICISCE*, 400–403.
213. Wang, X., L. Tang, H. Gao, and H. Liu. 2010. Discovering Overlapping Groups in Social Media. In *Proceedings of the IEEE International Conference on Data Mining*, 569–578.
214. Wang, Y. and M.S. Kankanhalli. 2015. Tweeting Cameras for Event Detection. In *Proceedings of the IW3C2 International Conference on World Wide Web*, 1231–1241.
215. Webster, A.A., C.T. Jones, M.H. Pinson, S.D. Voran, and S. Wolf. 1993. Objective Video Quality Assessment System based on Human Perception. In *Proceedings of the IS&T/SPIE's Symposium on Electronic Imaging: Science and Technology*, 15–26. International Society for Optics and Photonics.
216. Wei, C.Y., N. Dimitrova, and S.-F. Chang. 2004. Color-mood Analysis of Films based on Syntactic and Psychological Models. In *Proceedings of the IEEE International Conference on Multimedia and Expo*, 831–834.
217. Whissel, C. 1989. The Dictionary of Affect in Language. In *Emotion: Theory, Research and Experience. Vol. 4. The Measurement of Emotions*, ed. R. Plutchik and H. Kellerman, 113–131. New York: Academic.
218. Wu, L., L. Yang, N. Yu, and X.-S. Hua. 2009. Learning to Tag. In *Proceedings of the ACM World Wide Web Conference*, 361–370.

219. Xiao, J., W. Zhou, X. Li, M. Wang, and Q. Tian. 2012. Image Tag Re-ranking by Coupled Probability Transition. In *Proceedings of the ACM International Conference on Multimedia*, 849–852.
220. Xie, D., B. Qian, Y. Peng, and T. Chen. 2009. A Model of Job Scheduling with Deadline for Video-on-Demand System. In *Proceedings of the IEEE International Conference on Web Information Systems and Mining*, 661–668.
221. Xu, M., L.-Y. Duan, C. Xu, M. Kankanhalli, and Q. Tian. 2003. Event Detection in Basketball Video using Multiple Modalities. *Proceedings of the IEEE Joint Conference of the Fourth International Conference on Information, Communications and Signal Processing, and Fourth Pacific Rim Conference on Multimedia* 3: 1526–1530.
222. Xu, M., N.C. Maddage, C. Xu, M. Kankanhalli, and Q. Tian. 2003. Creating Audio Keywords for Event Detection in Soccer Video. In *Proceedings of the IEEE International Conference on Multimedia and Expo*, 2:II–281.
223. Yamamoto, N., J. Ogata, and Y. Ariki. 2003. Topic Segmentation and Retrieval System for Lecture Videos based on Spontaneous Speech Recognition. In *Proceedings of the INTERSPEECH*, 961–964.
224. Yang, H., M. Siebert, P. Luhne, H. Sack, and C. Meinel. 2011. Automatic Lecture Video Indexing using Video OCR Technology. In *Proceedings of the IEEE International Symposium on Multimedia*, 111–116.
225. Yang, Y.H., Y.C. Lin, Y.F. Su, and H.H. Chen. 2008. A Regression Approach to Music Emotion Recognition. *Proceedings of the IEEE Transactions on Audio, Speech, and Language Processing* 16(2): 448–457.
226. Ye, G., D. Liu, I.-H. Jhuo, and S.-F. Chang. 2012. Robust Late Fusion with Rank Minimization. In *Proceedings of the IEEE Conference on Computer Vision and Pattern Recognition*, 3021–3028.
227. Ye, Q., Q. Huang, W. Gao, and D. Zhao. 2005. Fast and Robust Text Detection in Images and Video Frames. *Proceedings of the Elsevier Image and Vision Computing* 23(6): 565–576.
228. Yin, Y., Z. Shen, L. Zhang, and R. Zimmermann. 2015. Spatial-temporal Tag Mining for Automatic Geospatial Video Annotation. *Proceedings of the ACM Transactions on Multimedia Computing, Communications, and Applications* 11(2): 29.
229. Yoon, S. and V. Pavlovic. 2014. Sentiment Flow for Video Interestingness Prediction. In *Proceedings of the Workshop on HuEvent at the ACM International Conference on Multimedia*, 29–34.
230. Yu, Y., K. Joe, V. Oria, F. Moerchen, J.S. Downie, and L. Chen. 2009. Multi-version Music Search using Acoustic Feature Union and Exact Soft Mapping. *Proceedings of the World Scientific International Journal of Semantic Computing* 3(02): 209–234.
231. Yu, Y., Z. Shen, and R. Zimmermann. 2012. Automatic Music Soundtrack Generation for Out-door Videos from Contextual Sensor Information. In *Proceedings of the ACM International Conference on Multimedia*, 1377–1378.
232. Zaharieva, M., M. Zeppelzauer, and C. Breiteneder. 2013. Automated Social Event Detection in Large Photo Collections. In *Proceedings of the ACM International Conference on Multimedia Retrieval*, 167–174.
233. Zhang, J., X. Liu, L. Zhuo, and C. Wang. 2015. Social Images Tag Ranking based on Visual Words in Compressed Domain. *Proceedings of the Elsevier Neurocomputing* 153: 278–285.
234. Zhang, J., S. Wang, and Q. Huang. 2015. Location-Based Parallel Tag Completion for Geo-tagged Social Image Retrieval. In *Proceedings of the ACM International Conference on Multimedia Retrieval*, 355–362.
235. Zhang, M., J. Wong, W. Tavanapong, J. Oh, and P. de Groen. 2004. Media Uploading Systems with Hard Deadlines. In *Proceedings of the Citeseer International Conference on Internet and Multimedia Systems and Applications*, 305–310.
236. Zhang, M., J. Wong, W. Tavanapong, J. Oh, and P. de Groen. 2008. Deadline-constrained Media Uploading Systems. *Proceedings of the Springer Multimedia Tools and Applications* 38(1): 51–74.

237. Zhang, W., J. Lin, X. Chen, Q. Huang, and Y. Liu. 2006. Video Shot Detection using Hidden Markov Models with Complementary Features. *Proceedings of the IEEE International Conference on Innovative Computing, Information and Control* 3: 593–596.
238. Zheng, L., V. Noroozi, and P.S. Yu. 2017. Joint Deep Modeling of Users and Items using Reviews for Recommendation. In *Proceedings of the ACM International Conference on Web Search and Data Mining*, 425–434.
239. Zhou, X.S. and T.S. Huang. 2000. CBIR: from Low-level Features to High-level Semantics. In *Proceedings of the International Society for Optics and Photonics Electronic Imaging*, 426–431.
240. Zhuang, J. and S.C. Hoi. 2011. A Two-view Learning Approach for Image Tag Ranking. In *Proceedings of the ACM International Conference on Web Search and Data Mining*, 625–634.
241. Zimmermann, R. and Y. Yu. 2013. Social Interactions over Geographic-aware Multimedia Systems. In *Proceedings of the ACM International Conference on Multimedia*, 1115–1116.
242. Shah, R.R. 2016. Multimodal-based Multimedia Analysis, Retrieval, and Services in Support of Social Media Applications. In *Proceedings of the ACM International Conference on Multimedia*, 1425–1429.
243. Shah, R.R. 2016. Multimodal Analysis of User-Generated Content in Support of Social Media Applications. In *Proceedings of the ACM International Conference in Multimedia Retrieval*, 423–426.
244. Yin, Y., R.R. Shah, and R. Zimmermann. 2016. A General Feature-based Map Matching Framework with Trajectory Simplification. In *Proceedings of the ACM SIGSPATIAL International Workshop on GeoStreaming*, 7.

Chapter 5
Soundtrack Recommendation for UGVs

Abstract Capturing videos anytime and anywhere, and then instantly sharing them online, has become a very popular activity. However, many outdoor user-generated videos (UGVs) lack a certain appeal because their soundtracks consist mostly of ambient background noise. Aimed at making UGVs more attractive, we introduce *ADVISOR*, a personalized video soundtrack recommendation system. We propose a fast and effective heuristic ranking approach based on heterogeneous late fusion by jointly considering three aspects: venue categories, visual scene, and user listening history. Specifically, we combine confidence scores, produced by SVM^{hmm} models constructed from geographic, visual, and audio features, to obtain different types of video characteristics. Our contributions are threefold. First, we predict scene moods from a real-world video dataset that was collected from users' daily outdoor activities. Second, we perform heuristic rankings to fuse the predicted confidence scores of multiple models, and third we customize the video soundtrack recommendation functionality to make it compatible with mobile devices. A series of extensive experiments confirm that our approach performs well and recommends appealing soundtracks for UGVs to enhance the viewing experience.

Keywords Soundtrack recommendation • User-generated videos • Scene understanding Music recommendation • Video understanding • Multimodal analysis • ADVISOR

5.1 Introduction

In the era of ubiquitous availability of mobile devices with wireless connectivity, user-generated videos (UGVs) have become popular since they can be easily acquired using most modern smartphones or tablets and are instantly available for sharing on social media websites (*e.g.*, YouTube, Vimeo, Dailymotion). A very large number of such videos are generated and shared on social media web sites every day. In addition, people enjoy listening to music online. Thus, various user-generated data of online activities (*e.g.*, sharing videos, listening to music) can be rich sources containing users' preferences. It is very interesting to extract activity related data with a user-centric point of view. Exploiting such data may be very

R. Shah, R. Zimmermann, *Multimodal Analysis of User-Generated Multimedia Content*, Socio-Affective Computing 6, DOI 10.1007/978-3-319-61807-4_5

beneficial to individual users, especially for preference-aware multimedia recommendations [241]. We consider location (*e.g.*, GPS information) and online listening histories as user-centric preference-aware activities. GPS information can be used in map matching [244] along with Foursquare to determine geo categories that describe the users' preferences. We categorize user activity logs from different data sources, correlate them with user preferences by using semantic concepts, *i.e.*, moods, and leverage them to complement recommendations for personal multimedia events. To enhance the appeal of a UGV for viewing and sharing, we have designed the ADVISOR system [187], which replaces the ambient background noise of a UGV with a soundtrack that matches both the video scenes and a user's preferences. A generated music video (the UGV with the recommended soundtrack) enhances the video viewing experience because it not only provides the visual experience but simultaneously renders music that matches the captured scenes and locations. We leverage multimodal information [242, 243] in our music video generate model since it is very much useful in addressing several social media analytics problems such as lecture videos segmentation [183, 184], news videos uploading [180], event understanding [182, 186], tag relevance computation [181, 185], and SMS/MMS based FAQ retrieval [189, 190]. ADVISOR can be used in many applications such as to recommend music for a slideshow of sensor-rich Flickr images or an outdoor UGV live streaming. All notations used in this chapter are listed in Table 5.1.

In terms of the target environment, this work mainly studies soundtrack recommendations for outdoor UGVs in places where different geo-categories such as beach, temple, *etc.*, would be relevant. Thus, may not work well for indoor scenes (*e.g.*, parties). The reader may imagine the following scenario: a mom brings her son outdoors where she records a video of the little boy playing on a beach and swimming in the sea. Subsequently, they would like to add music of their style to this video to make it more appealing. Since video and audio have quite different low-level features, they are linked via high-level semantic concepts, *i.e.*, moods in this work. As shown in Fig. 5.1, the ADVISOR system consists of two parts: an offline training and an online processing component. Offline a training dataset with geo-tagged videos is used to train SVM^{hmm} [2, 27, 75] models that map videos to mood tags. The online processing is further divided into two modules: a smartphone application and a server backend system. The smartphone application allows users to capture sensor-annotated videos.[1] Geographic contextual information (*i.e.*, geo-categories such as *Theme Park*, *Lake*, *Plaza* and others derived from[2]) captured by geo-sensors (GPS and compass), can serve as an important dimension to represent valuable semantic information of multimedia data while video frame content is often used in scene understanding. Hence, scene moods are embodied in both the geographic context and the video content. The sensor data streams for a UGV V are mapped to a geo feature G_V, and a visual feature F_V is calculated from

[1]We use the terms sensor-annotated videos and UGVs interchangeably in this book to refer to the same outdoor videos acquired by our custom Android application.

[2]www.foursquare.com

Table 5.1 Notations used in the Soundtrack Recommendation for UGVs chapter

Symbols	Meanings
$M_{Last.fm}$	The 20 most frequent mood tags of Last.fm
D_{GeoVid}	1213 UGVs that were captured using the GeoVid[a] application
D_{ISMIR}	An offline music dataset 729 of candidate songs in all main music genres from the ISMIR'04 genre classification dataset
$D_{Hollywood}$	A collection of 402 soundtracks from Hollywood movies of all main movie genres
V	A UGV
G_V	The geo-feature of the UGV V
F_V	The visual feature of the UGV V
A_V	The audio feature of the UGV V
m	The set of predicted mood tags
T	The set of most frequent mood tags
C	The set of predicted mood clusters
$prob(\bar{m})$	The likelihood of mood tag $\bar{m}$ in V
$L_t(\bar{m})$	A song list for mood tag $\bar{m}$
$Model$	A SVM^{hmm} learning model that predicts mood tags or clusters
M_F	Visual features based *Model* that predicts mood tags
M_G	Geo-features based *Model* that predicts mood tags
M_A	Audio features based *Model* that predicts mood clusters
M_{Cat}	The *Model* based the on concatenation of geo- and visual features that predicts mood tags
M_{GVC}	The *Model* that is constructed by the late fusion of M_G and M_F, and predicts mood clusters
M_{GVM}	The *Model* that is constructed by the late fusion of M_G and M_F, and predicts mood tags
M_{Eval}	The *Model* that is constructed by the late fusion of M_A and M_F, and predicts mood clusters
$\bar{s}$	A song from the soundtrack dataset of ISMIR'04
S_t	A song that selected as a soundtrack for the UGV V

[a]http://www.geovid.org

the video content. With the trained models, G_V and F_V are mapped to mood tags. Then, songs matching these mood tags are recommended. Among them, the songs matching a user's listening history are considered as user preference-aware.

In the ADVISOR system, first, we classify the 20 most frequent mood tags of Last.fm[3] ($M_{Last.fm}$) into four mood clusters (see Table 5.2 and Sect. 5.3.1.1 for more details) in mood space based on the intensities of energy and stress (see Fig. 5.6). We use these mood tags and mood clusters to generate ground truths for the collected music and video datasets.

Next, in order to effectively exploit multimodal (geo, visual and audio) features, we propose methods to predict moods for a UGV. We construct two offline learning models (see M_{GVM} and M_{GVC} in Fig. 5.1) which predict moods for the UGV based

[3]Last.fm is a popular music website.

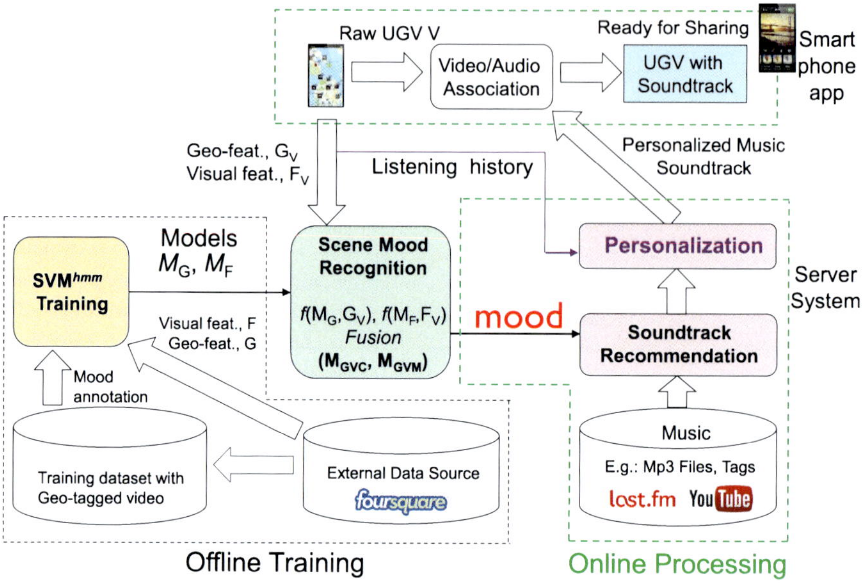

Fig. 5.1 System overview of soundtrack recommendations for UGVs with the ADVISOR system

Table 5.2 Four mood clusters

Mood cluster	Cluster type	Mood tags
$Cluster_1$	High Stress,	Angry, Quirky,
	High Energy	Aggressive
$Cluster_2$	Low Stress,	Fun, Playful, Happy,
	High Energy	Intense, Gay, Sweet
$Cluster_3$	Low Stress,	Calm, Sentimental, Quiet,
	Low Energy	Dreamy, Sleepy, Soothing
$Cluster_4$	High Stress,	Bittersweet, Depressing,
	Low Energy	Heavy, Melancholy, sad

on the late fusion of geo and visual features. Furthermore, we also construct an offline learning model (Fig. 5.2, M_{Eval}) based on the late fusion of visual and concatenated audio features (MFCC, mel-spectrum, and pitch [230]) to learn from the experience of experts who create professional *soundtracks* in Hollywood movies. We leverage this experience in the automatic selection of a matching soundtrack for the UGV using M_{Eval} (see Fig. 5.2). We deploy these models (M_{GVM}, M_{GVC} and M_{Eval}) in the backend system. The Android application first uploads its recorded sensor data and selected keyframes to the backend system for generating the music soundtrack for the UGV. Next, the backend system computes geo and visual features for the UGV and forwards these features to M_{GVM} and M_{GVC} to predict scene *mood tags* and *mood clusters*, respectively, for the UGV. Moreover, we also construct a novel heuristic method to retrieve a list of songs from an offline music database based on the predicted scene moods of the UGV. The soundtrack recommendation component of the backend system re-ranks a list of songs retrieved by the heuristic method based on user preferences and recommends them for the UGV (see Fig. 5.5). Next, the backend system determines the most

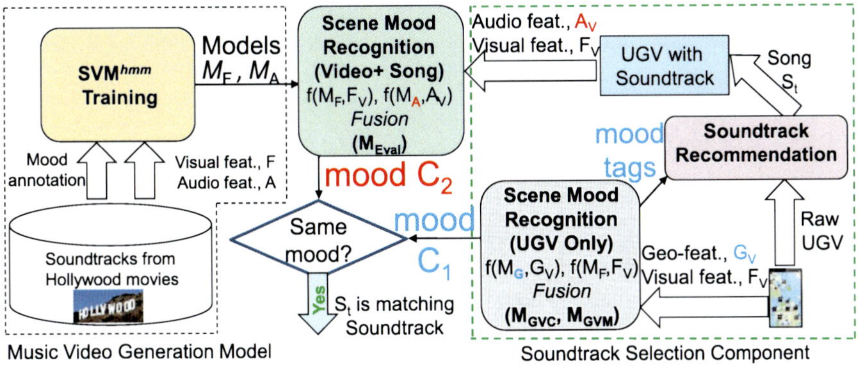

Fig. 5.2 Soundtrack selection process for UGVs in the ADVISOR system

appropriate song from the recommended list by comparing the characteristics of a composition of a selected song and the UGV with a *soundtrack* dataset of Hollywood movies of all movie genres using the learning model M_{Eval}. Finally, the Android application generates a music video using that song as a soundtrack for the UGV. The remaining parts of this chapter are organized as follows.

In Sect. 5.2, we describe the ADVISOR system. The evaluation results are presented in Sect. 5.3. Finally, we conclude the chapter with a summary in Sect. 5.4.

5.2 Music Video Generation

To generate a music video for a UGV, first predicts scene moods from the UGV using learning models described next in Sect. 5.2.1. The scene moods used in this study are the 20 most frequent mood tags of Last.fm, described in detail in Sect. 5.3.1.1. Next, the *soundtrack recommendation component* in the backend system recommends a list of songs, using a heuristic music retrieval method, described in Sect. 5.2.2. Finally, the *soundtrack selection component* selects the most appropriate song from the recommended list to generate the music video of the UGV, using a novel method, described in Sect. 5.2.3.

5.2.1 Scene Moods Prediction Models

In our custom Android recording app, a continuous stream of geo-sensor information is captured together with each video using GPS sensors. This sensor information is mapped to geo-categories such as *Concert Hall*, *Racetrack*, and others using the Foursquare API (see Sect. 1.4.1 for details). Then the geo-categories for a UGV V are mapped to a geo-feature G_V using the bag-of-word model. With the trained SVM^{hmm} model (M_G), mood tags C_G with geo-aware likelihood are generated. Furthermore, a visual feature F_V such as a color histogram is calculated from the video content. With the trained SVM^{hmm} model (M_F), mood tags C_F associated with visual-aware likelihood are generated. In the next step, the mood tags associated with location information and video content are combined by late fusion. Finally, mood tags with high likelihoods are regarded as scene moods of this video.

5.2.1.1 Geo and Visual Features

Based on the geo-information, a UGV is split into multiple segments with timestamps, with each segment representing a video scene. The geo-information (GPS location) for each video segment is mapped to geo-categories using APIs provided by Foursquare. The Foursquare API also provides distances of geo-categories on the queried GPS location, which describe the typical objects near the video scene in the video segments. We treat each geo-tag as a word and exploit the bag-of-words model [93] on a set of 317 different geo-tags in this study. Next, for each video segment, a geo-feature with 317 dimensions is computed from geo-tags with their score used as weights.

A color histogram [95, 188] with 64 dimensions is computed from each UGV video frame by dividing each component of RGB into four bins. Next, the UGV is divided into multiple continuously correlated parts (CCP), within each of which color histograms have high correlations. Specifically, starting with an initial frame, each subsequent frame is regarded as part of the same CCP if its correlation with the initial frame is above a pre-selected threshold. Next, a frame with its timestamp, which is most correlated with all the other frames in the same CCP, is regarded as a key-frame. Color histograms of key-frames are treated as visual features.

5.2.1.2 Scene Moods Classification Model

Wang et al. [209] classified emotions for a video using an SVM-based probabilistic inference machine. To arrange scenes depicting *fear*, *happiness* or *sadness*, Kang [85] used visual characteristics and camera motion with hidden Markov models (HMMs) at both the shot and scene levels. To effectively exploit multimodal features, late fusion techniques have been advocated in various applications and semantic video analysis [194, 195, 226]. These approaches inspired us to use SVM^{hmm} models based on the late fusion of various features of UGVs to learn the relationships between UGVs and scene moods. Table 5.3 shows the summary of all the SVM^{hmm} learning models used in this study.

To establish the relation between UGVs and their associated scene moods, we train several offline learning models with the *GeoVid dataset* as described later in Sect. 5.3.1.2. Experimental results in Sect. 5.3.2.1 confirm that a model based on late fusion outperforms other models in scene mood prediction. Therefore, we construct two learning models based on the late fusion of geo and visual features and refer to them as *emotion prediction models* in this study. A geo feature computed from geo-categories reflects the environmental atmosphere associated with moods and a color histogram computed from keyframes represents moods in

Table 5.3 The list of SVM^{hmm} models that are used in the ADVISOR system, where, G_V, F_V, and A_V represent the geo, visual and audio features, respectively

Model	Input-1	Input-2	Output
M_F	F_V	–	T
M_G	G_V	–	T
M_A	A_V	–	C
M_{GVC}	M_G	M_F	$C = f_1(M_G, M_F)$
M_{GVM}	M_G	M_F	$T = f_2(M_G, M_F)$
M_{Eval}	M_A	M_F	$C = f_3(M_A, M_F)$
M_{Cat}	G_V	F_V	$T = f_4(G_V, F_V)$

T and C denote the set of predicted mood tags and mood clusters, respectively. M_{GVC} and M_{GVM} are models constructed by the late fusion of M_G and M_F . M_{Eval} is constructed by the late fusion of M_A and M_F

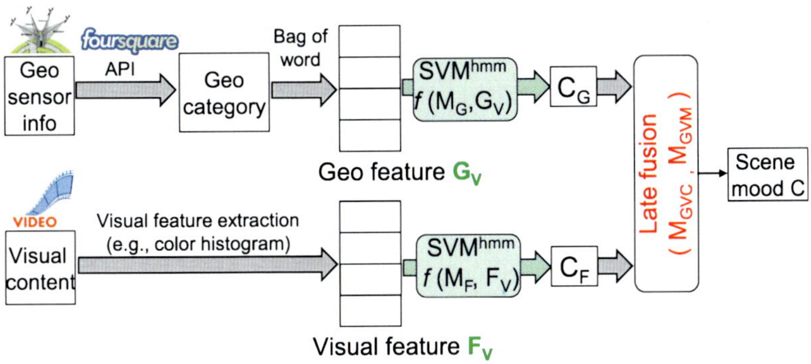

Fig. 5.3 Mood recognition from UGVs using M_{GVC} and M_{GVM} SVM^{hmm} models

the video content. Next, the sequence of geo-features and the sequence of visual features are synchronized based on their respective timestamps to train *emotion prediction models* using SVM^{hmm} method. Figure 5.3 shows the process of mood recognition from UGVs based on heterogeneous late fusion of SVM^{hmm} models constructed from geo and visual features. M_{GVC} and M_{GVM} are *emotion prediction models* trained with mood clusters and mood tags, respectively, as ground truths for the training dataset. Hence, M_{GVC} and M_{GVM} predict mood clusters and mood tags, respectively, for a UGV based on a heterogeneous late fusion of SVM^{hmm} models constructed from geographic and visual features.

5.2.1.3 Scene Moods Recognition

UGVs acquired by our Android application are enhanced with geo-information by using sensors such as GPS and compass. When a user requests soundtracks for a UGV, then the Android application determines timestamps for multiple video segments of the UGV with each segment representing a video scene based on

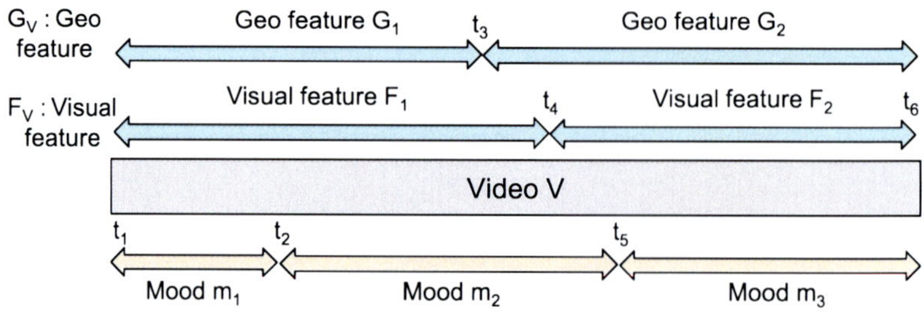

Fig. 5.4 The concatenation model M_{Cat} from Shah et al. [188]

geo-information of the UGV. Furthermore, the Android application extracts keyframes of the UGV based on timestamps of video segments and uploads them to the backend system along with the geo-information of video segments. The backend system computes geo and visual features of the UGV from the uploaded sensor information and keyframes. The SVM^{hmm} models, M_{GVC}, M_{GVM} and M_{Cat}, read the sequence of geo and visual features and recognize moods for the UGV. For example, M_{Cat} is trained with the concatenation of geo and visual features as described in the following sequence (see Fig. 5.4).

$$\langle V, G_1, F_1, m_1 \rangle, \langle V, G_1, F_1, m_2 \rangle, \langle V, G_2, F_1, m_2 \rangle, \ldots \quad (5.1)$$

In this specific example, in the emotion recognition step, when M_{Cat} is fed with geo features G_V and visual features F_V using $f_4(G_V; F_V)$, then it automatically predicts a set of scene mood tags $m = \{m_1; m_2; m_2; m_2; m_3\}$ for the UGV V.

5.2.2 *Music Retrieval Techniques*

We prepared an offline music dataset of candidate songs in all main music genres, with details described later in Sect. 5.3.1.3. We refer to this dataset as the *soundtrack dataset*. The next step in the ADVISOR system is to find music from the *soundtrack dataset* that matches with both the predicted mood tags and the user preferences. With the given mood tags, the soundtrack retrieval stage returns an initial song list L_1. For this task, we propose a novel music retrieval method. Many state-of-the-art methods for music retrieval use heuristic approaches [62, 115, 173, 197]. Such work inspired us to propose a heuristic method which retrieves a list of songs based on the predicted scene moods by M_{GVM} and M_{GVC}. We take the user's listening history as user preferences and calculate the correlation between audio features of songs in the initial list L_1 and the listening history. From the initial list, songs with high correlations are regarded as user specific songs L_2 and recommended to users as video soundtracks.

5.2.2.1 Heuristic Method for Soundtrack Retrieval

An improvement in mood tag prediction accuracy for a UGV is also an improvement in matching music retrieval because songs in the *soundtrack dataset* are organized in a hash table with as keys. However, retrieving songs based on only one mood tag suffers from subjectivity because the *mood clusters* prediction accuracy of M_{GVC} is much better than the *mood tags* prediction accuracy of M_{GVM} for a UGV (see Table 5.6). Since a song may have multiple mood tags, when the *emotion prediction models* predict multiple mood tags, a song may be matched with several tags. Therefore, we calculate the total score of each song to reduce this subjectivity issue and propose a heuristics based on a music retrieval method to rank all the predicted mood tags for the UGV and then normalize them as the likelihood to retrieve the final ranked list L_1 of N songs. Algorithm 5.1 describes this retrieval process and its *composition operation* is defined such that it outputs only those most frequent *mood tags* T from the list of mood tags predicted by $f_2(M_G, M_F)$ which belong to the most frequent *mood clusters* predicted by $f_1(M_G, M_F)$. Thus, the *composition operation* is defined by the following equation:

$$T = f_1(M_G, M_F) * f_2(M_G, M_F) \tag{5.2}$$

where T, G_V, F_V, f_1 and f_2 have the usual meaning, with details described in Table 5.3.

Algorithm 5.1 Heuristic based song retrieval procedure

---

procedure HEURISTICSONGSRETRIEVAL(H)
    INPUT: geo and visual features (G_V, F_V) of the UGV V
    OUTPUT: A ranked list of songs L_1 for the UGV V
    $T = f_1(M_G, M_F) * f_2(M_G, M_F)$
    $L = []$ ▷ Initialize with empty list.
    for each mood tag $\overline{m}$ in T **do**
        $prob(\overline{m}) = likelihood(\overline{m})$ ▷ Likelihood of mood tag $\overline{m}$.
        $L_t(\overline{m}) = songList(\overline{m})$ ▷ Song list for mood tag $\overline{m}$.
        $L = L \cup L_t(\overline{m})$ ▷ L has all unique songs.
    end for
▷ *isPrsnt* returns 1 if s is present in $L_t(\overline{m})$ else 0.
▷ $scr(\overline{s}, \overline{m})$ is the score of song $\overline{s}$ with mood tag $\overline{m}$.
    for each song $\overline{s}$ in L **do**
        $Score(\overline{s}) = 0$ ▷ Initialize song score.
        for each mood tag $\overline{m}$ in T **do**
            $Score(\overline{s}) += prob(\overline{m}) * scr(\overline{s}, \overline{m}) * isPrsnt(\overline{s}, L_t(\overline{m}))$
        end for
    end for
    $L_1 = sortSongScore(L)$ ▷ Sort songs.
    Return L_1 ▷ A ranked list of N songs.
end procedure

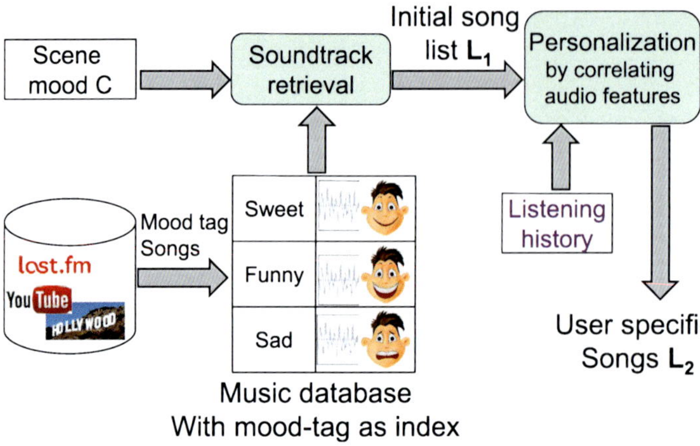

Fig. 5.5 Matching songs with a user's preferences

5.2.2.2 Post-Filtering with User Preferences

A new paradigm shift in music information retrieval (MIR) is currently creating a move from a system-centric perspective towards user-centric approaches. Therefore, addressing user-specific demands in music recommendation is receiving increased attention. User preference-aware music recommendations based on users' preferences observed from their listening history is very common. Music genres of the user's frequently listened to songs are treated as his/her listening preference and later used for the re-ranking of a list of songs L_1 recommended by the heuristics method. Our system extracts audio features including MFCC [172] and pitch from audio tracks of the user's frequently listened to songs. These features help in re-ranking the list of recommended songs L_1 by comparing the correlation coefficients of songs matching the genres preferred by the user, and then recommending a list of user preference-aware songs L_2 (see Fig. 5.5). Next, the *soundtrack selection component* automatically chooses the most appropriately matching song from L_2 and attaches it as the soundtrack to the UGV.

5.2.3 Automatic Music Video Generation Model

Wang et al. [209] concatenated audio and visual cues to form scene vectors which were sent to an SVM method to obtain high-level audio cues at the scene level. We propose a novel method to automatically select the most appropriate soundtrack from the list of songs L_2 recommended by our music retrieval system as described in the previous Sect. 5.2.2, to generate a *music video* from the UGV.

We use soundtracks of Hollywood movies in our system to select appropriate UGV soundtracks since music in Hollywood movies is designed to be emotional and hence is easier to associate with mood tags. Moreover, music used by Hollywood movies is generated by professionals, which ensures a good harmony with the movie content. Therefore, we learn from the experience of such experts using their professional *soundtracks* of Hollywood movies through a SVM^{hmm} learning model. We refer to the collection of such soundtracks as the *evaluation dataset*, with details described later in Sect. 5.3.1.4. We construct a *music video generation model* (M_{Eval}) using the training dataset of the *evaluation dataset*, which can predict mood clusters for any music video. We leverage this model to select the most appropriate soundtrack for the UGV. We construct M_{Eval} based on a heterogeneous of SVM^{hmm} models constructed from visual features such as a color histogram and audio features such as MFCC, mel-spectrum, and pitch. Similar to our findings with the learning model to predict *scene moods* based on the late fusion of geo and visual features of UGVs, we find that the learning model M_{Eval} based on the late fusion of visual features and concatenated MFCC, mel-spectrum and pitch audio features, also performs well.

Figure 5.2 shows the process of soundtrack selection for a UGV V. It consists of two components, first, music video generation model (M_{Eval}), and second, a soundtrack selection component. M_{Eval} maps visual features F_V and audio features A_V of the UGV with a soundtrack to mood clusters C_2, *i.e.*,$f_3(F_V, A_V)$ corresponds to mood clusters C_2 based on the late fusion of F_V and A_V. The soundtrack selection component compares moods (C_2 and C_1) of the UGV predicted by M_{Eval} and, M_{GVC} and M_{GVM}.

Algorithm 5.2 describes the process of the most appropriate soundtrack selection from the list of songs recommended by the heuristic method to generate the music video of the UGV. To automatically select the most appropriate soundtrack, we compute audio features of a selected song and visual features of the UGV and refer to this combination as the *prospective music video*. We compare the characteristics of the *prospective music video* with video songs of the *evaluation dataset* of many famous Hollywood movies. Next, we predict mood clusters (C) for the *prospective music video* using M_{Eval}. We treat the predicted mood clusters (C_1) of the UGV by M_{GVC} as ground truth for the UGV, since the mood cluster prediction accuracy of M_{GVC} is very good (see Sect. 5.3.2.1). Finally, if the most frequent mood clusters C_2 from C for the *prospective music video* is similar to the ground truth (C_1) of the UGV, then the selected song (S_t) is treated as the soundtrack and the music video of the UGV is generated. If both mood clusters are different, then we repeat the same process with the next song in the recommended list L_2. In the worst case, if none of the songs in the recommended list L_2 satisfies the above criteria then we repeat the same process with the second most frequent mood cluster from C, and so on.

Algorithm 5.2 Music video generation for a UGV

```
procedure MusicVideoGeneration(MV)
    INPUT: A UGV V by the Android application
    OUTPUT: A music video MV for V
    m = moodTags(V)                                  ▷ M_GVM predicts mood tags.
    C_1 = moodClusters(V)                            ▷ M_GVC predicts clusters.
    L_2 = HeuristicSongsRetrieval(m, C_1)
    F_V = visualFeatures(V)                          ▷ Compute visual features.
    for rank = 1 to numMoodCluster do
        for each song S_t in L_2 do
            a_1 = calcMFCC(S_t)                      ▷ MFCC feat.
            a_2 = calcMelSpec(S_t)                   ▷ Mel-spec feat.
            a_3 = calcPitch(S_t)                     ▷ Pitch feat.
                                                     ▷ Concatenate all audio features.
            A_V = concatenate(a_1, a_2, a_3)
            C = findMoodCluster(F_V, A_V)            ▷ using M_Eval.
            C_2 = mostFreqMoodCluster(rank, C)
                                                     ▷ Check for similar mood clusters
                                                     ▷ predicted by M_GVC and M_Eval.
            if C_2 == C_1 then
                                                     ▷ Android app generates music video.
                MV = generateMusicVideo(S_t, V)
                Return MV                            ▷ Music video for V.
            end if
        end for
    end for
end procedure
```

5.3 Evaluation

5.3.1 Dataset and Experimental Settings

The input dataset in our study consists of sensor-annotated (sensor-rich) videos acquired from a custom Android (or iOS) application running on smartphones. As described in Sect. 5.2, we train several learning models to generate a music video from a UGV. It is important to have good ground truths for the training and the testing dataset to train effective models for the system. However, due to the difference in age, occupation, gender, environment, cultural background and personality, music perception is highly subjective among users. Hence generating ground truths for the evaluation of various music mood classification algorithms are very challenging [91]. Furthermore, there is no standard music dataset with associated mood tags (ground truths) available due to the lack of an authoritative taxonomy of music moods and an associated audio dataset. Therefore, we prepare our datasets as follows to address the above issues.

5.3.1.1 Emotion Tag Space

Mood tags are important keywords in digital audio libraries and online music repositories for effective music retrieval. Furthermore, often, music experts refer to music as the finest *language of emotion*. Therefore it is very important to learn the relationship between music and emotions (mood tags) to build a robust ADVISOR. Some prior methods [91, 99] have described state-of-the-art classifications of mood tags into different emotion classes. The first type of approach is the *categorical approach* which classifies mood tags into emotion clusters such as *happy*, *sad*, *fear*, *anger*, and *tender*. Hevner [70] categorized 67 mood tags into eight mood clusters with similar emotions based on musical characteristics such as *pitch*, *mode*, *rhythm*, *tempo*, *melody*, and *harmony*. Thayer [200] proposed an energy-stress model, where the mood space is divided into four clusters such as *low energy / low stress*, *high energy / low stress*, *high energy/high stress*, and *low energy / high stress* (see Fig. 5.6). The second type of method is based on the *dimensional approach* to affect, which represents music samples along a two-dimensional emotion space (characterized by arousal and valence) as a set of points.

We consider the *categorical approach* of music mood classification to classify the mood tags used in this work. We extracted the 20 most frequent mood tags $M_{Last.fm}$ of Last.fm from the crawled dataset of 575,149 tracks with 6,814,068 tag annotations in all main music genres by Laurier et al. [99]. Last.fm is a music website with more than 30 million users, who have created a site-wide folksonomy of music through end-user tagging. We classified tags in $M_{Last.fm}$ into four mood clusters based on mood tag clustering introduced in earlier work [70, 171, 188]. Four mood clusters represent four quadrants of a 2-dimensional emotion plane with energy and stress characterized as its two dimensions (see Table 5.2).

However, emotion recognition is a very challenging task due to its cross-disciplinary nature and high subjectivity. Therefore experts have suggested the need for the use of multi-label emotion classification. Since the recommendation

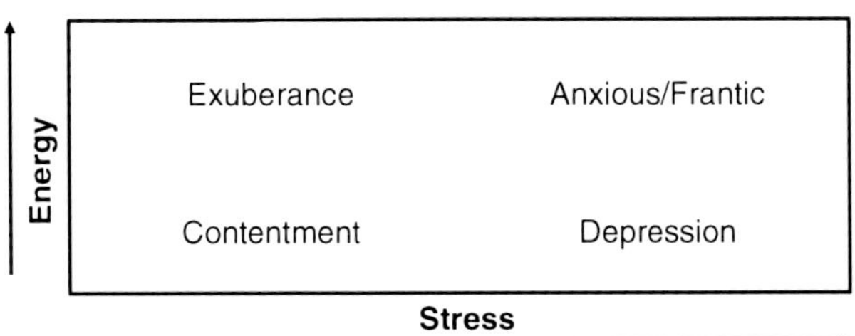

Fig. 5.6 Thayer's [200] model of moods

of music based on low-level mood tags can be very subjective, many earlier approaches [91, 225] on emotion classification and music recommendation are based on high-level mood clusters. Therefore, to calculate the annotator consistency, accuracy, and inter-annotator agreement, we compare annotations at four high-level *mood clusters* instead of the 20 low-level *mood tags* in this study. Moreover, we leverage the mood tags and mood clusters together to improve the scene mood prediction accuracy of ADVISOR.

5.3.1.2 GeoVid Dataset

To create an offline training model for the proposed framework of scene mood prediction of a UGV we utilized 1213 UGVs D_{GeoVid} which were captured during 8 months (4 March 2013 to 8 November 2013) using the GeoVid[4] application. These videos were captured with iPhone 4S and iPad 3 devices. The video resolution of all videos was 720×480 pixels, and their frame rate 24 frames per second. The minimum sampling rate for the location and orientation information was five samples per second (*i.e.*, a 200-millisecond sampling rate). In our case, we mainly focus on videos that contain additional information provided by sensors and we refer to these videos as *sensor-annotated videos*. The captured videos cover a diverse range of rich scenes across Singapore and we refer to this video collection as the *GeoVid dataset*.

Since emotion classification is highly subjective and can vary from person to person [91], generating ground truths for the evaluation of the various emotion classifications from video techniques are difficult. It is necessary to use some filtering mechanism to discard bad annotations. In the E6K music dataset for MIREX,[5] IMIRSEL assigns each music sample to three different evaluators for mood annotations. They then evaluate the quality of ground truths by the degree of agreement on the music samples. Only those annotations are considered as ground truths where the majority of evaluators selected the same mood cluster. Music experts resolve the ground truth of music samples for which all annotators select different mood clusters.

For the *GeoVid dataset*, we recruited 30 volunteers to annotate emotions (the mood tags listed in Table 5.2). First, we identified annotators who are consistent with their annotations by introducing redundancy. We repeated one of the videos in the initial sets of the annotation task with ten videos given to each of the evaluators. If any annotated mood tag belonged to a different mood cluster for a repeated video then this annotator's tags were discarded. Annotators passing this criterion were

[4]The GeoVid app and portal at http://www.geovid.org provide recorded videos annotated with location meta-data.

[5]The MIR Evaluation eXchange is an annual evaluation campaign for various MIR algorithms hosted by IMIRSEL (International MIR System Evaluation Lab) at the University of Illinois at Urbana-Champaign.

Table 5.4 Ground truth annotation statistics with three annotators per video segment

All different	Two the same	All the same
298	1293	710

selected for mood annotation of the *GeoVid dataset*. Furthermore, all videos of the *GeoVid dataset* were split into multiple segments with each segment representing a video scene, based on its geo-information and timestamps. For each video segment, we asked three randomly chosen evaluators to annotate one mood tag each after watching the UGV carefully. To reduce subjectivity and check the inter-annotator agreement of the three human evaluators for any video, we inspected whether the majority (at least two) of the evaluators chose mood tags that belonged to the same mood cluster. If the majority of evaluators annotated mood tags from the same mood cluster then that particular cluster and its associated mood tags were considered as ground truth for the UGV. Otherwise, the decision was resolved by music experts. Due to the subjectivity of music moods, we found that all three evaluators annotated different mood clusters for 298 segments during annotation for the *GeoVid dataset*, hence their ground truths were resolved by music experts (see Table 5.4).

5.3.1.3 Soundtrack Dataset

We prepared an offline music dataset D_{ISMIR} of candidate songs (729 songs altogether) in all main music genres such as *classical*, *electronic*, *jazz*, *metal*, *pop*, *punk*, *rock* and *world* from the ISMIR'04 genre classification dataset.[6] We refer to this dataset as the *soundtrack dataset* and we divided it into 15 *emotion annotation tasks (EAT)*. We recruited 30 annotators and assigned each EAT (with 48–50 songs) to two randomly chosen annotators and asked them to annotate one mood tag for each song. Each EAT had two randomly selected repetitive songs to check the annotation consistency of each human evaluator, *i.e.*, if the evaluator-chosen mood tags belonged to the same mood cluster for redundant songs then the evaluator was consistent; otherwise, the evaluator's annotations were discarded. Since the same set of EATs was assigned to two different annotators, their inter-annotator agreement is calculated by *Cohen's kappa coefficient* (k) [52]. This coefficient is considered to be a robust statistical measure of inter-annotator agreement and defined earlier in Sect. 4.3.2.

If $k = 1$ then both annotators for an EAT are in complete agreement while there is no agreement when $k = 0$. According to Schuller et al. [179], an agreement level with a k value of 0.40 and 0.44, respectively, for the music mood assessment with regard to valence and arousal, are considered to be *moderate to good*. Table 5.5

[6] ismir2004.ismir.net/genre contest/index.htm

Table 5.5 Summary of the emotion annotation tasks

Total number of songs	729
Pairs of annotators	15
Common songs per pair	48–50
κ: Maximum	0.67
κ: Minimum	0.29
κ: Mean	0.47
κ: Standard deviation	0.12

shows the summary of the mood annotation tasks for the *soundtrack dataset* with a mean k value of 0.47, which is considered to be *moderate to good* in music judgment. For four EATs, annotations were carried out again since evaluators for these EATs failed to fulfill the *annotation consistency* criteria.

For a fair comparison of music excerpts, samples were converted to a uniform format (22,050 Hz, 16 bits, and a mono channel PCM WAV) and normalized to the same volume level. Yang et al. [225] suggested to using 25-second music excerpts from around the segment middle to reduce the burden on evaluators. Therefore, we manually selected 25-second music excerpts from near the middle such that the mood was likely to be constant within the excerpt by avoiding drastic changes in musical characteristics. Furthermore, songs were organized in a hash structure with their mood tags as hash keys, so that ADVISOR was able to retrieve the relevant songs from the hash table with the predicted mood tags as keys. We then considered a sequence of the most frequent mood tags T predicted by the *emotion prediction model*, with details described in Sect. 5.2.1, for song retrievals.

The *soundtrack dataset* was stored in a database, indexed and used for soundtrack recommendation for UGVs. A song with ID s and k tags is described by a list of tag attributes and scores from $s;tag_1;\ scr_1$ to $s;tag_k;\ scr_k$, where tag_1 to tag_k are mood tags and scr_1 to scr_k are their corresponding scores. Tag attributes describe the relationship between mood tags and songs and are organized in a hash table where each bucket is associated with a mood tag. With the aforementioned song s as an example, its k tag attributes are separately stored in k buckets. Since a tag is common to all songs in the same bucket, it is sufficient to only store tuples consisting of song ID and tag score.

5.3.1.4 Evaluation Dataset

We collected 402 soundtracks $D_{Hollywood}$ from Hollywood movies of all main movie genres such as *action*, *comedy*, *romance*, *war*, *horror* and others. We refer to this video collection as the *evaluation dataset*. We manually selected 1-minute video segments from around the middle for each clip in the *evaluation dataset* such that the emotion was likely to be constant within that segment by avoiding drastic changes in scene and musical characteristics. We ignored segments having dialogues in a scene while selecting 1-minute excerpts. Since the segments in the *evaluation dataset* are professionally produced and their genres, lyrics, and context are known, emotions elicited by these segments are easy to determine. Mood clusters (listed in Table 5.2) were manually annotated for each segment based on its movie genre, lyrics, and context and treated as ground truth for the *evaluation dataset*.

5.3.2 Experimental Results

5.3.2.1 Scene Moods Prediction Accuracy

To investigate the relationship between *geo* and *visual features* to predict video scene moods for a UGV, we trained four SVM^{hmm} models and compared their accuracy. First, the *Geo model* (M_G) was trained with geo features only, second, the *Visual model* (M_F) was trained with visual features only and third, the *Concatenation model* (M_{Cat}) was trained with the concatenation of both geo and visual features (see Fig. 5.4). Finally, fourth, the *Late fusion models* (M_{GVM}; M_{GVC}) were trained by the late fusion of the first (M_G) and second (M_F) models.

We randomly divided videos in the *GeoVid dataset* into training and testing datasets with 80:20 and 70:30 ratios. The reason we divided the dataset into two ratios is that we wanted to investigate how the emotion prediction accuracies vary by changing the training and testing dataset ratios. We performed tenfold cross-validation experiments on various learning models, as described in Table 5.3, to compare their scene mood prediction accuracy for UGVs in the test dataset. We used three experimental settings. First, we trained all models from the training dataset with *mood tags* as ground truth and compared their scene mood prediction accuracy at the *mood tags level* (*i.e.*, whether the predicted *mood tags* and ground truth *mood tags* were the same). Second, we trained all models from the training dataset with *mood tags* as ground truth and compared their scene mood prediction accuracy at the *mood clusters level* (*i.e.*, whether the most frequent mood cluster of predicted *mood tags* and ground truth *mood tags* were the same).

Lastly, we trained all models from the training dataset with *mood clusters* as ground truth and compared their scene mood prediction accuracy at the *mood clusters level* (*i.e.*, whether the predicted *mood clusters* and ground truth *mood clusters* were the same). Our experiments confirm that the model based on the late fusion of geo and visual features outperforms the other three models. We noted that the scene mood prediction accuracy at the *mood tag level* does not perform well because the accuracy of the SVM classifier degrades as the number of classes increases. A comparison of the scene mood prediction accuracies for all four models is listed in Table 5.6. Particularly, M_{GVC} performs 30.83%, 13.93%, and 14.26% better than M_F, M_G, and M_{Cat}, respectively.

Table 5.6 Accuracies of emotion prediction models with tenfold cross validation for the following three experimental settings: (i) Exp-1: Model trained at mood *tags* level and predicted moods accuracy checked at mood *tags* level, (ii) Exp-2: Model trained at mood *tags* level and predicted moods accuracy checked at mood *cluster* level, and (iii) Exp-3: Model trained at mood *cluster* level and predicted moods accuracy checked at mood *cluster* level

Ratio type	Model	Exp-1 (%)	Exp-2 (%)	Exp-3 (%)	Feature dimension
70:30	M_F	18.87	52.62	64.63	64
	M_G	25.56	60.12	74.22	317
	M_{Cat}	24.47	60.79	73.52	381
	M_{GVM}	**37.18**	**76.42**	–	317
	M_{GVC}	–	–	**84.56**	317
80:20	M_F	17.76	51.65	63.93	64
	M_G	24.68	60.83	73.06	317
	M_{Cat}	25.97	61.96	71.97	381
	M_{GVM}	**34.86**	**75.95**	–	317
	M_{GVC}	–	–	**84.08**	317

5.3.2.2 Soundtrack Selection Accuracy

We randomly divided the *evaluation dataset* into training and testing datasets with an 80:20 ratio, and performed fivefold cross-validation experiments to calculate the scene mood prediction accuracy of M_{Eval} for UGVs in the test dataset. We performed two experiments. First, we trained M_{Eval} from the training set with *mood clusters* as ground truth and compared their scene mood prediction accuracy at the *mood clusters level* for UGVs in the test dataset of the *evaluation dataset* (*i.e.*, whether the predicted *mood clusters* and ground truth *mood clusters* matched). In the second experiment, we replaced the test dataset of the *evaluation dataset* with the same number of *music videos* generated by our system for randomly selected UGVs from the *GeoVid dataset*. The M_{Eval} maps visual features F and audio features A of a video V to mood clusters C, *i.e.*, $f_3(F, A)$ corresponds to mood clusters C based on the late fusion of F and A (see Fig. 5.2). An input vector (in time order) for M_{Eval} can be represented by the following sequence (see Fig. 5.7).

$$\langle F_1, A_1\rangle, \langle F_1, A_2\rangle, \langle F_2, A_2\rangle, \langle F_2, A_3\rangle, \ldots \quad (5.3)$$

M_{Eval} reads the above input vector and predicts mood clusters for it. Table 5.7 shows that the emotions (mood clusters) prediction accuracy (68.75%) of M_{Eval} for *music videos* is comparable to the emotion prediction accuracy at the scene level in movies by state-of-the-art approaches such as introduced by Soleymani et al. [196] (63.40%) and Wang et al. [209] (74.69%). To check the effectiveness of the ADVISOR system, we generated *music videos* for 80 randomly selected UGVs from the *GeoVid dataset* and predicted their mood clusters by M_{Eval} with 70.0% accuracy, which is again comparable to state-of-the-art algorithms for emotion prediction at the scene level in movies. The experimental results in Table 5.7

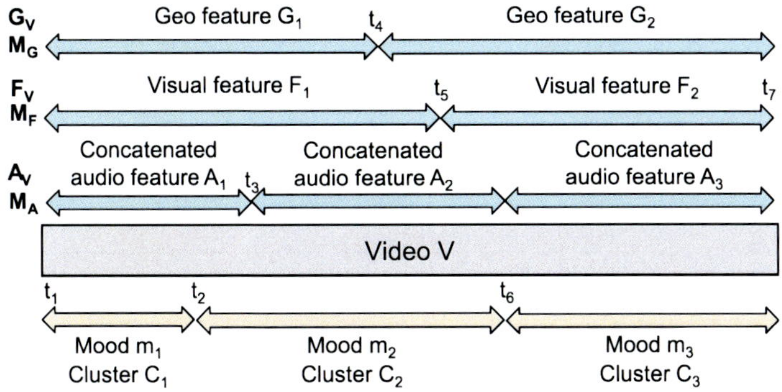

Fig. 5.7 Features to mood tags/clusters mapping

Table 5.7 Emotion classification accuracy of M_{Eval} with 5-fold cross validation. M_{Eval} is trained with 322 videos from the Evaluation dataset $D_{Hollywood}$

Experiment type	Number of test videos	Accuracy (in %)
Prediction on videos from $D_{Hollywood}$	80	68.75
Prediction on videos from D_{GeoVid}	80	70.00

confirm that ADVISOR effectively combines objective scene moods and music to recommend appealing soundtracks for UGVs.

5.3.3 *User Study*

Based on the techniques introduced earlier, we implemented the system to generate music videos for UGVs. All UGVs were single-shot clips with sensor metadata, acquired by our Android application designed specifically for recording sensor-annotated videos. We randomly selected five UGVs each for six different sites in Singapore as listed in Table 5.8, from a set of acquired videos. To judge whether the recommended songs capture the scene moods of videos, we recruited fifteen volunteers to assess the appropriateness and entertainment value of the music videos (UGVs with recommended songs). We asked every user to select one video for each site by choosing the most likely candidate that they themselves would have captured at that site. The predicted scene moods listed in Table 5.8 are the first three mood tags belonging to the most frequent mood cluster predicted by M_{GVC} for five videos at different sites. A soundtrack for all selected videos was generated using ADVISOR and users were asked to assign a score 1 (worst) to 5 (best) to the generated music videos. Finally, we calculated the average score of music videos for all sites. Table 5.8 summarizes the ratings and the most appropriate scene moods from a list of predicted for videos from the six sites mentioned above. The feedback from these volunteers was encouraging, indicating that our

Table 5.8 User study feedback (ratings) on a scale from 1 (worst) to 5 (best) from 15 volunteers

Video location	Predicted scene moods	1 2 3 4 5	Average rating
Cemetery	Melancholy, sad, sentimental	0 0 3 4 8	4.3
Clarke Quay	Fun, sweet, calm	0 2 5 7 1	3.5
Gardens by the bay	Soothing, fun, calm	0 3 3 9 0	3.4
Marina Bay Sands	Fun, playful	0 0 2 6 7	4.3
Siloso Beach	Happy, fun, quiet	0 0 1 6 8	4.5
Universal Studios	Fun, intense, happy, playful	0 2 5 5 3	3.6

technique achieves its goal of automatic music video generation to enhance the video viewing experience.

5.4 Summary

Our work represents one of the first attempts for user preference-aware video soundtrack generation. We categorize user activity logs from different data sources by using semantic concepts. This way, the correlation of preference-aware activities based on the categorization of user-generated heterogeneous data complements video soundtrack recommendations for individual users. The ADVISOR system automatically generates a matching soundtrack for a UGV in four steps. More specifically, first, a learning model based on the late fusion of geo and visual features recognizes scene moods in the UGV. Particularly, M_{GVC} predicts scene moods for the UGV since it performs better than all other models (*i.e.*, 30.83%, 13.93%, and 14.26% better than M_F, M_G, and M_{Cat}, respectively). Second, a novel heuristic method recommends a list of songs based on the predicted scene moods. Third, the soundtrack recommendation component re-ranks songs recommended by the heuristics method based on the user's listening history. Finally, our Android application generates a music video from the UGV by automatically selecting the most appropriate song using a learning model based on the late fusion of visual and concatenated audio features. Particularly, we use M_{Eval} to select the most suitable song since the emotion prediction accuracy (70.0%) of the generated soundtrack UGVs from D_{GeoVid} using M_{Eval} is comparable to the emotion prediction accuracy (68.8%) of soundtrack videos from $D_{Hollywood}$ of the Hollywood movies. Thus, the experimental results and our user study confirm that the ADVISOR system can effectively combine objective scene moods and individual music tastes to recommend appealing soundtracks for UGVs. In the future, each one of these steps could be further enhanced (see Chap. 8 for details).

References

1. Apple Denies Steve Jobs Heart Attack Report: "It Is Not True". http://www.businessinsider.com/2008/10/apple-s-steve-jobs-rushed-to-er-after-heart-attack-says-cnn-citizen-journalist/. October 2008. Online: Last Accessed Sept 2015.
2. SVM^{hmm}: Sequence Tagging with Structural Support Vector Machines. https://www.cs.cornell.edu/people/tj/svm light/svm hmm.html. August 2008. Online: Last Accessed May 2016.
3. NPTEL. 2009, December. http://www.nptel.ac.in. Online; Accessed Apr 2015.
4. iReport at 5: Nearly 900,000 contributors worldwide. http://www.niemanlab.org/2011/08/ireport-at-5-nearly-900000-contributors-worldwide/. August 2011. Online: Last Accessed Sept 2015.
5. Meet the million: 999,999 iReporters + you! http://www.ireport.cnn.com/blogs/ireport-blog/2012/01/23/meet-the-million-999999-ireporters-you. January 2012. Online: Last Accessed Sept 2015.
6. 5 Surprising Stats about User-generated Content. 2014, April. http://www.smartblogs.com/social-media/2014/04/11/6.-surprising-stats-about-user-generated-content/. Online: Last Accessed Sept 2015.
7. The Citizen Journalist: How Ordinary People are Taking Control of the News. 2015, June. http://www.digitaltrends.com/features/the-citizen-journalist-how-ordinary-people-are-taking-control-of-the-news/. Online: Last Accessed Sept 2015.
8. Wikipedia API. 2015, April. http://tinyurl.com/WikiAPI-AI. API: Last Accessed Apr 2015.
9. Apache Lucene. 2016, June. https://lucene.apache.org/core/. Java API: Last Accessed June 2016.
10. By the Numbers: 14 Interesting Flickr Stats. 2016, May. http://www.expandedramblings.com/index.php/flickr-stats/. Online: Last Accessed May 2016.
11. By the Numbers: 180+ Interesting Instagram Statistics (June 2016). 2016, June. http://www.expandedramblings.com/index.php/important-instagram-stats/. Online: Last Accessed July 2016.
12. Coursera. 2016, May. https://www.coursera.org/. Online: Last Accessed May 2016.
13. FourSquare API. 2016, June. https://developer.foursquare.com/. Last Accessed June 2016.
14. Google Cloud Vision API. 2016, December. https://cloud.google.com/vision/. Online: Last Accessed Dec 2016.
15. Google Forms. 2016, May. https://docs.google.com/forms/. Online: Last Accessed May 2016.
16. MIT Open Course Ware. 2016, May. http://www.ocw.mit.edu/. Online: Last Accessed May 2016.
17. Porter Stemmer. 2016, May. https://tartarus.org/martin/PorterStemmer/. Online: Last Accessed May 2016.
18. SenticNet. 2016, May. http://www.sentic.net/computing/. Online: Last Accessed May 2016.
19. Sentics. 2016, May. https://en.wiktionary.org/wiki/sentics. Online: Last Accessed May 2016.
20. VideoLectures.Net. 2016, May. http://www.videolectures.net/. Online: Last Accessed May, 2016.
21. YouTube Statistics. 2016, July. http://www.youtube.com/yt/press/statistics.html. Online: Last Accessed July, 2016.
22. Abba, H.A., S.N.M. Shah, N.B. Zakaria, and A.J. Pal. 2012. Deadline based performance evaluation of job scheduling algorithms. In *Proceedings of the IEEE International Conference on Cyber-Enabled Distributed Computing and Knowledge Discovery*, 106–110.
23. Achanta, R.S., W.-Q. Yan, and M.S. Kankanhalli. 2006. Modeling Intent for Home Video Repurposing. *Proceedings of the IEEE MultiMedia* 45(1): 46–55.
24. Adcock, J., M. Cooper, A. Girgensohn, and L. Wilcox. 2005. Interactive Video Search Using Multilevel Indexing. In *Proceedings of the Springer Image and Video Retrieval*, 205–214.

25. Agarwal, B., S. Poria, N. Mittal, A. Gelbukh, and A. Hussain. 2015. Concept-level Sentiment Analysis with Dependency-based Semantic Parsing: A Novel Approach. In *Proceedings of the Springer Cognitive Computation*, 1–13.
26. Aizawa, K., D. Tancharoen, S. Kawasaki, and T. Yamasaki. 2004. Efficient Retrieval of Life Log based on Context and Content. In *Proceedings of the ACM Workshop on Continuous Archival and Retrieval of Personal Experiences*, 22–31.
27. Altun, Y., I. Tsochantaridis, and T. Hofmann. 2003. Hidden Markov Support Vector Machines. In *Proceedings of the International Conference on Machine Learning*, 3–10.
28. Anderson, A., K. Ranghunathan, and A. Vogel. 2008. Tagez: Flickr Tag Recommendation. In *Proceedings of the Association for the Advancement of Artificial Intelligence*.
29. Atrey, P.K., A. El Saddik, and M.S. Kankanhalli. 2011. Effective Multimedia Surveillance using a Human-centric Approach. *Proceedings of the Springer Multimedia Tools and Applications* 51(2): 697–721.
30. Barnard, K., P. Duygulu, D. Forsyth, N. De Freitas, D.M. Blei, and M.I. Jordan. 2003. Matching Words and Pictures. *Proceedings of the Journal of Machine Learning Research* 3: 1107–1135.
31. Basu, S., Y. Yu, V.K. Singh, and R. Zimmermann. 2016. Videopedia: Lecture Video Recommendation for Educational Blogs Using Topic Modeling. In *Proceedings of the Springer International Conference on Multimedia Modeling*, 238–250.
32. Basu, S., Y. Yu, and R. Zimmermann. 2016. Fuzzy Clustering of Lecture Videos Based on Topic Modeling. In *Proceedings of the IEEE International Workshop on Content-Based Multimedia Indexing*, 1–6.
33. Basu, S., R. Zimmermann, K.L. OHalloran, S. Tan, and K. Marissa. 2015. Performance Evaluation of Students Using Multimodal Learning Systems. In *Proceedings of the Springer International Conference on Multimedia Modeling*, 135–147.
34. Beeferman, D., A. Berger, and J. Lafferty. 1999. Statistical Models for Text Segmentation. *Proceedings of the Springer Machine Learning* 34(1–3): 177–210.
35. Bernd, J., D. Borth, C. Carrano, J. Choi, B. Elizalde, G. Friedland, L. Gottlieb, K. Ni, R. Pearce, D. Poland, et al. 2015. Kickstarting the Commons: The YFCC100M and the YLI Corpora. In *Proceedings of the ACM Workshop on Community-Organized Multimodal Mining: Opportunities for Novel Solutions*, 1–6.
36. Bhatt, C.A., and M.S. Kankanhalli. 2011. Multimedia Data Mining: State of the Art and Challenges. *Proceedings of the Multimedia Tools and Applications* 51(1): 35–76.
37. Bhatt, C.A., A. Popescu-Belis, M. Habibi, S. Ingram, S. Masneri, F. McInnes, N. Pappas, and O. Schreer. 2013. Multi-factor Segmentation for Topic Visualization and Recommendation: the MUST-VIS System. In *Proceedings of the ACM International Conference on Multimedia*, 365–368.
38. Bhattacharjee, S., W.C. Cheng, C.-F. Chou, L. Golubchik, and S. Khuller. 2000. BISTRO: A Framework for Building Scalable Wide-Area Upload Applications. *Proceedings of the ACM SIGMETRICS Performance Evaluation Review* 28(2): 29–35.
39. Cambria, E., J. Fu, F. Bisio, and S. Poria. 2015. AffectiveSpace 2: Enabling Affective Intuition for Concept-level Sentiment Analysis. In *Proceedings of the AAAI Conference on Artificial Intelligence*, 508–514.
40. Cambria, E., A. Livingstone, and A. Hussain. 2012. The Hourglass of Emotions. In *Proceedings of the Springer Cognitive Behavioural Systems*, 144–157.
41. Cambria, E., D. Olsher, and D. Rajagopal. 2014. SenticNet 3: A Common and Common-sense Knowledge Base for Cognition-driven Sentiment Analysis. In *Proceedings of the AAAI Conference on Artificial Intelligence*, 1515–1521.
42. Cambria, E., S. Poria, R. Bajpai, and B. Schuller. 2016. SenticNet 4: A Semantic Resource for Sentiment Analysis based on Conceptual Primitives. In *Proceedings of the International Conference on Computational Linguistics (COLING)*, 2666–2677.

43. Cambria, E., S. Poria, F. Bisio, R. Bajpai, and I. Chaturvedi. 2015. The CLSA Model: A Novel Framework for Concept-Level Sentiment Analysis. In *Proceedings of the Springer Computational Linguistics and Intelligent Text Processing*, 3–22.
44. Cambria, E., S. Poria, A. Gelbukh, and K. Kwok. 2014. Sentic API: A Common-sense based API for Concept-level Sentiment Analysis. *CEUR Workshop Proceedings* 144: 19–24.
45. Cao, J., Z. Huang, and Y. Yang. 2015. Spatial-aware Multimodal Location Estimation for Social Images. In *Proceedings of the ACM Conference on Multimedia Conference*, 119–128.
46. Chakraborty, I., H. Cheng, and O. Javed. 2014. Entity Centric Feature Pooling for Complex Event Detection. In *Proceedings of the Workshop on HuEvent at the ACM International Conference on Multimedia*, 1–5.
47. Che, X., H. Yang, and C. Meinel. 2013. Lecture Video Segmentation by Automatically Analyzing the Synchronized Slides. In *Proceedings of the ACM International Conference on Multimedia*, 345–348.
48. Chen, B., J. Wang, Q. Huang, and T. Mei. 2012. Personalized Video Recommendation through Tripartite Graph Propagation. In *Proceedings of the ACM International Conference on Multimedia*, 1133–1136.
49. Chen, S., L. Tong, and T. He. 2011. Optimal Deadline Scheduling with Commitment. In *Proceedings of the IEEE Annual Allerton Conference on Communication, Control, and Computing*, 111–118.
50. Chen, W.-B., C. Zhang, and S. Gao. 2012. Segmentation Tree based Multiple Object Image Retrieval. In *Proceedings of the IEEE International Symposium on Multimedia*, 214–221.
51. Chen, Y., and W.J. Heng. 2003. Automatic Synchronization of Speech Transcript and Slides in Presentation. *Proceedings of the IEEE International Symposium on Circuits and Systems* 2: 568–571.
52. Cohen, J. 1960. A Coefficient of Agreement for Nominal Scales. *Proceedings of the Durham Educational and Psychological Measurement* 20(1): 37–46.
53. Cristani, M., A. Pesarin, C. Drioli, V. Murino, A. Rodà,M. Grapulin, and N. Sebe. 2010. Toward an Automatically Generated Soundtrack from Low-level Cross-modal Correlations for Automotive Scenarios. In *Proceedings of the ACM International Conference on Multimedia*, 551–560.
54. Dang-Nguyen, D.-T., L. Piras, G. Giacinto, G. Boato, and F.G. De Natale. 2015. A Hybrid Approach for Retrieving Diverse Social Images of Landmarks. In *Proceedings of the IEEE International Conference on Multimedia and Expo (ICME)*, 1–6.
55. Fabro, M. Del, A. Sobe, and L. Böszörmenyi. 2012. Summarization of Real-life Events Based on Community-contributed Content. In *Proceedings of the International Conferences on Advances in Multimedia*, 119–126.
56. Du, L., W.L. Buntine, and M. Johnson. 2013. Topic Segmentation with a Structured Topic Model. In *Proceedings of the North American Chapter of the Association for Computational Linguistics: Human Language Technologies*, 190–200.
57. Fan, Q., K. Barnard, A. Amir, A. Efrat, and M. Lin. 2006. Matching Slides to Presentation Videos using SIFT and Scene Background Matching. In *Proceedings of the ACM International Conference on Multimedia*, 239–248.
58. Filatova, E. and V. Hatzivassiloglou. 2004. Event-Based Extractive Summarization. In *Proceedings of the ACL Workshop on Summarization*, 104–111.
59. Firan, C.S., M. Georgescu, W. Nejdl, and R. Paiu. 2010. Bringing Order to Your Photos: Event-driven Classification of Flickr Images Based on Social Knowledge. In *Proceedings of the ACM International Conference on Information and Knowledge Management*, 189–198.
60. Gao, S., C. Zhang, and W.-B. Chen. 2012. An Improvement of Color Image Segmentation through Projective Clustering. In *Proceedings of the IEEE International Conference on Information Reuse and Integration*, 152–158.
61. Garg, N. and I. Weber. 2008. Personalized, Interactive Tag Recommendation for Flickr. In *Proceedings of the ACM Conference on Recommender Systems*, 67–74.

62. Ghias, A., J. Logan, D. Chamberlin, and B.C. Smith. 1995. Query by Humming: Musical Information Retrieval in an Audio Database. In *Proceedings of the ACM International Conference on Multimedia*, 231–236.
63. Golder, S.A., and B.A. Huberman. 2006. Usage Patterns of Collaborative Tagging Systems. *Proceedings of the Journal of Information Science* 32(2): 198–208.
64. Gozali, J.P., M.-Y. Kan, and H. Sundaram. 2012. Hidden Markov Model for Event Photo Stream Segmentation. In *Proceedings of the IEEE International Conference on Multimedia and Expo Workshops*, 25–30.
65. Guo, Y., L. Zhang, Y. Hu, X. He, and J. Gao. 2016. Ms-celeb-1m: Challenge of recognizing one million celebrities in the real world. *Proceedings of the Society for Imaging Science and Technology Electronic Imaging* 2016(11): 1–6.
66. Hanjalic, A., and L.-Q. Xu. 2005. Affective Video Content Representation and Modeling. *Proceedings of the IEEE Transactions on Multimedia* 7(1): 143–154.
67. Haubold, A. and J.R. Kender. 2005. Augmented Segmentation and Visualization for Presentation Videos. In *Proceedings of the ACM International Conference on Multimedia*, 51–60.
68. Healey, J.A., and R.W. Picard. 2005. Detecting Stress during Real-world Driving Tasks using Physiological Sensors. *Proceedings of the IEEE Transactions on Intelligent Transportation Systems* 6(2): 156–166.
69. Hefeeda, M., and C.-H. Hsu. 2010. On Burst Transmission Scheduling in Mobile TV Broadcast Networks. *Proceedings of the IEEE/ACM Transactions on Networking* 18(2): 610–623.
70. Hevner, K. 1936. Experimental Studies of the Elements of Expression in Music. *Proceedings of the American Journal of Psychology* 48: 246–268.
71. Hochbaum, D.S. 1996. Approximating Covering and Packing Problems: Set Cover, Vertex Cover, Independent Set, and related Problems. In *Proceedings of the PWS Approximation algorithms for NP-hard problems*, 94–143.
72. Hong, R., J. Tang, H.-K. Tan, S. Yan, C. Ngo, and T.-S. Chua. 2009. Event Driven Summarization for Web Videos. In *Proceedings of the ACM SIGMM Workshop on Social Media*, 43–48.
73. P. ITU-T Recommendation. 1999. Subjective Video Quality Assessment Methods for Multimedia Applications.
74. Jiang, L., A.G. Hauptmann, and G. Xiang. 2012. Leveraging High-level and Low-level Features for Multimedia Event Detection. In *Proceedings of the ACM International Conference on Multimedia*, 449–458.
75. Joachims, T., T. Finley, and C.-N. Yu. 2009. Cutting-plane Training of Structural SVMs. *Proceedings of the Machine Learning Journal* 77(1): 27–59.
76. Johnson, J., L. Ballan, and L. Fei-Fei. 2015. Love Thy Neighbors: Image Annotation by Exploiting Image Metadata. In *Proceedings of the IEEE International Conference on Computer Vision*, 4624–4632.
77. Johnson, J., A. Karpathy, and L. Fei-Fei. 2015. Densecap: Fully Convolutional Localization Networks for Dense Captioning. *In Proceedings of the arXiv preprint arXiv:1511.07571.*
78. Jokhio, F., A. Ashraf, S. Lafond, I. Porres, and J. Lilius. 2013. Prediction-Based dynamic resource allocation for video transcoding in cloud computing. In *Proceedings of the IEEE International Conference on Parallel, Distributed and Network-Based Processing*, 254–261.
79. Kaminskas, M., I. Fernández-Tobías, F. Ricci, and I. Cantador. 2014. Knowledge-Based Identification of Music Suited for Places of Interest. *Proceedings of the Springer Information Technology & Tourism* 14(1): 73–95.
80. Kaminskas, M. and F. Ricci. 2011. Location-adapted Music Recommendation using Tags. In *Proceedings of the Springer User Modeling, Adaption and Personalization*, 183–194.
81. Kan, M.-Y. 2001. Combining Visual Layout and Lexical Cohesion Features for Text Segmentation. *In Proceedings of the Citeseer.*
82. Kan, M.-Y. 2003. *Automatic Text Summarization as Applied to Information Retrieval*. PhD thesis, Columbia University.

83. Kan, M.-Y., J.L. Klavans, and K.R. McKeown. 1998. Linear Segmentation and Segment Significance. *In Proceedings of the arXiv preprint cs/9809020*.
84. Kan, M.-Y., K.R. McKeown, and J.L. Klavans. 2001. Applying Natural Language Generation to Indicative Summarization. *Proceedings of the ACL European Workshop on Natural Language Generation* 8: 1–9.
85. Kang, H.B. 2003. Affective Content Detection using HMMs. In *Proceedings of the ACM International Conference on Multimedia*, 259–262.
86. Kang, Y.-L., J.-H. Lim, M.S. Kankanhalli, C.-S. Xu, and Q. Tian. 2004. Goal Detection in Soccer Video using Audio/Visual Keywords. *Proceedings of the IEEE International Conference on Image Processing* 3: 1629–1632.
87. Kang, Y.-L., J.-H. Lim, Q. Tian, and M.S. Kankanhalli. 2003. Soccer Video Event Detection with Visual Keywords. *Proceedings of the Joint Conference of International Conference on Information, Communications and Signal Processing, and Pacific Rim Conference on Multimedia* 3: 1796–1800.
88. Kankanhalli, M.S., and T.-S. Chua. 2000. Video Modeling using Strata-Based Annotation. *Proceedings of the IEEE MultiMedia* 7(1): 68–74.
89. Kennedy, L., M. Naaman, S. Ahern, R. Nair, and T. Rattenbury. 2007. How Flickr Helps us Make Sense of the World: Context and Content in Community-Contributed Media Collections. In *Proceedings of the ACM International Conference on Multimedia*, 631–640.
90. Kennedy, L.S., S.-F. Chang, and I.V. Kozintsev. 2006. To Search or to Label?: Predicting the Performance of Search-Based Automatic Image Classifiers. In *Proceedings of the ACM International Workshop on Multimedia Information Retrieval*, 249–258.
91. Kim, Y.E., E.M. Schmidt, R. Migneco, B.G. Morton, P. Richardson, J. Scott, J.A. Speck, and D. Turnbull. 2010. Music Emotion Recognition: A State of the Art Review. In *Proceedings of the International Society for Music Information Retrieval*, 255–266.
92. Klavans, J.L., K.R. McKeown, M.-Y. Kan, and S. Lee. 1998. Resources for Evaluation of Summarization Techniques. *In Proceedings of the arXiv preprint cs/9810014*.
93. Ko, Y. 2012. A Study of Term Weighting Schemes using Class Information for Text Classification. In *Proceedings of the ACM Special Interest Group on Information Retrieval*, 1029–1030.
94. Kort, B., R. Reilly, and R.W. Picard. 2001. An Affective Model of Interplay between Emotions and Learning: Reengineering Educational Pedagogy-Building a Learning Companion. *Proceedings of the IEEE International Conference on Advanced Learning Technologies* 1: 43–47.
95. Kucuktunc, O., U. Gudukbay, and O. Ulusoy. 2010. Fuzzy Color Histogram-Based Video Segmentation. *Proceedings of the Computer Vision and Image Understanding* 114(1): 125–134.
96. Kuo, F.-F., M.-F. Chiang, M.-K. Shan, and S.-Y. Lee. 2005. Emotion-Based Music Recommendation by Association Discovery from Film Music. In *Proceedings of the ACM International Conference on Multimedia*, 507–510.
97. Lacy, S., T. Atwater, X. Qin, and A. Powers. 1988. Cost and Competition in the Adoption of Satellite News Gathering Technology. *Proceedings of the Taylor & Francis Journal of Media Economics* 1(1): 51–59.
98. Lambert, P., W. De Neve, P. De Neve, I. Moerman, P. Demeester, and R. Van de Walle. 2006. Rate-distortion performance of H. 264/AVC compared to state-of-the-art video codecs. *Proceedings of the IEEE Transactions on Circuits and Systems for Video Technology* 16(1): 134–140.
99. Laurier, C., M. Sordo, J. Serrà, and P. Herrera. 2009. Music Mood Representations from Social Tags. In *Proceedings of the International Society for Music Information Retrieval*, 381–386.
100. Li, C.T. and M.K. Shan. 2007. Emotion-Based Impressionism Slideshow with Automatic Music Accompaniment. In *Proceedings of the ACM International Conference on Multimedia*, 839–842.

101. Li, J., and J.Z. Wang. 2008. Real-Time Computerized Annotation of Pictures. *Proceedings of the IEEE Transactions on Pattern Analysis and Machine Intelligence* 30(6): 985–1002.
102. Li, X., C.G. Snoek, and M. Worring. 2009. Learning Social Tag Relevance by Neighbor Voting. *Proceedings of the IEEE Transactions on Multimedia* 11(7): 1310–1322.
103. Li, X., T. Uricchio, L. Ballan, M. Bertini, C.G. Snoek, and A.D. Bimbo. 2016. Socializing the Semantic Gap: A Comparative Survey on Image Tag Assignment, Refinement, and Retrieval. *Proceedings of the ACM Computing Surveys (CSUR)* 49(1): 14.
104. Li, Z., Y. Huang, G. Liu, F. Wang, Z.-L. Zhang, and Y. Dai. 2012. Cloud Transcoder: Bridging the Format and Resolution Gap between Internet Videos and Mobile Devices. In *Proceedings of the ACM International Workshop on Network and Operating System Support for Digital Audio and Video*, 33–38.
105. Liang, C., Y. Guo, and Y. Liu. 2008. Is Random Scheduling Sufficient in P2P Video Streaming? In *Proceedings of the IEEE International Conference on Distributed Computing Systems*, 53–60. IEEE.
106. Lim, J.-H., Q. Tian, and P. Mulhem. 2003. Home Photo Content Modeling for Personalized Event-Based Retrieval. *Proceedings of the IEEE MultiMedia* 4: 28–37.
107. Lin, M., M. Chau, J. Cao, and J.F. Nunamaker Jr. 2005. Automated Video Segmentation for Lecture Videos: A Linguistics-Based Approach. *Proceedings of the IGI Global International Journal of Technology and Human Interaction* 1(2): 27–45.
108. Liu, C.L., and J.W. Layland. 1973. Scheduling Algorithms for Multiprogramming in a Hard-real-time Environment. *Proceedings of the ACM Journal of the ACM* 20(1): 46–61.
109. Liu, D., X.-S. Hua, L. Yang, M. Wang, and H.-J. Zhang. 2009. Tag Ranking. In *Proceedings of the ACM World Wide Web Conference*, 351–360.
110. Liu, T., C. Rosenberg, and H.A. Rowley. 2007. Clustering Billions of Images with Large Scale Nearest Neighbor Search. In *Proceedings of the IEEE Winter Conference on Applications of Computer Vision*, 28–28.
111. Liu, X. and B. Huet. 2013. Event Representation and Visualization from Social Media. In *Proceedings of the Springer Pacific-Rim Conference on Multimedia*, 740–749.
112. Liu, Y., D. Zhang, G. Lu, and W.-Y. Ma. 2007. A Survey of Content-Based Image Retrieval with High-level Semantics. *Proceedings of the Elsevier Pattern Recognition* 40(1): 262–282.
113. Livingston, S., and D.A.V. Belle. 2005. The Effects of Satellite Technology on Newsgathering from Remote Locations. *Proceedings of the Taylor & Francis Political Communication* 22(1): 45–62.
114. Long, R., H. Wang, Y. Chen, O. Jin, and Y. Yu. 2011. Towards Effective Event Detection, Tracking and Summarization on Microblog Data. In *Proceedings of the Springer Web-Age Information Management*, 652–663.
115. L. Lu, H. You, and H. Zhang. 2001. A New Approach to Query by Humming in Music Retrieval. In *Proceedings of the IEEE International Conference on Multimedia and Expo*, 22–25.
116. Lu, Y., H. To, A. Alfarrarjeh, S.H. Kim, Y. Yin, R. Zimmermann, and C. Shahabi. 2016. GeoUGV: User-generated Mobile Video Dataset with Fine Granularity Spatial Metadata. In *Proceedings of the ACM International Conference on Multimedia Systems*, 43.
117. Mao, J., W. Xu, Y. Yang, J. Wang, Z. Huang, and A. Yuille. 2014. Deep Captioning with Multimodal Recurrent Neural Networks (M-RNN). *In Proceedings of the arXiv preprint arXiv:1412.6632*.
118. Matusiak, K.K. 2006. Towards User-Centered Indexing in Digital Image Collections. *Proceedings of the OCLC Systems & Services: International Digital Library Perspectives* 22(4): 283–298.
119. McDuff, D., R. El Kaliouby, E. Kodra, and R. Picard. 2013. Measuring Voter's Candidate Preference Based on Affective Responses to Election Debates. In *Proceedings of the IEEE Humaine Association Conference on Affective Computing and Intelligent Interaction*, 369–374.

120. McKeown, K.R., J.L. Klavans, and M.-Y. Kan. Method and System for Topical Segmentation, Segment Significance and Segment Function, 29 2002. US Patent 6,473,730.
121. Mezaris, V., A. Scherp, R. Jain, M. Kankanhalli, H. Zhou, J. Zhang, L. Wang, and Z. Zhang. 2011. Modeling and Rrepresenting Events in Multimedia. In *Proceedings of the ACM International Conference on Multimedia*, 613–614.
122. Mezaris, V., A. Scherp, R. Jain, and M.S. Kankanhalli. 2014. Real-life Events in Multimedia: Detection, Representation, Retrieval, and Applications. *Proceedings of the Springer Multimedia Tools and Applications* 70(1): 1–6.
123. Miller, G., and C. Fellbaum. 1998. *Wordnet: An Electronic Lexical Database*. Cambridge, MA: MIT Press.
124. Miller, G.A. 1995. WordNet: A Lexical Database for English. *Proceedings of the Communications of the ACM* 38(11): 39–41.
125. Moxley, E., J. Kleban, J. Xu, and B. Manjunath. 2009. Not All Tags are Created Equal: Learning Flickr Tag Semantics for Global Annotation. In *Proceedings of the IEEE International Conference on Multimedia and Expo*, 1452–1455.
126. Mulhem, P., M.S. Kankanhalli, J. Yi, and H. Hassan. 2003. Pivot Vector Space Approach for Audio-Video Mixing. *Proceedings of the IEEE MultiMedia* 2: 28–40.
127. Naaman, M. 2012. Social Multimedia: Highlighting Opportunities for Search and Mining of Multimedia Data in Social Media Applications. *Proceedings of the Springer Multimedia Tools and Applications* 56(1): 9–34.
128. Natarajan, P., P.K. Atrey, and M. Kankanhalli. 2015. Multi-Camera Coordination and Control in Surveillance Systems: A Survey. *Proceedings of the ACM Transactions on Multimedia Computing, Communications, and Applications* 11(4): 57.
129. Nayak, M.G. 2004. *Music Synthesis for Home Videos*. PhD thesis.
130. Neo, S.-Y., J. Zhao, M.-Y. Kan, and T.-S. Chua. 2006. Video Retrieval using High Level Features: Exploiting Query Matching and Confidence-Based Weighting. In *Proceedings of the Springer International Conference on Image and Video Retrieval*, 143–152.
131. Ngo, C.-W., F. Wang, and T.-C. Pong. 2003. Structuring Lecture Videos for Distance Learning Applications. In *Proceedings of the IEEE International Symposium on Multimedia Software Engineering*, 215–222.
132. Nguyen, V.-A., J. Boyd-Graber, and P. Resnik. 2012. SITS: A Hierarchical Nonparametric Model using Speaker Identity for Topic Segmentation in Multiparty Conversations. In *Proceedings of the Annual Meeting of the Association for Computational Linguistics*, 78–87.
133. Nwana, A.O. and T. Chen. 2016. Who Ordered This?: Exploiting Implicit User Tag Order Preferences for Personalized Image Tagging. *In Proceedings of the arXiv preprint arXiv:1601.06439*.
134. Papagiannopoulou, C. and V. Mezaris. 2014. Concept-Based Image Clustering and Summarization of Event-related Image Collections. In *Proceedings of the Workshop on HuEvent at the ACM International Conference on Multimedia*, 23–28.
135. Park, M.H., J.H. Hong, and S.B. Cho. 2007. Location-Based Recommendation System using Bayesian User's Preference Model in Mobile Devices. In *Proceedings of the Springer Ubiquitous Intelligence and Computing*, 1130–1139.
136. Petkos, G., S. Papadopoulos, V. Mezaris, R. Troncy, P. Cimiano, T. Reuter, and Y. Kompatsiaris. 2014. Social Event Detection at MediaEval: a Three-Year Retrospect of Tasks and Results. In *Proceedings of the Workshop on Social Events in Web Multimedia at ACM International Conference on Multimedia Retrieval*.
137. Pevzner, L., and M.A. Hearst. 2002. A Critique and Improvement of an Evaluation Metric for Text Segmentation. *Proceedings of the Computational Linguistics* 28(1): 19–36.
138. Picard, R.W., and J. Klein. 2002. Computers that Recognise and Respond to User Emotion: Theoretical and Practical Implications. *Proceedings of the Interacting with Computers* 14(2): 141–169.

139. Picard, R.W., E. Vyzas, and J. Healey. 2001. Toward machine emotional intelligence: Analysis of affective physiological state. *Proceedings of the IEEE Transactions on Pattern Analysis and Machine Intelligence* 23(10): 1175–1191.
140. Poisson, S.D. and C.H. Schnuse. 1841. Recherches Sur La Pprobabilité Des Jugements En Mmatieré Criminelle Et En Matieré Civile. Meyer.
141. Poria, S., E. Cambria, R. Bajpai, and A. Hussain. 2017. A Review of Affective Computing: From Unimodal Analysis to Multimodal Fusion. *Proceedings of the Elsevier Information Fusion* 37: 98–125.
142. Poria, S., E. Cambria, and A. Gelbukh. 2016. Aspect Extraction for Opinion Mining with a Deep Convolutional Neural Network. *Proceedings of the Elsevier Knowledge-Based Systems* 108: 42–49.
143. Poria, S., E. Cambria, A. Gelbukh, F. Bisio, and A. Hussain. 2015. Sentiment Data Flow Analysis by Means of Dynamic Linguistic Patterns. *Proceedings of the IEEE Computational Intelligence Magazine* 10(4): 26–36.
144. Poria, S., E. Cambria, and A.F. Gelbukh. 2015. Deep Convolutional Neural Network Textual Features and Multiple Kernel Learning for Utterance-level Multimodal Sentiment Analysis. In *Proceedings of the EMNLP*, 2539–2544.
145. Poria, S., E. Cambria, D. Hazarika, N. Mazumder, A. Zadeh, and L.-P. Morency. 2017. Context-Dependent Sentiment Analysis in User-Generated Videos. *In Proceedings of the Association for Computational Linguistics.*
146. Poria, S., E. Cambria, D. Hazarika, and P. Vij. 2016. A Deeper Look into Sarcastic Tweets using Deep Convolutional Neural Networks. In *Proceedings of the International Conference on Computational Linguistics (COLING).*
147. Poria, S., E. Cambria, N. Howard, G.-B. Huang, and A. Hussain. 2016. Fusing Audio Visual and Textual Clues for Sentiment Analysis from Multimodal Content. *Proceedings of the Elsevier Neurocomputing* 174: 50–59.
148. Poria, S., E. Cambria, N. Howard, and A. Hussain. 2015. Enhanced SenticNet with Affective Labels for Concept-Based Opinion Mining: Extended Abstract. In *Proceedings of the International Joint Conference on Artificial Intelligence.*
149. Poria, S., E. Cambria, A. Hussain, and G.-B. Huang. 2015. Towards an Intelligent Framework for Multimodal Affective Data Analysis. *Proceedings of the Elsevier Neural Networks* 63: 104–116.
150. Poria, S., E. Cambria, L.-W. Ku, C. Gui, and A. Gelbukh. 2014. A Rule-Based Approach to Aspect Extraction from Product Reviews. In *Proceedings of the Second Workshop on Natural Language Processing for Social Media (SocialNLP)*, 28–37.
151. Poria, S., I. Chaturvedi, E. Cambria, and F. Bisio. 2016. Sentic LDA: Improving on LDA with Semantic Similarity for Aspect-Based Sentiment Analysis. In *Proceedings of the IEEE International Joint Conference on Neural Networks (IJCNN)*, 4465–4473.
152. Poria, S., I. Chaturvedi, E. Cambria, and A. Hussain. 2016. Convolutional MKL Based Multimodal Emotion Recognition and Sentiment Analysis. In *Proceedings of the IEEE International Conference on Data Mining (ICDM)*, 439–448.
153. Poria, S., A. Gelbukh, B. Agarwal, E. Cambria, and N. Howard. 2014. Sentic Demo: A Hybrid Concept-level Aspect-Based Sentiment Analysis Toolkit. In *Proceedings of the ESWC.*
154. Poria, S., A. Gelbukh, E. Cambria, D. Das, and S. Bandyopadhyay. 2012. Enriching SenticNet Polarity Scores Through Semi-Supervised Fuzzy Clustering. In *Proceedings of the IEEE International Conference on Data Mining Workshops (ICDMW)*, 709–716.
155. Poria, S., A. Gelbukh, E. Cambria, A. Hussain, and G.-B. Huang. 2014. EmoSenticSpace: A Novel Framework for Affective Common-sense Reasoning. *Proceedings of the Elsevier Knowledge-Based Systems* 69: 108–123.
156. Poria, S., A. Gelbukh, E. Cambria, P. Yang, A. Hussain, and T. Durrani. 2012. Merging SenticNet and WordNet-Affect Emotion Lists for Sentiment Analysis. *Proceedings of the IEEE International Conference on Signal Processing (ICSP)* 2: 1251–1255.

157. Poria, S., A. Gelbukh, A. Hussain, S. Bandyopadhyay, and N. Howard. 2013. Music Genre Classification: A Semi-Supervised Approach. In *Proceedings of the Springer Mexican Conference on Pattern Recognition*, 254–263.
158. Poria, S., N. Ofek, A. Gelbukh, A. Hussain, and L. Rokach. 2014. Dependency Tree-Based Rules for Concept-level Aspect-Based Sentiment Analysis. In *Proceedings of the Springer Semantic Web Evaluation Challenge*, 41–47.
159. Poria, S., H. Peng, A. Hussain, N. Howard, and E. Cambria. 2017. Ensemble Application of Convolutional Neural Networks and Multiple Kernel Learning for Multimodal Sentiment Analysis. *In Proceedings of the Elsevier Neurocomputing.*
160. Pye, D., N.J. Hollinghurst, T.J. Mills, and K.R. Wood. 1998. Audio-visual Segmentation for Content-Based Retrieval. In *Proceedings of the International Conference on Spoken Language Processing.*
161. Qiao, Z., P. Zhang, C. Zhou, Y. Cao, L. Guo, and Y. Zhang. 2014. *Event Recommendation in Event-Based Social Networks.*
162. Raad, E.J. and R. Chbeir. 2014. Foto2Events: From Photos to Event Discovery and Linking in Online Social Networks. In *Proceedings of the IEEE Big Data and Cloud Computing*, 508–515, .
163. Radsch, C.C. 2013. The Revolutions will be Blogged: Cyberactivism and the 4th Estate in Egypt. Doctoral Disseration. American University.
164. Rae, A., B. Sigurbjörnssön, and R. van Zwol. 2010. Improving Tag Recommendation using Social Networks. In *Proceedings of the Adaptivity, Personalization and Fusion of Heterogeneous Information*, 92–99.
165. Rahmani, H., B. Piccart, D. Fierens, and H. Blockeel. 2010. Three Complementary Approaches to Context Aware Movie Recommendation. In *Proceedings of the ACM Workshop on Context-Aware Movie Recommendation*, 57–60.
166. Rattenbury, T., N. Good, and M. Naaman. 2007. Towards Automatic Extraction of Event and Place Semantics from Flickr Tags. In *Proceedings of the ACM Special Interest Group on Information Retrieval.*
167. Rawat, Y. and M. S. Kankanhalli. 2016. ConTagNet: Exploiting User Context for Image Tag Recommendation. In *Proceedings of the ACM International Conference on Multimedia*, 1102–1106.
168. Repp, S., A. Groß, and C. Meinel. 2008. Browsing within Lecture Videos Based on the Chain Index of Speech Transcription. *Proceedings of the IEEE Transactions on Learning Technologies* 1(3): 145–156.
169. Repp, S. and C. Meinel. 2006. Semantic Indexing for Recorded Educational Lecture Videos. In *Proceedings of the IEEE International Conference on Pervasive Computing and Communications Workshops*, 5.
170. Repp, S., J. Waitelonis, H. Sack, and C. Meinel. 2007. Segmentation and Annotation of Audiovisual Recordings Based on Automated Speech Recognition. In *Proceedings of the Springer Intelligent Data Engineering and Automated Learning*, 620–629.
171. Russell, J.A. 1980. A Circumplex Model of Affect. *Proceedings of the Journal of Personality and Social Psychology* 39: 1161–1178.
172. Sahidullah, M., and G. Saha. 2012. Design, Analysis and Experimental Evaluation of Block Based Transformation in MFCC Computation for Speaker Recognition. *Proceedings of the Speech Communication* 54: 543–565.
173. Salamon, J., J. Serra, and E. Gomez. 2013. Tonal Representations for Music Retrieval: From Version Identification to Query-by-Humming. In *Proceedings of the Springer International Journal of Multimedia Information Retrieval* 2(1): 45–58.
174. Schedl, M. and D. Schnitzer. 2014. Location-Aware Music Artist Recommendation. In *Proceedings of the Springer MultiMedia Modeling*, 205–213.
175. M. Schedl and F. Zhou. 2016. Fusing Web and Audio Predictors to Localize the Origin of Music Pieces for Geospatial Retrieval. In *Proceedings of the Springer European Conference on Information Retrieval*, 322–334.

176. Scherp, A., and V. Mezaris. 2014. Survey on Modeling and Indexing Events in Multimedia. *Proceedings of the Springer Multimedia Tools and Applications* 70(1): 7–23.
177. Scherp, A., V. Mezaris, B. Ionescu, and F. De Natale. 2014. HuEvent '14: Workshop on Human-Centered Event Understanding from Multimedia. In *Proceedings of the ACM International Conference on Multimedia*, 1253–1254.
178. Schmitz, P. 2006. Inducing Ontology from Flickr Tags. In *Proceedings of the Collaborative Web Tagging Workshop at ACM World Wide Web Conference*, vol 50.
179. Schuller, B., C. Hage, D. Schuller, and G. Rigoll. 2010. Mister DJ, Cheer Me Up!: Musical and Textual Features for Automatic Mood Classification. *Proceedings of the Journal of New Music Research* 39(1): 13–34.
180. Shah, R.R., M. Hefeeda, R. Zimmermann, K. Harras, C.-H. Hsu, and Y. Yu. 2016. NEWSMAN: Uploading Videos over Adaptive Middleboxes to News Servers In Weak Network Infrastructures. In *Proceedings of the Springer International Conference on Multimedia Modeling*, 100–113.
181. Shah, R.R., A. Samanta, D. Gupta, Y. Yu, S. Tang, and R. Zimmermann. 2016. PROMPT: Personalized User Tag Recommendation for Social Media Photos Leveraging Multimodal Information. In *Proceedings of the ACM International Conference on Multimedia*, 486–492.
182. Shah, R.R., A.D. Shaikh, Y. Yu, W. Geng, R. Zimmermann, and G. Wu. 2015. EventBuilder: Real-time Multimedia Event Summarization by Visualizing Social Media. In *Proceedings of the ACM International Conference on Multimedia*, 185–188.
183. Shah, R.R., Y. Yu, A.D. Shaikh, S. Tang, and R. Zimmermann. 2014. ATLAS: Automatic Temporal Segmentation and Annotation of Lecture Videos Based on Modelling Transition Time. In *Proceedings of the ACM International Conference on Multimedia*, 209–212.
184. Shah, R.R., Y. Yu, A.D. Shaikh, and R. Zimmermann. 2015. TRACE: A Linguistic-Based Approach for Automatic Lecture Video Segmentation Leveraging Wikipedia Texts. In *Proceedings of the IEEE International Symposium on Multimedia*, 217–220.
185. Shah, R.R., Y. Yu, S. Tang, S. Satoh, A. Verma, and R. Zimmermann. 2016. Concept-Level Multimodal Ranking of Flickr Photo Tags via Recall Based Weighting. In *Proceedings of the MMCommon's Workshop at ACM International Conference on Multimedia*, 19–26.
186. Shah, R.R., Y. Yu, A. Verma, S. Tang, A.D. Shaikh, and R. Zimmermann. 2016. Leveraging Multimodal Information for Event Summarization and Concept-level Sentiment Analysis. In *Proceedings of the Elsevier Knowledge-Based Systems*, 102–109.
187. Shah, R.R., Y. Yu, and R. Zimmermann. 2014. ADVISOR: Personalized Video Soundtrack Recommendation by Late Fusion with Heuristic Rankings. In *Proceedings of the ACM International Conference on Multimedia*, 607–616.
188. Shah, R.R., Y. Yu, and R. Zimmermann. 2014. User Preference-Aware Music Video Generation Based on Modeling Scene Moods. In *Proceedings of the ACM International Conference on Multimedia Systems*, 156–159.
189. Shaikh, A.D., M. Jain, M. Rawat, R.R. Shah, and M. Kumar. 2013. Improving Accuracy of SMS Based FAQ Retrieval System. In *Proceedings of the Springer Multilingual Information Access in South Asian Languages*, 142–156.
190. Shaikh, A.D., R.R. Shah, and R. Shaikh. 2013. SMS Based FAQ Retrieval for Hindi, English and Malayalam. In *Proceedings of the ACM Forum on Information Retrieval Evaluation*, 9.
191. Shamma, D.A., R. Shaw, P.L. Shafton, and Y. Liu. 2007. Watch What I Watch: Using Community Activity to Understand Content. In *Proceedings of the ACM International Workshop on Multimedia Information Retrieval*, 275–284.
192. Shaw, B., J. Shea, S. Sinha, and A. Hogue. 2013. Learning to Rank for Spatiotemporal Search. In *Proceedings of the ACM International Conference on Web Search and Data Mining*, 717–726.
193. Sigurbjörnsson, B. and R. Van Zwol. 2008. Flickr Tag Recommendation Based on Collective Knowledge. In *Proceedings of the ACM World Wide Web Conference*, 327–336.

194. Snoek, C.G., M. Worring, and A.W. Smeulders. 2005. Early versus Late Fusion in Semantic Video Analysis. In *Proceedings of the ACM International Conference on Multimedia*, 399–402.
195. Snoek, C.G., M. Worring, J.C. Van Gemert, J.-M. Geusebroek, and A.W. Smeulders. 2006. The Challenge Problem for Automated Detection of 101 Semantic Concepts in Multimedia. In *Proceedings of the ACM International Conference on Multimedia*, 421–430.
196. Soleymani, M., J.J.M. Kierkels, G. Chanel, and T. Pun. 2009. A Bayesian Framework for Video Affective Representation. In *Proceedings of the IEEE International Conference on Affective Computing and Intelligent Interaction and Workshops*, 1–7.
197. Stober, S., and A. . Nürnberger. 2013. Adaptive Music Retrieval – A State of the Art. *Proceedings of the Springer Multimedia Tools and Applications* 65(3): 467–494.
198. Stoyanov, V., N. Gilbert, C. Cardie, and E. Riloff. 2009. Conundrums in Noun Phrase Coreference Resolution: Making Sense of the State-of-the-art. In *Proceedings of the ACL International Joint Conference on Natural Language Processing of the Asian Federation of Natural Language Processing*, 656–664.
199. Stupar, A. and S. Michel. 2011. Picasso: Automated Soundtrack Suggestion for Multimodal Data. In *Proceedings of the ACM Conference on Information and Knowledge Management*, 2589–2592.
200. Thayer, R.E. 1989. *The Biopsychology of Mood and Arousal*. New York: Oxford University Press.
201. Thomee, B., B. Elizalde, D.A. Shamma, K. Ni, G. Friedland, D. Poland, D. Borth, and L.-J. Li. 2016. YFCC100M: The New Data in Multimedia Research. *Proceedings of the Communications of the ACM* 59(2): 64–73.
202. Tirumala, A., F. Qin, J. Dugan, J. Ferguson, and K. Gibbs. 2005. Iperf: The TCP/UDP Bandwidth Measurement Tool. http://dast.nlanr.net/Projects/Iperf/
203. Torralba, A., R. Fergus, and W.T. Freeman. 2008. 80 Million Tiny Images: A Large Data set for Nonparametric Object and Scene Recognition. *Proceedings of the IEEE Transactions on Pattern Analysis and Machine Intelligence* 30(11): 1958–1970.
204. Toutanova, K., D. Klein, C.D. Manning, and Y. Singer. 2003. Feature-Rich Part-of-Speech Tagging with a Cyclic Dependency Network. In *Proceedings of the North American Chapter of the Association for Computational Linguistics on Human Language Technology*, 173–180.
205. Toutanova, K. and C.D. Manning. 2000. Enriching the Knowledge Sources used in a Maximum Entropy Part-of-Speech Tagger. In *Proceedings of the Annual Meeting of the Association for Computational Linguistics*, 63–70.
206. Utiyama, M. and H. Isahara. 2001. A Statistical Model for Domain-Independent Text Segmentation. In *Proceedings of the Annual Meeting on Association for Computational Linguistics*, 499–506.
207. Vishal, K., C. Jawahar, and V. Chari. 2015. Accurate Localization by Fusing Images and GPS Signals. In *Proceedings of the IEEE Computer Vision and Pattern Recognition Workshops*, 17–24.
208. Wang, C., F. Jing, L. Zhang, and H.-J. Zhang. 2008. Scalable Search-Based Image Annotation. *Proceedings of the Springer Multimedia Systems* 14(4): 205–220.
209. Wang, H.L., and L.F. Cheong. 2006. Affective Understanding in Film. *Proceedings of the IEEE Transactions on Circuits and Systems for Video Technology* 16(6): 689–704.
210. Wang, J., J. Zhou, H. Xu, T. Mei, X.-S. Hua, and S. Li. 2014. Image Tag Refinement by Regularized Latent Dirichlet Allocation. *Proceedings of the Elsevier Computer Vision and Image Understanding* 124: 61–70.
211. Wang, P., H. Wang, M. Liu, and W. Wang. 2010. An Algorithmic Approach to Event Summarization. In *Proceedings of the ACM Special Interest Group on Management of Data*, 183–194.
212. Wang, X., Y. Jia, R. Chen, and B. Zhou. 2015. Ranking User Tags in Micro-Blogging Website. In *Proceedings of the IEEE ICISCE*, 400–403.

213. Wang, X., L. Tang, H. Gao, and H. Liu. 2010. Discovering Overlapping Groups in Social Media. In *Proceedings of the IEEE International Conference on Data Mining*, 569–578.
214. Wang, Y. and M.S. Kankanhalli. 2015. Tweeting Cameras for Event Detection. In *Proceedings of the IW3C2 International Conference on World Wide Web*, 1231–1241.
215. Webster, A.A., C.T. Jones, M.H. Pinson, S.D. Voran, and S. Wolf. 1993. Objective Video Quality Assessment System Based on Human Perception. In *Proceedings of the IS&T/SPIE's Symposium on Electronic Imaging: Science and Technology*, 15–26. International Society for Optics and Photonics.
216. Wei, C.Y., N. Dimitrova, and S.-F. Chang. 2004. Color-Mood Analysis of Films Based on Syntactic and Psychological Models. In *Proceedings of the IEEE International Conference on Multimedia and Expo*, 831–834.
217. Whissel, C. 1989. The Dictionary of Affect in Language. In *Emotion: Theory, Research and Experience. Vol. 4. The Measurement of Emotions*, ed. R. Plutchik and H. Kellerman, 113–131. New York: Academic.
218. Wu, L., L. Yang, N. Yu, and X.-S. Hua. 2009. Learning to Tag. In *Proceedings of the ACM World Wide Web Conference*, 361–370.
219. Xiao, J., W. Zhou, X. Li, M. Wang, and Q. Tian. 2012. Image Tag Re-ranking by Coupled Probability Transition. In *Proceedings of the ACM International Conference on Multimedia*, 849–852.
220. Xie, D., B. Qian, Y. Peng, and T. Chen. 2009. A Model of Job Scheduling with Deadline for Video-on-Demand System. In *Proceedings of the IEEE International Conference on Web Information Systems and Mining*, 661–668.
221. Xu, M., L.-Y. Duan, C. Xu, M. Kankanhalli, and Q. Tian. 2003. Event Detection in Basketball Video using Multiple Modalities. *Proceedings of the IEEE Joint Conference of the Fourth International Conference on Information, Communications and Signal Processing, and Fourth Pacific Rim Conference on Multimedia* 3: 1526–1530.
222. Xu, M., N.C. Maddage, C. Xu, M. Kankanhalli, and Q. Tian. 2003. Creating Audio Keywords for Event Detection in Soccer Video. In *Proceedings of the IEEE International Conference on Multimedia and Expo*, 2:II–281.
223. Yamamoto, N., J. Ogata, and Y. Ariki. 2003. Topic Segmentation and Retrieval System for Lecture Videos Based on Spontaneous Speech Recognition. In *Proceedings of the INTERSPEECH*, 961–964.
224. Yang, H., M. Siebert, P. Luhne, H. Sack, and C. Meinel. 2011. Automatic Lecture Video Indexing using Video OCR Technology. In *Proceedings of the IEEE International Symposium on Multimedia*, 111–116.
225. Yang, Y.H., Y.C. Lin, Y.F. Su, and H.H. Chen. 2008. A Regression Approach to Music Emotion Recognition. *Proceedings of the IEEE Transactions on Audio, Speech, and Language Processing* 16(2): 448–457.
226. Ye, G., D. Liu, I.-H. Jhuo, and S.-F. Chang. 2012. Robust Late Fusion with Rank Minimization. In *Proceedings of the IEEE Conference on Computer Vision and Pattern Recognition*, 3021–3028.
227. Ye, Q., Q. Huang, W. Gao, and D. Zhao. 2005. Fast and Robust Text Detection in Images and Video Frames. *Proceedings of the Elsevier Image and Vision Computing* 23(6): 565–576.
228. Yin, Y., Z. Shen, L. Zhang, and R. Zimmermann. 2015. Spatial temporal Tag Mining for Automatic Geospatial Video Annotation. *Proceedings of the ACM Transactions on Multimedia Computing, Communications, and Applications* 11(2): 29.
229. Yoon, S. and V. Pavlovic. 2014. Sentiment Flow for Video Interestingness Prediction. In *Proceedings of the Workshop on HuEvent at the ACM International Conference on Multimedia*, 29–34.
230. Yu, Y., K. Joe, V. Oria, F. Moerchen, J.S. Downie, and L. Chen. 2009. Multiversion Music Search using Acoustic Feature Union and Exact Soft Mapping. *Proceedings of the World Scientific International Journal of Semantic Computing* 3(02): 209–234.

231. Yu, Y., Z. Shen, and R. Zimmermann. 2012. Automatic Music Soundtrack Generation for Outdoor Videos from Contextual Sensor Information. In *Proceedings of the ACM International Conference on Multimedia*, 1377–1378.
232. Zaharieva, M., M. Zeppelzauer, and C. Breiteneder. 2013. Automated Social Event Detection in Large Photo Collections. In *Proceedings of the ACM International Conference on Multimedia Retrieval*, 167–174.
233. Zhang, J., X. Liu, L. Zhuo, and C. Wang. 2015. Social Images Tag Ranking Based on Visual Words in Compressed Domain. *Proceedings of the Elsevier Neurocomputing* 153: 278–285.
234. Zhang, J., S. Wang, and Q. Huang. 2015. Location-Based Parallel Tag Completion for Geo-tagged Social Image Retrieval. In *Proceedings of the ACM International Conference on Multimedia Retrieval*, 355–362.
235. Zhang, M., J. Wong, W. Tavanapong, J. Oh, and P. de Groen. 2004. Media Uploading Systems with Hard Deadlines. In *Proceedings of the Citeseer International Conference on Internet and Multimedia Systems and Applications*, 305–310.
236. Zhang, M., J. Wong, W. Tavanapong, J. Oh, and P. de Groen. 2008. Deadline-constrained Media Uploading Systems. *Proceedings of the Springer Multimedia Tools and Applications* 38(1): 51–74.
237. Zhang, W., J. Lin, X. Chen, Q. Huang, and Y. Liu. 2006. Video Shot Detection using Hidden Markov Models with Complementary Features. *Proceedings of the IEEE International Conference on Innovative Computing, Information and Control* 3: 593–596.
238. Zheng, L., V. Noroozi, and P.S. Yu. 2017. Joint Deep Modeling of Users and Items using Reviews for Recommendation. In *Proceedings of the ACM International Conference on Web Search and Data Mining*, 425–434.
239. Zhou, X.S. and T.S. Huang. 2000. CBIR: from Low-level Features to High-level Semantics. In *Proceedings of the International Society for Optics and Photonics Electronic Imaging*, 426–431.
240. Zhuang, J. and S.C. Hoi. 2011. A Two-view Learning Approach for Image Tag Ranking. In *Proceedings of the ACM International Conference on Web Search and Data Mining*, 625–634.
241. Zimmermann, R. and Y. Yu. 2013. Social Interactions over Geographic-aware Multimedia Systems. In *Proceedings of the ACM International Conference on Multimedia*, 1115–1116.
242. Shah, R.R. 2016. Multimodal-based Multimedia Analysis, Retrieval, and Services in Support of Social Media Applications. In *Proceedings of the ACM International Conference on Multimedia*, 1425–1429.
243. Shah, R.R. 2016. Multimodal Analysis of User-Generated Content in Support of Social Media Applications. In *Proceedings of the ACM International Conference in Multimedia Retrieval*, 423–426.
244. Yin, Y., R.R. Shah, and R. Zimmermann. 2016. A General Feature-based Map Matching Framework with Trajectory Simplification. In *Proceedings of the ACM SIGSPATIAL International Workshop on GeoStreaming*, 7.

Chapter 6
Lecture Video Segmentation

Abstract In multimedia-based e-learning systems, the accessibility and searchability of most lecture video content is still insufficient due to the unscripted and spontaneous speech of the speakers. Thus, it is very desirable to enable people to navigate and access specific topics within lecture videos by performing an automatic topic-wise video segmentation. This problem becomes even more challenging when the quality of such lecture videos is not sufficiently high. To this end, we first present the ATLAS system that has two main novelties: (i) a SVM^{hmm} model is proposed to learn temporal transition cues and (ii) a fusion scheme is suggested to combine transition cues extracted from heterogeneous information of lecture videos. Subsequently, considering that contextual information is very useful in determining knowledge structures, we present the TRACE system to automatically perform such a segmentation based on a linguistic approach using Wikipedia texts. TRACE has two main contributions: (i) the extraction of a novel linguistic-based Wikipedia feature to segment lecture videos efficiently, and (ii) the investigation of the late fusion of video segmentation results derived from state-of-the-art algorithms. Specifically for the late fusion, we combine confidence scores produced by the models constructed from visual, transcriptional, and Wikipedia features. According to our experiments on lecture videos from VideoLectures.NET and NPTEL, the proposed algorithms in the ATLAS and TRACE systems segment knowledge structures more accurately compared to existing state-of-the-art algorithms.

Keywords Lecture videos segmentation • Segment boundaries detection • Video understanding • Multimodal analysis • ATLAS • TRACE

6.1 Introduction

A large volume of digital lecture videos has accumulated on the web due to the ubiquitous availability of digital cameras and affordable network infrastructures. Lecture videos are now also frequently streamed in e-learning applications. However, a significant number of old (but important) lecture videos with low visual quality from well-known speakers (experts) is also commonly part of such databases. Therefore, it is essential to perform an efficient and fast topic boundary

R. Shah, R. Zimmermann, *Multimodal Analysis of User-Generated Multimedia Content*, Socio-Affective Computing 6, DOI 10.1007/978-3-319-61807-4_6

detection that works robustly even with low-quality videos. However, an automatic topic-wise indexing and content-based retrieval of appropriate information from a large collection of lecture videos is very challenging due to the following reasons. First, transcripts/SRTs (subtitle resource tracks) of lecture videos contain repetitions, mistakes, and rephrasings. Second, the low visual quality of a lecture video may be challenging for topic boundary detection. Third, the camera may in many parts of a video focus on the speaker instead of the, *e.g.*, whiteboard. Hence, the topic-wise segmentation of a lecture video into smaller cohesive intervals is a highly necessary approach to enable an easy search of the desired pieces of information. Moreover, an automatic segmentation of lecture videos is highly desired because of the high cost of manual video segmentation. All notations used in this chapter are listed in Table 6.1.

State-of-the-art methods of automatic lecture video segmentation are based on the analysis of visual content, speech signals, and transcripts. However, most earlier approaches perform an analysis of only one of these modalities. Hence, the late fusion of the results of these analyses has been largely unexplored for the segmentation of lecture videos. Furthermore, none of the above approaches consistently yields the best segmentation results for all lecture videos due to unclear topic boundaries, varying video qualities, and the subjectiveness inherent in the transcripts of lecture videos. Since multimodal information has shown great importance in addressing different multimedia analytics problems [242, 243], we leverage knowledge structures from different modalities to address the lecture video segmentation problem. Interestingly, the segment boundaries derived from the different modalities (*e.g.*, video content, speech, and SRT) are highly correlated. Therefore, it is desirable to investigate the idea of late-fusing results from multiple state-of-the-art lecture video segmentation algorithms. Note that the topic boundaries derived from different modalities have different granularity. For instance, the topic boundaries derived from visual content are mostly the shot changes. Thus, often several such boundaries result in false positive topic boundaries and even miss several actual topic boundaries. Similarly, the topic boundaries derived from speech transcript are mostly the coherent blocks of words (say, a window of 120 words). However, the drawback of such topic boundaries is that they are of fixed sizes, that is not often the case in real-time. Furthermore, the topic boundaries derived from audio content are mostly the long pauses. However, similar to topic boundaries derived from visual content, such boundaries consist several false positive cases. Thus, we want to investigate the effect of fusing segment boundaries derived from different modalities (Fig. 6.1).

To solve the problem of automatic lecture video segmentation, we present the ATLAS system which stands for *automatic temporal segmentation and annotation of lecture videos based on modeling transition time*. We follow the theme of this book [182, 186–188, 244], i.e., multimodal analysis of user-generated content in our solution to this problem [180, 181, 185, 189, 190]. ATLAS first predicts temporal transitions (TT_1) using supervised learning on video content. Specifically, a color histogram of a keyframe at each shot boundary is used as a visual feature to represent the slide transition in the video content. The relationship between the visual features and the transition time of a slide is established with a training dataset of lecture videos from VideoLectures.NET, using a machine-learning SVM^{hmm}

Table 6.1 Notations used in the lecture video segmentation chapter

Symbols	Meanings
TT_1	Temporal transitions predicted using supervised learning on video content
TT_2	Temporal transitions derived using text (transcripts and/or slides) analysis using an N-gram based language model
$\bar{\chi}$	Similarity threshold for lecture video segmentation
$D_{LectureVideo.Net}$	The Videolectures.NET dataset
D_{NPTEL}	The NPTEL dataset
TS_{ATLAS}	Test set for the ATLAS system
N_{slides}	The number of slides in a PPT of a lecture video
$precision$	The precision of lecture video segmentation
$recall$	The recall of lecture video segmentation
PTT	A set of predicted transition time
PTT_i	i^{th} predicted transition time from PTT
ATT	A set of Actual transition time
ATT_j	j^{th} actual transition time from ATT
td	Time difference between PTT_i and the nearest ATT_j
Υ	The number of (PTT_i, ATT_j) pairs
$\|PTT\|$	The number of PTT_i, i.e., the cardinality of PTT
$\|PTT\|$	The number of ATT_j, i.e., the cardinality of $AT\ T$
H	The HTML file for PPT of a lecture video
T_{Text}	The list of title text from PPT of a lecture video
TF	Term frequency
$\mathcal{N}$	N-gram count
$\mathcal{B}$	A text block in SRT
W_k	Weight for different N-gram count k
$\bar{t}_k$	An N-gram token
$TF(\bar{t}_k \| \mathcal{B}_i)$	The TF of an N-gram token $\bar{t}_k$ in a block $\mathcal{B}_i$
$TF(\bar{t}_k \| SRT)$	The TF of the token $\bar{t}_k$ in the SRT file
b_S	A block of 120 words from SRT
b_W	The block of texts corresponding to a Wikipedia topic
B_W	The list of Wikipedia blocks
B_S	The list of SRT blocks
$\mathcal{I}$	Linguistic feature vector for the SRT window b_S
$\mathcal{J}$	Linguistic feature vector for the SRT window adjacent to b_S
$\alpha(b_w, b_s)$	Cosine similarity between blocks b_W and b_S
f_W	A Wikipedia feature vector
f_S	An SRT feature vector for the block b_S
$\bar{S}$	SRT text
S_S	List of segment boundaries (SB) derived from SRT analysis
S_V	List of SB derived from Visual analysis
S_W	List of SB derived from Wikipedia analysis
S_F	List of SB derived from the fusion of different modalities

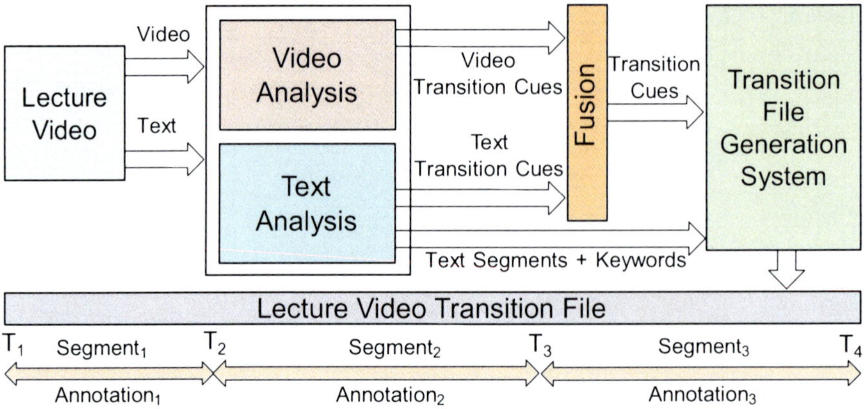

Fig. 6.1 System framework of the ATLAS system

technique. The SVM^{hmm} model predicts temporal transitions for a lecture video. In the next step, temporal transitions (TT_2) are derived from text (transcripts and slides) analysis using an N-gram based language model. Finally, TT_1 and TT_2 are fused by our algorithm to obtain a list of transition times for lecture videos. Moreover, text annotations corresponding to these temporal segments are determined by assigning the most frequent N-gram token of the SRT block under consideration (and similar to the N-gram token of slide titles, if available). Furthermore, our solution can help in recommending similar content to the users using text annotations as keywords for searching. Our initial experiments have confirmed that the ATLAS system recommends reasonable temporal segmentations for lecture videos. In this way, the proposed ATLAS system improves the automatic temporal segmentation of lecture videos so that online learning becomes much easier and users can search sections within a lecture video.

A specific topic of interest is often discussed in only a few minutes of a long lecture video recording. Therefore, the information requested by a user may be buried within a long video that is stored along with thousands of others. It is of ten relatively easy to find the relevant lecture video in an archive, but then the main challenge is to find the proper position within that video. Our goal is to produce a semantically meaningful segmentation of lecture videos appropriate for information retrieval in e-learning systems. Specifically, we target the lecture videos whose video qualities are not sufficiently high to allow robust visual segmentation. A large collection of lecture videos presents a unique set of challenges to a search system designer. SRT does not always provide an accurate index of segment boundaries corresponding to the visual content. Moreover, the performance of semantic extraction techniques based on visual content is often inadequate for segmentation and search tasks. Therefore, we postulate that a crowdsourced knowledge base such as Wikipedia can be very helpful in automatic lecture video segmentation since it provides several semantic contexts to analyze and divide lecture videos more accurately. To solve this problem, we propose the TRACE system which employs a linguistic-based approach for automatic lecture video segmentation using Wikipedia text.

The target lecture videos for TRACE are mainly videos whose video and/or SRT quality are not sufficiently good for segmenting videos automatically. We propose a novel approach to determine segment boundaries by matching blocks of SRT and Wikipedia texts of the topics of a lecture video. An overview of the method is as follows. First, we create feature vectors for Wikipedia blocks (one block for one Wikipedia topic) and SRT blocks (120 words in one SRT block) based on noun phrases in the entire Wikipedia texts. Next, we compute the similarity between a Wikipedia block and an SRT block using cosine similarity. Finally, the SRT block which has both the maximum cosine similarity and is above a similarity threshold $\overline{\chi}$ is considered as a segment boundary corresponding to the Wikipedia block. Empirical results in Sect. 6.3 confirm our intuition. To the best of our knowledge, this work is the first to attempt to segment lecture videos by leveraging a crowdsourced knowledge base such as Wikipedia. Moreover, combining Wikipedia with other segmenting techniques also shows significant improvements in the *recall* measure. Therefore, the segment boundaries computed from SRT using state-of-the-art [107] methods is further improved by refining these results using Wikipedia features. TRACE also works well for the detection of topic boundaries when only Wikipedia texts and the SRT of the lecture videos are available. Generally, the length of lecture videos ranges from 30 min to 2 h, and computing the visual and audio features is a very time-consuming process. Since TRACE is based on a linguistic approach, it does not require the computation of visual and audio features from video content and audio signals, respectively. Therefore, the TRACE system is scalable and executes very fast.

We use a supervised learning technique on video content and linguistic features with SRT, inspired by state-of-the-art methods to compute segment boundaries from video content [183] and SRT [107], respectively. Next, we compare these results with segment boundaries derived from our proposed method by leveraging Wikipedia texts [184]. To compute segment boundaries from SRT, we employ the linguistic method suggested in the state-of-the-art work by Lin et al. [107]. They used a noun phrase as a content-based feature, but other discourse-based features such as cue phrases are also employed as linguistic features to represent the topic transitions in SRT (see Sect. 6.2.3 for details). A color histogram of keyframes at each shot boundary is used as a visual feature to represent the slide transition in the video content to determine segment boundaries from the video content [183]. The relationship between the visual features and the segment boundary of a slide transition is established with a training dataset of lecture videos from VideoLectures.NET, using a machine-learning SVM^{hmm} technique. The SVM^{hmm} model predicts segment boundaries for lecture videos (see Sect. 6.2.1 for details).

Our systems are time-efficient and scale well to large repositories of lecture videos since both ATLAS and TRACE can determine segment boundaries offline rather than at search time. Results from experiments confirm that our systems recommend segment boundaries more accurately than existing state-of-the-art [107, 183] approaches on lecture video segmentation. We also investigated the effects of a late fusion of the segment boundaries determined from the different modalities such as visual, SRT, and Wikipedia content. We found that the proposed TRACE system improves the automatic temporal segmentation of lecture videos which facilitates online learning and users can accurately search sections within lecture videos.

The remaining parts of this chapter are organized as follows. In Sect. 6.2, we describe the ATLAS and TRACE systems. The evaluation results are presented in Sect. 6.3. Finally, we conclude the chapter with a summary in Sect. 6.4.

6.2 Lecture Video Segmentation

Our systems have several novel components which together form its innovative contributions (see Figs. 6.4 and 6.2 for the system frameworks). The ATLAS system performs the temporal segmentation and annotation of a lecture video in three steps. First, transition cues are predicted from the visual content, using supervised learning described in Sect. 6.2.1. Second, transition cues are computed from the available texts using an N-gram based language model described in Sect. 6.2.3. Finally, transition cues derived from the previous steps are fused to compute the final temporal transitions and annotations with text, as described in Sect. 6.2.5.

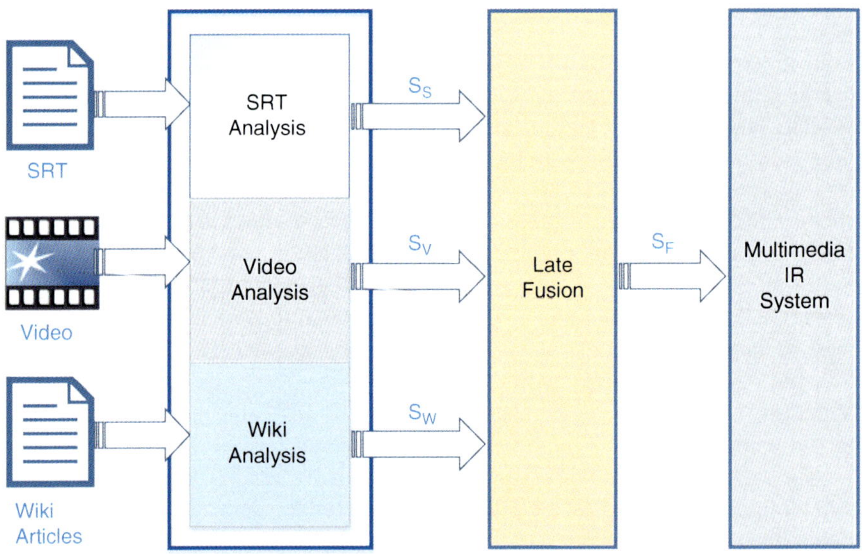

Fig. 6.2 Architecture for the late fusion of the segment boundaries derived from different modalities such as video content, SRT, and Wikipedia texts

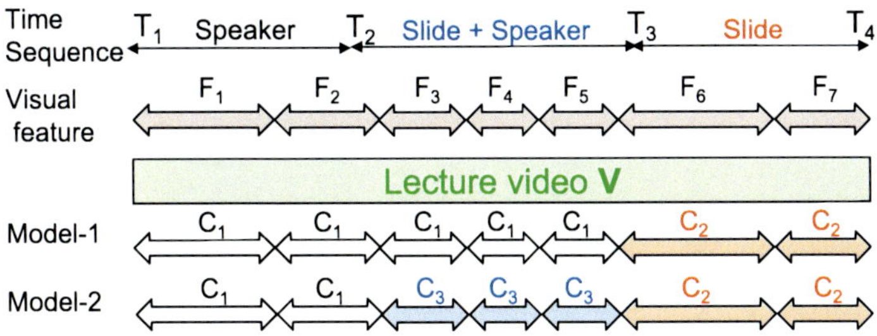

Fig. 6.3 Slide transition models

6.2.1 *Prediction of Video Transition Cues Using Supervised Learning*

A lecture video is composed of several shots combined with cuts and gradual transitions. Kucuktunc et al. [95] proposed a video segmentation approach based on fuzzy color histograms, which detects shot-boundaries. Therefore, we train two machine learning models using an SVM^{hmm} [75] technique by exploiting the color histograms (64-D) of keyframes to detect the slide transitions automatically in a lecture video. As described in the later Sect. 6.3.1, we use lecture videos (V_T) with known transition times as the test set and the remaining in the dataset as the training set. We employ human annotators to annotate ground truths for lecture videos in the training set (see Fig. 6.3 for an illustration of the annotation with both models).

First, an SVM^{hmm} model M_1 is trained with two classes C_1 and C_2. Class C_2 represents the segment of a lecture video when only a slideshow is visible (or the slideshow covers a major fraction of a frame), and class C_1 represents the remaining part of the video (see Model-1 in Fig. 6.3). Therefore, whenever a transition occurs from a sequence of classes C_1 (*i.e.*, from speaker only or, both speaker and slide) to C_2 (*i.e.*, slideshow only), it indicates a temporal transition with high probability in the majority of cases. However, we find that this model detects very few transitions (less than five transitions only) for some videos. We notice that there are mainly three reasons for this issue, first, when lecture videos are recorded with a single shot, second, when the transition occurs from a speaker to a slideshow but the speaker is still visible in the frame most of the time, and third, when the transition occurs between two slides only.

To resolve the above issues, we train another SVM^{hmm} model M_2 by adding an other class C_3, which represents the part of a video when a slideshow and a speaker are both visible. We use this model to predict transitions from only those videos for which M_1 predicted very few transitions. We do not use this model for all videos due to two reasons. First, the classification accuracy of M_1 is better than that of M_2 when there is a clear transition from C_1 to C_2. Second, we want to focus on only those videos which exhibit most of their transitions from C_1 to C_3 throughout the video (this is the reason M_1 was predicting very few transitions). Hence, a transition from a sequence of classes C_1 to C_3 is considered a slide transition for such kind of videos.

6.2.2 Computation of Text Transition Cues Using $\mathcal{N}$-Gram Based Language Model

6.2.2.1 Preparation

In the preparation step, we convert slides (a PDF file) of a lecture video to an HTML file using Adobe Acrobat software. However, this can be done with any other proprietary or open source software as well. The benefit of converting the PDF to an HTML file is that we obtain the text from slides along with their positions and font sizes, which are very important cues to determine the title of slides.

6.2.2.2 Title/Sub-Title Text Extraction

Algorithm 6.1 extracts titles/sub-titles from the HTML file derived from slides, which represent most of the slide titles of lecture videos accurately. A small variation of this algorithm produces the textual content of a slide by extracting the text between two consecutive title texts.

Algorithm 6.1 Title/sub-title text extraction from slides

1: **procedure** TITLEOFSLIDES
2: INPUT: A HTML file for slides (H)
3: OUTPUT: A list of title text T_{Text}
4: extractFontFreq(H, $fontList$, $freq$) ▷ this function finds all font and their frequency counts in slides.
5: $titleFontSize$ = findTitleFontSize($fontList$, $freq$) ▷ this function determines the font size of the title of slides.
6: $numSlides$ = findNumSlides($titleFontSize$) ▷ this function calculates the approx number of slides.
7: T_{Text} = findTitleText($titleFontSize$, $position$) ▷ this function determines the text for titles of all slides which located in top 1/3 of vertically or 2/3 of horizontally in slides.
8: **end procedure**

6.2.2.3 Transition Time Recommendation from SRT File

We employ an $\mathcal{N}$-gram based language model to calculate the relevance score R for every block of 30 tokens from an SRT file. We use a hash map to keep track of all $\mathcal{N}$-gram tokens and their respective term frequencies (TF). The relevance score is defined by the following equation:

$$R(\mathcal{B}_i) = \sum_{j=1}^{\mathcal{N}} \sum_{k=1}^{n} W_j * w(\bar{t}_k), \tag{6.1}$$

$$\text{and}\, w(\bar{t}_k) = \frac{TF(\bar{t}_k|\mathcal{B}_i) * \log(TF(\bar{t}_k|SRT) + 1)}{\log \frac{TF(\bar{t}_k|SRT)+1}{TF(\bar{t}_k|\mathcal{B}_i)}}, \tag{6.2}$$

where, $TF(\bar{t}_k|\mathcal{B}_i)$ is the TF of an $\mathcal{N}$-gram token $\bar{t}_k$ in a block $\mathcal{B}_i$ and $TF(\bar{t}_k|SRT)$ is the TF of the token $\bar{t}_k$ in the SRT file. $\mathcal{N}$ is the $\mathcal{N}$-gram count (we consider up to $\mathcal{N} = 3$, *i.e.*, trigram), W_j is the weight for different $\mathcal{N}$-gram counts such that the sum of all W_j is equal to one, and n is the number of unique tokens in the block $\mathcal{B}_i$. We place more importance to a higher order N-gram count by assigning high values to W_j in the relevance score equation.

If slides of a lecture video are available, then we calculate the approximate number of slides (N_{slides}) using the Algorithm 6.1. We consider the N_{slides} number of SRT blocks with the highest relevance scores to determine transitions using text analysis. We infer the start time of these blocks from the hashmap and designate them as the temporal transitions derived from the available texts.

6.2.3 *Computation of SRT Segment Boundaries Using a Linguistic-Based Approach*

Lin et al. [107] proposed a new video segmentation approach that used natural language processing techniques such as noun phrases extraction and utilized lexical knowledge sources such as WordNet. They used multiple linguistic-based segmentation features, including content-based features such as noun phrases and discourse based features such as cue phrases. They found that the noun phrase feature is salient in automatic lecture video segmentation. We implemented this state-of-the-art work [107] based on NLP techniques mentioned above to compute segment boundaries from a lecture video. We used Reconcile [198] to compute noun phrases from the available SRT texts. To compute the part of speech (POS) tags, we used the Stanford POS Tagger [204, 205] (see Sect. 1.4.5 for details). We used Porter stemmer [17] for stemming words. As suggested in the work [107], we used the block size of 120 words and shifted the window by 20 words every time. Subsequently, we computed cosine similarities between feature vectors of adjacent

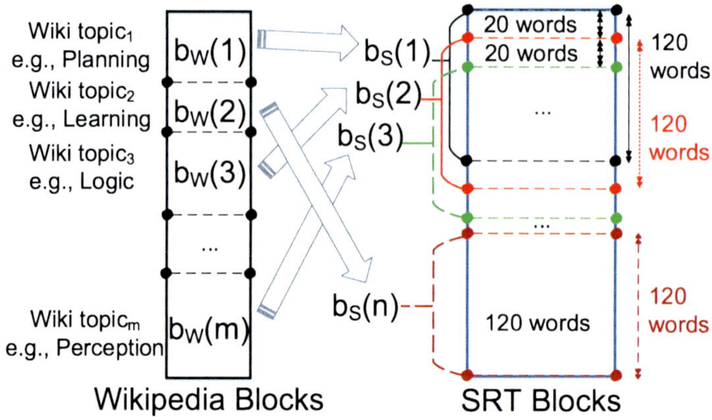

Fig. 6.4 Architecture for segment boundary detection using Wikipedia

windows by the standard formula $(\mathscr{I} \cdot \mathscr{J})/(\| \mathscr{I} \| + \| \mathscr{J} \|)$.$\mathscr{I}$ and $\mathscr{J}$ are the linguistic feature vectors for the adjacent SRT windows b_S. $\|\mathscr{I}\|$ and $\| \mathscr{J} \|$ are magnitude of the feature vectors.

6.2.4 Computation of Wikipedia Segment Boundaries

TRACE performs the temporal segmentation of a lecture video by leveraging SRT and Wikipedia texts using linguistic features. Figure 6.4 shows the system framework for segment boundaries detection from SRT using the proposed linguistic based method which leverages the Wikipedia texts of subjects. We assume that the subject (*e.g.*, Artificial Intelligence) of the lecture video is known. We used the Wikipedia API to find the related Wikipedia articles. Since a Wikipedia article consists of many topics, we parse the Wikipedia article to get texts of different topics. We refer to the block of texts corresponding to a Wikipedia topic as b_W. We determine the POS tags for Wikipedia texts and SRT of the lecture video. Next, we find a block b_S of 120 words from SRT which matches closely with the Wikipedia block b_W for each topic in Wikipedia texts. Specifically, first, we create a Wikipedia feature vector f_W for each Wikipedia topic and an SRT feature vector $\mathrm{f_S}$ for each SRT block b_S of 120 words based on noun phrases in the entire Wikipedia texts. Next, we compute the cosine similarity $\alpha(b_W, b_S)$ between a Wikipedia block b_W and all SRT blocks b_S. An SRT block with maximum cosine similarity is considered as a match for the given Wikipedia block b_W. We consider match only if the cosine similarity is above some threshold δ. Algorithm 6.2 describes the procedure for determining the segment boundaries using SRT and Wikipedia texts.

Algorithm 6.2 Computation of lecture video segments using SRT and Wikipedia texts

procedure WIKIPEDIABASEDTEMPORALSEGMENTATION
    INPUT: SRT $\overline{\mathscr{S}}$ for a lecture video
    OUTPUT: A list of segment boundaries S_W
    P = getWikiPages("*Artificial Inteligence*") ▷ Get Wikipedia page
    T = getTopics(P) ▷ Get topics from Wikipedia pages
    B_W = getWikiBlocks(P, T) ▷ List of Wikipedia blocks
    B_S = getSRTBlocks($\overline{\mathscr{S}}$, 120) ▷ List of SRT blocks
    for for each b_W in B_W **do** ▷ b_W is a Wikipedia block
        f_w = getWikiFeatVector(b_W) ▷ Wiki feature vector
        $\overline{\alpha} = 0$ ▷ Initialize cosine similarity value
        for for each b_S in B_S **do** ▷ b_S is a SRT block
            f_S = getSRTFeatVector(b_S) ▷ SRT feature vector
            $\alpha(b_W, b_S)$ = getCosineSimilarity(f_W, f_S)
            if $(\alpha(b_W, b_S) > \overline{\alpha}) \wedge (\alpha(b_W, b_S) > \delta)$ **then** ▷ Select b_S from SRT which has the highest cosine similarity with b_W
                $\overline{\alpha} = \alpha(b_W, b_S)$ ▷ $\overline{\alpha}$ is max. cosine similarity
                $b_{S'} = b_S$ ▷ $b_{S'}$ is a boundary block in SRT
            end if
        end for
        t_s = getStartTime($b_{s'}$) ▷ Get start time of the block $b_{S'}$
        S_W = addToSegmentBoundaries(t_s) ▷ S_W is a list of derived segment boundaries by leveraging Wikipedia text
    end for
    S_W = sortList(S_W) ▷ S_W is sorted in ascending order
end procedure

6.2.5 *Transition File Generation*

In our ATLAS system, we fuse the temporal transitions derived from the visual content and the speech transcript file by replacing two transitions less than 10 s apart by their average transitions time and keeping the remaining transitions as the final temporal transitions for the lecture video. Next, we compare $\mathcal{N}$-gram tokens of blocks corresponding to the final temporal transitions and calculate their similarity with $\mathcal{N}$-gram tokens derived from the title of slides. We assign the most similar $\mathcal{N}$-gram token of a block B_i as a text annotation A for a temporal segment which consists of B_i. If slides of lecture videos are not available, then an $\mathcal{N}$-gram token with high TF is assigned as a text annotation for the lecture segment.

In our TRACE system, we propose a novel method to compute the segment boundaries derived from a lecture video by leveraging the Wikipedia texts of the lecture video's subject. Next, we perform an empirical investigation of the late fusion of segment boundaries derived from state-of-the-art methods. Figure 6.2

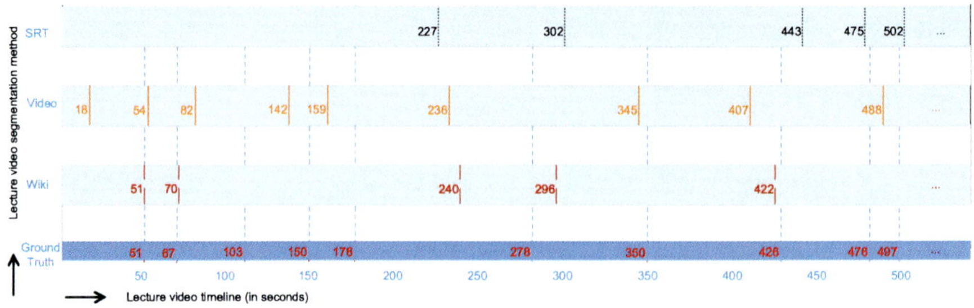

Fig. 6.5 The visualization of segment boundaries derived from different modalities

shows the system framework for the late fusion of the segment boundaries derived from different modalities. First, the segment boundaries are computed from SRT using the state-of-the-art work [107]. Second, the segment boundaries of the lecture video are predicted from the visual content using the supervised learning method described in the state-of-the-art work [37, 183]. Third, the segment boundaries are computed by leveraging Wikipedia texts using the proposed method (see Sect. 6.2.4). Finally, the segment boundaries are derived from the previous steps are fused as described in the earlier work [183] to compute the fused segment boundaries. The results of the late fusion in the TRACE system are summarized in Table 6.5. Figure 6.5 shows first a few segment boundaries derived from different modalities for a lecture video[1] from the test dataset.

6.3 Evaluation

6.3.1 Dataset and Experimental Settings

We used 133 videos with several metadata annotations such as speech transcriptions (SRT), slides, transition details (ground truths), *etc*., from the VideoLectures. NET and NPTEL. Specifically, we collected 65 videos[2] of different subjects from VideoLectures.NET. We evaluated the ATLAS system on VideoLectures.NET dataset $D_{LectureVideo.Net}$ by considering its 17 videos into the test set TS_{ATLAS} and rest of the videos into the train set. Furthermore, we collected 68 videos belonging to the *Artificial Intelligence* D_{NPTEL} course from NPTEL [3]. We evaluated the TRACE system using both $D_{LectureVideo.Net}$ and D_{NPTEL} datasets. We added the videos of D_{NPTEL} to test set and the videos of $D_{LectureVideo.Net}$ to train various models. Most of the videos in D_{NPTEL} are the old low-quality videos since the target videos for the TRACE system is the mainly old lecture videos in low video

[1]http://nptel.ac.in/courses/106105077/1

[2]This dataset is released as part of ACM International Conference on Multimedia Grand Challenge 2014. URL: http://acmmm.org/2014/docs/mm˙gc/MediaMixer.pdf

qualities. NPTEL and VideoLectures.NET provide transition files which contain details of all transitions for lecture videos. Therefore, details in the transition file are treated as ground truth for the lecture video segmentation task. We used the Wikipedia API [8] to determine texts for different courses and topics.

6.3.2 Results from the ATLAS System

The ATLAS system determines the temporal transitions and the corresponding annotations of lecture videos, with details described earlier in Sects. 6.2.1, 6.2.3, and 6.2.5. To evaluate the effectiveness of our approach, we compute precision, recall, and F1 scores for each video in TS_{ATLAS}. However, for a few videos in the test set, precision, recall, and F1 scores are very low because our SVM^{hmm} models are not able to detect transitions in lecture videos if lectures are recorded with a single shot, or without zoom in, zoom out, or when the slide transitions occur between two slides without any other change in the background. For example, precision and recall for the lecture video *cd07_eco_thu* are *zero*, since only a speaker is visible in the whole video except for a few seconds at the end when both the speaker and a slide consisting of an image with similar color as the background are visible. Therefore, for videos in which our machine learning techniques are not able to detect transitions, we determine transitions from analyzing the speech transcripts (and the text from slides if available) using an *N*-gram based language model as described in the earlier Sect. 6.2.3.

For an evaluation of the temporal segmentation, we connect one predicted transition time (PTT_i) with only one nearest actual transition time (ATT_j) from the provided transition files. It is possible that some PTT_i is not connected with any ATT_j and vice versa, as shown in Fig. 6.6. For example, PTT_4 and PTT_N are not connected with any actual transition time in *ATT*. Similarly, ATT_5 and ATT_6 are not connected with any predicted transition time in *PTT*. We refer to these PTT_i and ATT_j as *ExtraPTT* and *MissedATT*, respectively. We compute the score for each (PTT_i, ATT_j) pair based on the time difference between them, by employing a relaxed approach as depicted in Fig. 6.6 because it is very difficult to predict the same transition time at the granularity of seconds. Therefore, to evaluate the accuracy of the temporal segmentation, we use the following equations to compute precision and recall, and then compute F1 score using the standard formulas.

$$precision = \frac{\sum_{k=1}^{\Upsilon} score(PTT_i, ATT_j)}{|\ ATT\ |} \tag{6.3}$$

$$recall = \frac{\sum_{k=1}^{\Upsilon} score(PTT_i, ATT_j)}{|\ PTT\ |} \tag{6.4}$$

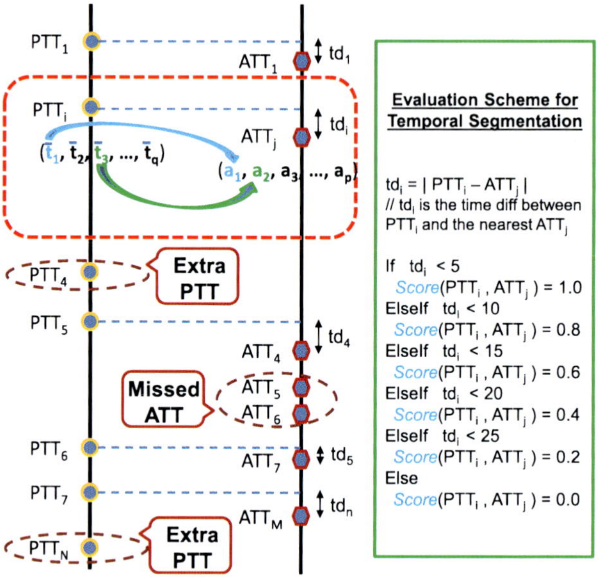

Fig. 6.6 The mapping of PTT, ATT and their respective text to calculate precision, recall and F1 scores

$$F1 = (2 \times \text{precision} \times \text{recall})/(\text{precision} + \text{recall}) \tag{6.5}$$

where $|ATT|$ is the cardinality of ATT, $|PTT|$ is the cardinality of $PTT\,s$ and Υ is the number of (PTT_i, ATT_j) pairs.

Tables 6.2, 6.3, and 6.4 show the precision, recall and F1 scores for the temporal segmentation of lecture videos, (I) when visual transition cues are predicted by our SVM^{hmm} models, (II) when text transition cues are predicted by our $\mathcal{N}$- gram based approach, and (III) when the visual transition cues are fused with text transition cues, respectively. Furthermore, it shows that the proposed scheme (III), improves the average *recall* much and the average *F1* slightly, compared with the other two schemes. Therefore, the transition cues determined from the text analysis are also very helpful, especially when the supervised learning fails to detect temporal transitions.

6.3.3 Results from the TRACE System

The TRACE system determines the segment boundaries of a lecture video with details described earlier in Sect. 6.2. Precision, recall, and F1 scores are important measures to examine the effectiveness of any systems in information retrieval. Similar to earlier work [183], we computed precision, recall, and F1 scores for each

Table 6.2 Evaluation of temporal segmentation based on visual features

Video name	Segmentation accuracy with visual transition cues (I)		
	Precision	Recall	F1 score
sparsemethods_01	0.536	0.728	0.618
scholkopf_kernel_01	0.434	0.451	0.442
denberghe_convex_01	0.573	0.487	0.526
bartok_games	0.356	0.246	0.291
abernethy_learning	0.511	0.192	0.279
agarwal_fgc	0.478	0.287	0.358
abernethy_strategy	0.600	0.235	0.338
cd07_eco_thu	0	0	–
szathmary_eol	0.545	0.988	0.702
2011_agarwal_model	0.350	0.088	0.140
2010_agarwal_itl	0.571	0.174	0.267
leskovec_mlg_01	0.492	0.451	0.471
taylor_kmsvm_01	0.650	0.325	0.433
green_bayesian_01	0.473	0.492	0.483
icml08_agarwal_mpg	0.200	0.012	0.023
nonparametrics_01	0.384	0.571	0.459
bubeck_games	0.655	0.465	0.543
Overall score	**0.459**	**0.364**	**0.375**

video in D_{NPTEL} to evaluate the effectiveness of our approach. For a few videos in D_{NPTEL}, these scores are very low due to the following reasons: (i) if a lecture video is recorded with a single shot, (ii) when the slide transitions occur between two slides alone, and (iii) if the video quality of the lecture video is low. Therefore, it is desirable to leverage crowdsourced knowledge bases such as Wikipedia. Specifically, it is advantageous to use Wikipedia features for videos in which machine learning techniques are not able to detect the segment boundaries since the video quality of such videos is not sufficiently high for the analysis. Moreover, it is desirable to investigate the fusion of the boundary segmentation results derived from different modalities. Therefore, we implemented the state-of-the-art methods of lecture video segmentation based on SRT [107] and video content analysis [37, 183].

For an evaluation of the lecture video segmentation, we computed precision, recall, and F-measure (F1 score) using the standard formula used for the ATLAS system. Similar to earlier work [107], we considered a perfect match if *PTT* and *ATT* are at most 30 s apart, and partial match if *PTT* and *ATT* are at most 120 s apart. We computed the score for each (*PTT*, *ATT*) pair based on the time difference between them by employing a staircase function as follows:

Table 6.3 Evaluation of temporal segmentation based on SRT features

Video name	Segmentation accuracy with text transition cues (II)		
	Precision	Recall	F1 score
sparsemethods_01	0.245	0.185	0.211
scholkopf_kernel_01	0.186	0.255	0.219
denberghe_convex_01	0.397	0.296	0.339
bartok_games	0.156	0.938	0.268
abernethy_learning	0.340	0.625	0.441
agarwal_fgc	0.440	0.367	0.400
abernethy_strategy	0.518	0.496	0.507
cd07_eco_thu	0.166	0.154	0.160
szathmary_eol	0.109	0.225	0.147
2011_agarwal_model	0.366	0.331	0.348
2010_agarwal_itl	0.371	0.339	0.354
leskovec_mlg_01	0.356	0.251	0.294
taylor_kmsvm_01	0.260	0.232	0.245
green_bayesian_01	0.362	0.353	0.357
icml08_agarwal_mpg	0.363	0.352	0.357
nonparametrics_01	0.231	0.331	0.272
bubeck_games	0.280	0.452	0.347
Overall score	**0.303**	**0.363**	**0.310**

Table 6.4 Evaluation of temporal segmentation based on fusion

Video name	Segmentation accuracy with fused transition cues (III)		
	Precision	Recall	F1 score
sparsemethods_01	0.393	0.638	0.486
scholkopf_kernel_01	0.258	0.506	0.341
denberghe_convex_01	0.452	0.496	0.473
bartok_games	0.169	0.831	0.281
abernethy_learning	0.379	0.600	0.465
agarwal_fgc	0.358	0.393	0.375
abernethy_strategy	0.500	0.435	0.465
cd07_eco_thu	0.183	0.154	0.167
szathmary_eol	0.307	0.825	0.447
2011_agarwal_model	0.366	0.331	0.348
2010_agarwal_itl	0.320	0.348	0.333
leskovec_mlg_01	0.397	0.419	0.408
taylor_kmsvm_01	0.391	0.489	0.435
green_bayesian_01	0.339	0.539	0.416
icml08_agarwal_mpg	0.500	0.121	0.190
nonparametrics_01	0.301	0.584	0.397
bubeck_games	0.379	0.574	0.456
Overall score	**0.352**	**0.487**	**0.381**

Table 6.5 Evaluation of the TRACE system [184] that introduced Wikipedia (Wiki, in short) for lecture video segmentations

Sr. No.	Segmentation method	Average precision	Average recall	Average F1 score
1	Visual [183]	0.360247	0.407794	0.322243
2	SRT [107]	0.348466	0.630344	0.423925
3	Visual [183] + SRT [107]	0.372229	0.578942	0.423925
4	Wikipedia [184]	**0.452257**	0.550133	**0.477073**
5	Visual [183] + Wikipedia [184]	0.396253	0.577951	0.436109
6	SRT [107] + Wikipedia [184]	0.388168	0.62403	0.455365
7	Visual [183] + SRT [107] + Wiki [184]	0.386877	**0.630717**	0.4391

Results in rows 1, 2, and 3 correspond to state-of-the-arts that derive segment boundaries from visual content [183] and speech transcript (SRT) [107]

$$score(PTT, ATT) = \begin{cases} 1.0, & \text{if } distance(PTT, ATT) \leq 30 \\ 0.5, & \text{else if } distance(PTT, ATT) \leq 120 \\ 0, & \text{otherwise.} \end{cases} \tag{6.6}$$

Table 6.5 shows the precision, recall and F1 scores of the lecture video segmentation for the TRACE system, state-of-the-art work, and their late fusion. We evaluated the segment boundaries computed from the video content and SRT using the state-of-the-art work. Moreover, we evaluated the segment boundaries computed from Wikipedia texts using our proposed method. Next, we evaluate the performance of the late fusion of the segment boundaries determined from different approaches. Experimental results show that our proposed scheme to determine segment boundaries by leveraging Wikipedia texts results in the highest precision and F1 scores. Specifically, the segment boundaries derived from the Wikipedia knowledge base outperforms state-of-the-arts regarding precision, *i.e.*, 25.54% and 29.78% better than approaches when only visual content [183] and speech transcript [107] are used in segment boundaries detection from lecture videos, respectively. Moreover, the segment boundaries derived from the Wikipedia knowledge base outperforms state-of-the-arts regarding F1 score, *i.e.*, 48.04% and 12.53% better than approaches when only visual content [183] and speech transcript [107] are used in segment boundaries detection from lecture videos, respectively. Furthermore, when we performed the late fusion of all approaches, then it results in the highest recall value. Therefore, the segment boundaries determined from the Wikipedia texts and its late fusion with other approaches are also very helpful, especially when the state-of-the-art methods based on the visual content and SRT fails to detect lecture video segmentations.

6.4 Summary

The proposed ATLAS and TRACE systems provide a novel and time-efficient way to automatically determine the segment boundaries of a lecture video by leveraging multimodal content such as visual content, SRT texts, and Wikipedia texts. To the best of our knowledge, our work is the first attempt to compute segment boundaries using crowdsourced knowledge base such as Wikipedia. We further investigated their fusion with the segment boundaries determined from the visual content and SRT of a lecture video using the state-of-the-art work. First, we determine the segment boundaries using visual content, SRT, and Wikipedia texts. Next, we perform a late fusion to determine the fused segment boundaries for the lecture video. Experimental results confirm that the TRACE system (*i.e.*, the segment boundaries derived from the Wikipedia knowledge base) can effectively segment the lecture video to facilitate the accessibility and traceability within its content although the video quality is not sufficiently high. Specifically, TRACE outperforms the segment boundaries detection based on only visual content [183] by 25.54% and 48.04% in terms of precision and F1 score, respectively. Moreover, it outperforms the segment boundaries detection based on only speech transcript [107] by 29.78% and 12.53% in terms of precision and F1 score, respectively. Finally, the fusion of segment boundaries derived from visual content, speech transcript, and Wikipedia knowledge base results in the highest recall score. Chapter 8 describes our future work that we plan to pursue. Specifically, we want to develop a SmartTutor systems that can give tuition to students based on their learning speeds, capabilities, and interests. That is, it can adaptively mold its teaching, style, content, and language to give the best tuition to its students. SmartTutor can have the capabilities of the ATLAS and TRACE systems to perform topic boundaries detection automatically. So that it can automatically get the video segments which are required for its students. Moreover, the capability of SmartTutor can be extended to develop a browsing tool for use and evaluation by students.

References

1. Apple Denies Steve Jobs Heart Attack Report: "It Is Not True". http://www.businessinsider.com/2008/10/apple-s-steve-jobs-rushed-to-er-after-heart-attack-says-cnn-citizen-journalist/. October 2008. Online: Last Accessed Sept 2015.
2. SVM^{hmm}: Sequence Tagging with Structural Support Vector Machines. https://www.cs.cornell.edu/people/tj/svm light/svm hmm.html. August 2008. Online: Last Accessed May 2016.
3. NPTEL. 2009, December. http://www.nptel.ac.in. Online; Accessed Apr 2015.
4. iReport at 5: Nearly 900,000 contributors worldwide. http://www.niemanlab.org/2011/08/ireport-at-5-nearly-900000-contributors-worldwide/. August 2011. Online: Last Accessed Sept 2015.

5. Meet the million: 999,999 iReporters + you! http://www.ireport.cnn.com/blogs/ireport-blog/2012/01/23/meet-the-million-999999-ireporters-you. January 2012. Online: Last Accessed Sept 2015.
6. 5 Surprising Stats about User-generated Content. 2014, April. http://www.smartblogs.com/social-media/2014/04/11/6.-surprising-stats-about-user-generated-content/. Online: Last Accessed Sept 2015.
7. The Citizen Journalist: How Ordinary People are Taking Control of the News. 2015, June. http://www.digitaltrends.com/features/the-citizen-journalist-how-ordinary-people-are-taking-control-of-the-news/. Online: Last Accessed Sept 2015.
8. Wikipedia API. 2015, April. http://tinyurl.com/WikiAPI-AI. API: Last Accessed Apr 2015.
9. Apache Lucene. 2016, June. https://lucene.apache.org/core/. Java API: Last Accessed June 2016.
10. By the Numbers: 14 Interesting Flickr Stats. 2016, May. http://www.expandedramblings.com/index.php/flickr-stats/. Online: Last Accessed May 2016.
11. By the Numbers: 180+ Interesting Instagram Statistics (June 2016). 2016, June. http://www.expandedramblings.com/index.php/important-instagram-stats/. Online: Last Accessed July 2016.
12. Coursera. 2016, May. https://www.coursera.org/. Online: Last Accessed May 2016.
13. FourSquare API. 2016, June. https://developer.foursquare.com/. Last Accessed June 2016.
14. Google Cloud Vision API. 2016, December. https://cloud.google.com/vision/. Online: Last Accessed Dec 2016.
15. Google Forms. 2016, May. https://docs.google.com/forms/. Online: Last Accessed May 2016.
16. MIT Open Course Ware. 2016, May. http://www.ocw.mit.edu/. Online: Last Accessed May 2016.
17. Porter Stemmer. 2016, May. https://tartarus.org/martin/PorterStemmer/. Online: Last Accessed May 2016.
18. SenticNet. 2016, May. http://www.sentic.net/computing/. Online: Last Accessed May 2016.
19. Sentics. 2016, May. https://en.wiktionary.org/wiki/sentics. Online: Last Accessed May 2016.
20. VideoLectures.Net. 2016, May. http://www.videolectures.net/. Online: Last Accessed May, 2016.
21. YouTube Statistics. 2016, July. http://www.youtube.com/yt/press/statistics.html. Online: Last Accessed July, 2016.
22. Abba, H.A., S.N.M. Shah, N.B. Zakaria, and A.J. Pal. 2012. Deadline based performance evaluation of job scheduling algorithms. In *Proceedings of the IEEE International Conference on Cyber-Enabled Distributed Computing and Knowledge Discovery*, 106–110.
23. Achanta, R.S., W.-Q. Yan, and M.S. Kankanhalli. 2006. Modeling Intent for Home Video Repurposing. *Proceedings of the IEEE MultiMedia* 45(1): 46–55.
24. Adcock, J., M. Cooper, A. Girgensohn, and L. Wilcox. 2005. Interactive Video Search Using Multilevel Indexing. In *Proceedings of the Springer Image and Video Retrieval*, 205–214.
25. Agarwal, B., S. Poria, N. Mittal, A. Gelbukh, and A. Hussain. 2015. Concept-level Sentiment Analysis with Dependency-based Semantic Parsing: A Novel Approach. In *Proceedings of the Springer Cognitive Computation,* 1–13.
26. Aizawa, K., D. Tancharoen, S. Kawasaki, and T. Yamasaki. 2004. Efficient Retrieval of Life Log based on Context and Content. In *Proceedings of the ACM Workshop on Continuous Archival and Retrieval of Personal Experiences*, 22–31.
27. Altun, Y., I. Tsochantaridis, and T. Hofmann. 2003. Hidden Markov Support Vector Machines. In *Proceedings of the International Conference on Machine Learning*, 3–10.
28. Anderson, A., K. Ranghunathan, and A. Vogel. 2008. Tagez: Flickr Tag Recommendation. In *Proceedings of the Association for the Advancement of Artificial Intelligence*.
29. Atrey, P.K., A. El Saddik, and M.S. Kankanhalli. 2011. Effective Multimedia Surveillance using a Human-centric Approach. *Proceedings of the Springer Multimedia Tools and Applications* 51(2): 697–721.

30. Barnard, K., P. Duygulu, D. Forsyth, N. De Freitas, D.M. Blei, and M.I. Jordan. 2003. Matching Words and Pictures. *Proceedings of the Journal of Machine Learning Research* 3: 1107–1135.
31. Basu, S., Y. Yu, V.K. Singh, and R. Zimmermann. 2016. Videopedia: Lecture Video Recommendation for Educational Blogs Using Topic Modeling. In *Proceedings of the Springer International Conference on Multimedia Modeling*, 238–250.
32. Basu, S., Y. Yu, and R. Zimmermann. 2016. Fuzzy Clustering of Lecture Videos Based on Topic Modeling. In *Proceedings of the IEEE International Workshop on Content-Based Multimedia Indexing*, 1–6.
33. Basu, S., R. Zimmermann, K.L. OHalloran, S. Tan, and K. Marissa. 2015. Performance Evaluation of Students Using Multimodal Learning Systems. In *Proceedings of the Springer International Conference on Multimedia Modeling*, 135–147.
34. Beeferman, D., A. Berger, and J. Lafferty. 1999. Statistical Models for Text Segmentation. *Proceedings of the Springer Machine Learning* 34(1–3): 177–210.
35. Bernd, J., D. Borth, C. Carrano, J. Choi, B. Elizalde, G. Friedland, L. Gottlieb, K. Ni, R. Pearce, D. Poland, et al. 2015. Kickstarting the Commons: The YFCC100M and the YLI Corpora. In *Proceedings of the ACM Workshop on Community-Organized Multimodal Mining: Opportunities for Novel Solutions*, 1–6.
36. Bhatt, C.A., and M.S. Kankanhalli. 2011. Multimedia Data Mining: State of the Art and Challenges. *Proceedings of the Multimedia Tools and Applications* 51(1): 35–76.
37. Bhatt, C.A., A. Popescu-Belis, M. Habibi, S. Ingram, S. Masneri, F. McInnes, N. Pappas, and O. Schreer. 2013. Multi-factor Segmentation for Topic Visualization and Recommendation: the MUST-VIS System. In *Proceedings of the ACM International Conference on Multimedia*, 365–368.
38. Bhattacharjee, S., W.C. Cheng, C.-F. Chou, L. Golubchik, and S. Khuller. 2000. BISTRO: A Framework for Building Scalable Wide-Area Upload Applications. *Proceedings of the ACM SIGMETRICS Performance Evaluation Review* 28(2): 29–35.
39. Cambria, E., J. Fu, F. Bisio, and S. Poria. 2015. AffectiveSpace 2: Enabling Affective Intuition for Concept-level Sentiment Analysis. In *Proceedings of the AAAI Conference on Artificial Intelligence*, 508–514.
40. Cambria, E., A. Livingstone, and A. Hussain. 2012. The Hourglass of Emotions. In *Proceedings of the Springer Cognitive Behavioural Systems*, 144–157.
41. Cambria, E., D. Olsher, and D. Rajagopal. 2014. SenticNet 3: A Common and Common-sense Knowledge Base for Cognition-driven Sentiment Analysis. In *Proceedings of the AAAI Conference on Artificial Intelligence*, 1515–1521.
42. Cambria, E., S. Poria, R. Bajpai, and B. Schuller. 2016. SenticNet 4: A Semantic Resource for Sentiment Analysis based on Conceptual Primitives. In *Proceedings of the International Conference on Computational Linguistics (COLING)*, 2666–2677.
43. Cambria, E., S. Poria, F. Bisio, R. Bajpai, and I. Chaturvedi. 2015. The CLSA Model: A Novel Framework for Concept-Level Sentiment Analysis. In *Proceedings of the Springer Computational Linguistics and Intelligent Text Processing*, 3–22.
44. Cambria, E., S. Poria, A. Gelbukh, and K. Kwok. 2014. Sentic API: A Common-sense based API for Concept-level Sentiment Analysis. *CEUR Workshop Proceedings* 144: 19–24.
45. Cao, J., Z. Huang, and Y. Yang. 2015. Spatial-aware Multimodal Location Estimation for Social Images. In *Proceedings of the ACM Conference on Multimedia Conference*, 119–128.
46. Chakraborty, I., H. Cheng, and O. Javed. 2014. Entity Centric Feature Pooling for Complex Event Detection. In *Proceedings of the Workshop on HuEvent at the ACM International Conference on Multimedia*, 1–5.
47. Che, X., H. Yang, and C. Meinel. 2013. Lecture Video Segmentation by Automatically Analyzing the Synchronized Slides. In *Proceedings of the ACM International Conference on Multimedia*, 345–348.

48. Chen, B., J. Wang, Q. Huang, and T. Mei. 2012. Personalized Video Recommendation through Tripartite Graph Propagation. In *Proceedings of the ACM International Conference on Multimedia*, 1133–1136.
49. Chen, S., L. Tong, and T. He. 2011. Optimal Deadline Scheduling with Commitment. In *Proceedings of the IEEE Annual Allerton Conference on Communication, Control, and Computing*, 111–118.
50. Chen, W.-B., C. Zhang, and S. Gao. 2012. Segmentation Tree based Multiple Object Image Retrieval. In *Proceedings of the IEEE International Symposium on Multimedia*, 214–221.
51. Chen, Y., and W.J. Heng. 2003. Automatic Synchronization of Speech Transcript and Slides in Presentation. *Proceedings of the IEEE International Symposium on Circuits and Systems* 2: 568–571.
52. Cohen, J. 1960. A Coefficient of Agreement for Nominal Scales. *Proceedings of the Durham Educational and Psychological Measurement* 20(1): 37–46.
53. Cristani, M., A. Pesarin, C. Drioli, V. Murino, A. Rodà,M. Grapulin, and N. Sebe. 2010. Toward an Automatically Generated Soundtrack from Low-level Cross-modal Correlations for Automotive Scenarios. In *Proceedings of the ACM International Conference on Multimedia*, 551–560.
54. Dang-Nguyen, D.-T., L. Piras, G. Giacinto, G. Boato, and F.G. De Natale. 2015. A Hybrid Approach for Retrieving Diverse Social Images of Landmarks. In *Proceedings of the IEEE International Conference on Multimedia and Expo (ICME)*, 1–6.
55. Fabro, M. Del, A. Sobe, and L. Böszörmenyi. 2012. Summarization of Real-life Events Based on Community-contributed Content. In *Proceedings of the International Conferences on Advances in Multimedia*, 119–126.
56. Du, L., W.L. Buntine, and M. Johnson. 2013. Topic Segmentation with a Structured Topic Model. In *Proceedings of the North American Chapter of the Association for Computational Linguistics: Human Language Technologies*, 190–200.
57. Fan, Q., K. Barnard, A. Amir, A. Efrat, and M. Lin. 2006. Matching Slides to Presentation Videos using SIFT and Scene Background Matching. In *Proceedings of the ACM International Conference on Multimedia*, 239–248.
58. Filatova, E. and V. Hatzivassiloglou. 2004. Event-Based Extractive Summarization. In *Proceedings of the ACL Workshop on Summarization*, 104–111.
59. Firan, C.S., M. Georgescu, W. Nejdl, and R. Paiu. 2010. Bringing Order to Your Photos: Event-driven Classification of Flickr Images Based on Social Knowledge. In *Proceedings of the ACM International Conference on Information and Knowledge Management*, 189–198.
60. Gao, S., C. Zhang, and W.-B. Chen. 2012. An Improvement of Color Image Segmentation through Projective Clustering. In *Proceedings of the IEEE International Conference on Information Reuse and Integration*, 152–158.
61. Garg, N. and I. Weber. 2008. Personalized, Interactive Tag Recommendation for Flickr. In *Proceedings of the ACM Conference on Recommender Systems*, 67–74.
62. Ghias, A., J. Logan, D. Chamberlin, and B.C. Smith. 1995. Query by Humming: Musical Information Retrieval in an Audio Database. In *Proceedings of the ACM International Conference on Multimedia*, 231–236.
63. Golder, S.A., and B.A. Huberman. 2006. Usage Patterns of Collaborative Tagging Systems. *Proceedings of the Journal of Information Science* 32(2): 198–208.
64. Gozali, J.P., M.-Y. Kan, and H. Sundaram. 2012. Hidden Markov Model for Event Photo Stream Segmentation. In *Proceedings of the IEEE International Conference on Multimedia and Expo Workshops*, 25–30.
65. Guo, Y., L. Zhang, Y. Hu, X. He, and J. Gao. 2016. Ms-celeb-1m: Challenge of recognizing one million celebrities in the real world. *Proceedings of the Society for Imaging Science and Technology Electronic Imaging* 2016(11): 1–6.
66. Hanjalic, A., and L.-Q. Xu. 2005. Affective Video Content Representation and Modeling. *Proceedings of the IEEE Transactions on Multimedia* 7(1): 143–154.

67. Haubold, A. and J.R. Kender. 2005. Augmented Segmentation and Visualization for Presentation Videos. In *Proceedings of the ACM International Conference on Multimedia*, 51–60.
68. Healey, J.A., and R.W. Picard. 2005. Detecting Stress during Real-world Driving Tasks using Physiological Sensors. *Proceedings of the IEEE Transactions on Intelligent Transportation Systems* 6(2): 156–166.
69. Hefeeda, M., and C.-H. Hsu. 2010. On Burst Transmission Scheduling in Mobile TV Broadcast Networks. *Proceedings of the IEEE/ACM Transactions on Networking* 18(2): 610–623.
70. Hevner, K. 1936. Experimental Studies of the Elements of Expression in Music. *Proceedings of the American Journal of Psychology* 48: 246–268.
71. Hochbaum, D.S.. 1996. Approximating Covering and Packing Problems: Set Cover, Vertex Cover, Independent Set, and related Problems. In *Proceedings of the PWS Approximation algorithms for NP-hard problems*, 94–143.
72. Hong, R., J. Tang, H.-K. Tan, S. Yan, C. Ngo, and T.-S. Chua. 2009. Event Driven Summarization for Web Videos. In *Proceedings of the ACM SIGMM Workshop on Social Media*, 43–48.
73. P. ITU-T Recommendation. 1999. Subjective Video Quality Assessment Methods for Multimedia Applications.
74. Jiang, L., A.G. Hauptmann, and G. Xiang. 2012. Leveraging High-level and Low-level Features for Multimedia Event Detection. In *Proceedings of the ACM International Conference on Multimedia*, 449–458.
75. Joachims, T., T. Finley, and C.-N. Yu. 2009. Cutting-plane Training of Structural SVMs. *Proceedings of the Machine Learning Journal* 77(1): 27–59.
76. Johnson, J., L. Ballan, and L. Fei-Fei. 2015. Love Thy Neighbors: Image Annotation by Exploiting Image Metadata. In *Proceedings of the IEEE International Conference on Computer Vision*, 4624–4632.
77. Johnson, J., A. Karpathy, and L. Fei-Fei. 2015. Densecap: Fully Convolutional Localization Networks for Dense Captioning. *In Proceedings of the arXiv preprint arXiv:1511.07571*.
78. Jokhio, F., A. Ashraf, S. Lafond, I. Porres, and J. Lilius. 2013. Prediction-Based dynamic resource allocation for video transcoding in cloud computing. In *Proceedings of the IEEE International Conference on Parallel, Distributed and Network-Based Processing*, 254–261.
79. Kaminskas, M., I. Fernández-Tobías, F. Ricci, and I. Cantador. 2014. Knowledge-Based Identification of Music Suited for Places of Interest. *Proceedings of the Springer Information Technology & Tourism* 14(1): 73–95.
80. Kaminskas, M. and F. Ricci. 2011. Location-adapted Music Recommendation using Tags. In *Proceedings of the Springer User Modeling, Adaption and Personalization*, 183–194.
81. Kan, M.-Y.. 2001. Combining Visual Layout and Lexical Cohesion Features for Text Segmentation. *In Proceedings of the Citeseer*.
82. Kan, M.-Y.. 2003. *Automatic Text Summarization as Applied to Information Retrieval*. PhD thesis, Columbia University.
83. Kan, M.-Y., J.L. Klavans, and K.R. McKeown. 1998. Linear Segmentation and Segment Significance. *In Proceedings of the arXiv preprint cs/9809020*.
84. Kan, M.-Y., K.R. McKeown, and J.L. Klavans. 2001. Applying Natural Language Generation to Indicative Summarization. *Proceedings of the ACL European Workshop on Natural Language Generation* 8: 1–9.
85. Kang, H.B.. 2003. Affective Content Detection using HMMs. In *Proceedings of the ACM International Conference on Multimedia*, 259–262.
86. Kang, Y.-L., J.-H. Lim, M.S. Kankanhalli, C.-S. Xu, and Q. Tian. 2004. Goal Detection in Soccer Video using Audio/Visual Keywords. *Proceedings of the IEEE International Conference on Image Processing* 3: 1629–1632.
87. Kang, Y.-L., J.-H. Lim, Q. Tian, and M.S. Kankanhalli. 2003. Soccer Video Event Detection with Visual Keywords. *Proceedings of the Joint Conference of International Conference on*

Information, Communications and Signal Processing, and Pacific Rim Conference on Multimedia 3: 1796–1800.
88. Kankanhalli, M.S., and T.-S. Chua. 2000. Video Modeling using Strata-Based Annotation. *Proceedings of the IEEE MultiMedia* 7(1): 68–74.
89. Kennedy, L., M. Naaman, S. Ahern, R. Nair, and T. Rattenbury. 2007. How Flickr Helps us Make Sense of the World: Context and Content in Community-Contributed Media Collections. In *Proceedings of the ACM International Conference on Multimedia*, 631–640.
90. Kennedy, L.S., S.-F. Chang, and I.V. Kozintsev. 2006. To Search or to Label?: Predicting the Performance of Search-Based Automatic Image Classifiers. In *Proceedings of the ACM International Workshop on Multimedia Information Retrieval*, 249–258.
91. Kim, Y.E., E.M. Schmidt, R. Migneco, B.G. Morton, P. Richardson, J. Scott, J.A. Speck, and D. Turnbull. 2010. Music Emotion Recognition: A State of the Art Review. In *Proceedings of the International Society for Music Information Retrieval*, 255–266.
92. Klavans, J.L., K.R. McKeown, M.-Y. Kan, and S. Lee. 1998. Resources for Evaluation of Summarization Techniques. *In Proceedings of the arXiv preprint cs/9810014*.
93. Ko, Y.. 2012. A Study of Term Weighting Schemes using Class Information for Text Classification. In *Proceedings of the ACM Special Interest Group on Information Retrieval*, 1029–1030.
94. Kort, B., R. Reilly, and R.W. Picard. 2001. An Affective Model of Interplay between Emotions and Learning: Reengineering Educational Pedagogy-Building a Learning Companion. *Proceedings of the IEEE International Conference on Advanced Learning Technologies* 1: 43–47.
95. Kucuktunc, O., U. Gudukbay, and O. Ulusoy. 2010. Fuzzy Color Histogram-Based Video Segmentation. *Proceedings of the Computer Vision and Image Understanding* 114(1): 125–134.
96. Kuo, F.-F., M.-F. Chiang, M.-K. Shan, and S.-Y. Lee. 2005. Emotion-Based Music Recommendation by Association Discovery from Film Music. In *Proceedings of the ACM International Conference on Multimedia*, 507–510.
97. Lacy, S., T. Atwater, X. Qin, and A. Powers. 1988. Cost and Competition in the Adoption of Satellite News Gathering Technology. *Proceedings of the Taylor & Francis Journal of Media Economics* 1(1): 51–59.
98. Lambert, P., W. De Neve, P. De Neve, I. Moerman, P. Demeester, and R. Van de Walle. 2006. Rate-distortion performance of H. 264/AVC compared to state-of-the-art video codecs. *Proceedings of the IEEE Transactions on Circuits and Systems for Video Technology* 16(1): 134–140.
99. Laurier, C., M. Sordo, J. Serrà, and P. Herrera. 2009. Music Mood Representations from Social Tags. In *Proceedings of the International Society for Music Information Retrieval*, 381–386.
100. Li, C.T. and M.K. Shan. 2007. Emotion-Based Impressionism Slideshow with Automatic Music Accompaniment. In *Proceedings of the ACM International Conference on Multimedia*, 839–842.
101. Li, J., and J.Z. Wang. 2008. Real-Time Computerized Annotation of Pictures. *Proceedings of the IEEE Transactions on Pattern Analysis and Machine Intelligence* 30(6): 985–1002.
102. Li, X., C.G. Snoek, and M. Worring. 2009. Learning Social Tag Relevance by Neighbor Voting. *Proceedings of the IEEE Transactions on Multimedia* 11(7): 1310–1322.
103. Li, X., T. Uricchio, L. Ballan, M. Bertini, C.G. Snoek, and A.D. Bimbo. 2016. Socializing the Semantic Gap: A Comparative Survey on Image Tag Assignment, Refinement, and Retrieval. *Proceedings of the ACM Computing Surveys (CSUR)* 49(1): 14.
104. Li, Z., Y. Huang, G. Liu, F. Wang, Z.-L. Zhang, and Y. Dai. 2012. Cloud Transcoder: Bridging the Format and Resolution Gap between Internet Videos and Mobile Devices. In *Proceedings of the ACM International Workshop on Network and Operating System Support for Digital Audio and Video*, 33–38.

105. Liang, C., Y. Guo, and Y. Liu. 2008. Is Random Scheduling Sufficient in P2P Video Streaming? In *Proceedings of the IEEE International Conference on Distributed Computing Systems*, 53–60. IEEE.
106. Lim, J.-H., Q. Tian, and P. Mulhem. 2003. Home Photo Content Modeling for Personalized Event-Based Retrieval. *Proceedings of the IEEE MultiMedia* 4: 28–37.
107. Lin, M., M. Chau, J. Cao, and J.F. Nunamaker Jr. 2005. Automated Video Segmentation for Lecture Videos: A Linguistics-Based Approach. *Proceedings of the IGI Global International Journal of Technology and Human Interaction* 1(2): 27–45.
108. Liu, C.L., and J.W. Layland. 1973. Scheduling Algorithms for Multiprogramming in a Hard-real-time Environment. *Proceedings of the ACM Journal of the ACM* 20(1): 46–61.
109. Liu, D., X.-S. Hua, L. Yang, M. Wang, and H.-J. Zhang. 2009. Tag Ranking. In *Proceedings of the ACM World Wide Web Conference*, 351–360.
110. Liu, T., C. Rosenberg, and H.A. Rowley. 2007. Clustering Billions of Images with Large Scale Nearest Neighbor Search. In *Proceedings of the IEEE Winter Conference on Applications of Computer Vision*, 28–28.
111. Liu, X. and B. Huet. 2013. Event Representation and Visualization from Social Media. In *Proceedings of the Springer Pacific-Rim Conference on Multimedia*, 740–749.
112. Liu, Y., D. Zhang, G. Lu, and W.-Y. Ma. 2007. A Survey of Content-Based Image Retrieval with High-level Semantics. *Proceedings of the Elsevier Pattern Recognition* 40(1): 262–282.
113. Livingston, S., and D.A.V. Belle. 2005. The Effects of Satellite Technology on Newsgathering from Remote Locations. *Proceedings of the Taylor & Francis Political Communication* 22(1): 45–62.
114. Long, R., H. Wang, Y. Chen, O. Jin, and Y. Yu. 2011. Towards Effective Event Detection, Tracking and Summarization on Microblog Data. In *Proceedings of the Springer Web-Age Information Management*, 652–663.
115. L. Lu, H. You, and H. Zhang. 2001. A New Approach to Query by Humming in Music Retrieval. In *Proceedings of the IEEE International Conference on Multimedia and Expo*, 22–25.
116. Lu, Y., H. To, A. Alfarrarjeh, S.H. Kim, Y. Yin, R. Zimmermann, and C. Shahabi. 2016. GeoUGV: User-generated Mobile Video Dataset with Fine Granularity Spatial Metadata. In *Proceedings of the ACM International Conference on Multimedia Systems*, 43.
117. Mao, J., W. Xu, Y. Yang, J. Wang, Z. Huang, and A. Yuille. 2014. Deep Captioning with Multimodal Recurrent Neural Networks (M-RNN). *In Proceedings of the arXiv preprint arXiv:1412.6632*.
118. Matusiak, K.K. 2006. Towards User-Centered Indexing in Digital Image Collections. *Proceedings of the OCLC Systems & Services: International Digital Library Perspectives* 22(4): 283–298.
119. McDuff, D., R. El Kaliouby, E. Kodra, and R. Picard. 2013. Measuring Voter's Candidate Preference Based on Affective Responses to Election Debates. In *Proceedings of the IEEE Humaine Association Conference on Affective Computing and Intelligent Interaction*, 369–374.
120. McKeown, K.R., J.L. Klavans, and M.-Y. Kan. Method and System for Topical Segmentation, Segment Significance and Segment Function, 29 2002. US Patent 6,473,730.
121. Mezaris, V., A. Scherp, R. Jain, M. Kankanhalli, H. Zhou, J. Zhang, L. Wang, and Z. Zhang. 2011. Modeling and Rrepresenting Events in Multimedia. In *Proceedings of the ACM International Conference on Multimedia*, 613–614.
122. Mezaris, V., A. Scherp, R. Jain, and M.S. Kankanhalli. 2014. Real-life Events in Multimedia: Detection, Representation, Retrieval, and Applications. *Proceedings of the Springer Multimedia Tools and Applications* 70(1): 1–6.
123. Miller, G., and C. Fellbaum. 1998. *Wordnet: An Electronic Lexical Database*. Cambridge, MA: MIT Press.
124. Miller, G.A. 1995. WordNet: A Lexical Database for English. *Proceedings of the Communications of the ACM* 38(11): 39–41.

125. Moxley, E., J. Kleban, J. Xu, and B. Manjunath. 2009. Not All Tags are Created Equal: Learning Flickr Tag Semantics for Global Annotation. In *Proceedings of the IEEE International Conference on Multimedia and Expo*, 1452–1455.
126. Mulhem, P., M.S. Kankanhalli, J. Yi, and H. Hassan. 2003. Pivot Vector Space Approach for Audio-Video Mixing. *Proceedings of the IEEE MultiMedia* 2: 28–40.
127. Naaman, M. 2012. Social Multimedia: Highlighting Opportunities for Search and Mining of Multimedia Data in Social Media Applications. *Proceedings of the Springer Multimedia Tools and Applications* 56(1): 9–34.
128. Natarajan, P., P.K. Atrey, and M. Kankanhalli. 2015. Multi-Camera Coordination and Control in Surveillance Systems: A Survey. *Proceedings of the ACM Transactions on Multimedia Computing, Communications, and Applications* 11(4): 57.
129. Nayak, M.G. 2004. *Music Synthesis for Home Videos*. PhD thesis.
130. Neo, S.-Y., J. Zhao, M.-Y. Kan, and T.-S. Chua. 2006. Video Retrieval using High Level Features: Exploiting Query Matching and Confidence-Based Weighting. In *Proceedings of the Springer International Conference on Image and Video Retrieval*, 143–152.
131. Ngo, C.-W., F. Wang, and T.-C. Pong. 2003. Structuring Lecture Videos for Distance Learning Applications. In *Proceedings of the IEEE International Symposium on Multimedia Software Engineering*, 215–222.
132. Nguyen, V.-A., J. Boyd-Graber, and P. Resnik. 2012. SITS: A Hierarchical Nonparametric Model using Speaker Identity for Topic Segmentation in Multiparty Conversations. In *Proceedings of the Annual Meeting of the Association for Computational Linguistics*, 78–87.
133. Nwana, A.O. and T. Chen. 2016. Who Ordered This?: Exploiting Implicit User Tag Order Preferences for Personalized Image Tagging. *In Proceedings of the arXiv preprint arXiv:1601.06439*.
134. Papagiannopoulou, C. and V. Mezaris. 2014. Concept-Based Image Clustering and Summarization of Event-related Image Collections. In *Proceedings of the Workshop on HuEvent at the ACM International Conference on Multimedia*, 23–28.
135. Park, M.H., J.H. Hong, and S.B. Cho. 2007. Location-Based Recommendation System using Bayesian User's Preference Model in Mobile Devices. In *Proceedings of the Springer Ubiquitous Intelligence and Computing*, 1130–1139.
136. Petkos, G., S. Papadopoulos, V. Mezaris, R. Troncy, P. Cimiano, T. Reuter, and Y. Kompatsiaris. 2014. Social Event Detection at MediaEval: a Three-Year Retrospect of Tasks and Results. In *Proceedings of the Workshop on Social Events in Web Multimedia at ACM International Conference on Multimedia Retrieval*.
137. Pevzner, L., and M.A. Hearst. 2002. A Critique and Improvement of an Evaluation Metric for Text Segmentation. *Proceedings of the Computational Linguistics* 28(1): 19–36.
138. Picard, R.W., and J. Klein. 2002. Computers that Recognise and Respond to User Emotion: Theoretical and Practical Implications. *Proceedings of the Interacting with Computers* 14(2): 141–169.
139. Picard, R.W., E. Vyzas, and J. Healey. 2001. Toward machine emotional intelligence: Analysis of affective physiological state. *Proceedings of the IEEE Transactions on Pattern Analysis and Machine Intelligence* 23(10): 1175–1191.
140. Poisson, S.D. and C.H. Schnuse. 1841. Recherches Sur La Pprobabilité Des Jugements En Mmatieré Criminelle Et En Matieré Civile. Meyer.
141. Poria, S., E. Cambria, R. Bajpai, and A. Hussain. 2017. A Review of Affective Computing: From Unimodal Analysis to Multimodal Fusion. *Proceedings of the Elsevier Information Fusion* 37: 98–125.
142. Poria, S., E. Cambria, and A. Gelbukh. 2016. Aspect Extraction for Opinion Mining with a Deep Convolutional Neural Network. *Proceedings of the Elsevier Knowledge-Based Systems* 108: 42–49.
143. Poria, S., E. Cambria, A. Gelbukh, F. Bisio, and A. Hussain. 2015. Sentiment Data Flow Analysis by Means of Dynamic Linguistic Patterns. *Proceedings of the IEEE Computational Intelligence Magazine* 10(4): 26–36.

144. Poria, S., E. Cambria, and A.F. Gelbukh. 2015. Deep Convolutional Neural Network Textual Features and Multiple Kernel Learning for Utterance-level Multimodal Sentiment Analysis. In *Proceedings of the EMNLP*, 2539–2544.
145. Poria, S., E. Cambria, D. Hazarika, N. Mazumder, A. Zadeh, and L.-P. Morency. 2017. Context-Dependent Sentiment Analysis in User-Generated Videos. *In Proceedings of the Association for Computational Linguistics*.
146. Poria, S., E. Cambria, D. Hazarika, and P. Vij. 2016. A Deeper Look into Sarcastic Tweets using Deep Convolutional Neural Networks. In *Proceedings of the International Conference on Computational Linguistics (COLING)*.
147. Poria, S., E. Cambria, N. Howard, G.-B. Huang, and A. Hussain. 2016. Fusing Audio Visual and Textual Clues for Sentiment Analysis from Multimodal Content. *Proceedings of the Elsevier Neurocomputing* 174: 50–59.
148. Poria, S., E. Cambria, N. Howard, and A. Hussain. 2015. Enhanced SenticNet with Affective Labels for Concept-Based Opinion Mining: Extended Abstract. In *Proceedings of the International Joint Conference on Artificial Intelligence*.
149. Poria, S., E. Cambria, A. Hussain, and G.-B. Huang. 2015. Towards an Intelligent Framework for Multimodal Affective Data Analysis. *Proceedings of the Elsevier Neural Networks* 63: 104–116.
150. Poria, S., E. Cambria, L.-W. Ku, C. Gui, and A. Gelbukh. 2014. A Rule-Based Approach to Aspect Extraction from Product Reviews. In *Proceedings of the Second Workshop on Natural Language Processing for Social Media (SocialNLP)*, 28–37.
151. Poria, S., I. Chaturvedi, E. Cambria, and F. Bisio. 2016. Sentic LDA: Improving on LDA with Semantic Similarity for Aspect-Based Sentiment Analysis. In *Proceedings of the IEEE International Joint Conference on Neural Networks (IJCNN)*, 4465–4473.
152. Poria, S., I. Chaturvedi, E. Cambria, and A. Hussain. 2016. Convolutional MKL Based Multimodal Emotion Recognition and Sentiment Analysis. In *Proceedings of the IEEE International Conference on Data Mining (ICDM)*, 439–448.
153. Poria, S., A. Gelbukh, B. Agarwal, E. Cambria, and N. Howard. 2014. Sentic Demo: A Hybrid Concept-level Aspect-Based Sentiment Analysis Toolkit. In *Proceedings of the ESWC*.
154. Poria, S., A. Gelbukh, E. Cambria, D. Das, and S. Bandyopadhyay. 2012. Enriching SenticNet Polarity Scores Through Semi-Supervised Fuzzy Clustering. In *Proceedings of the IEEE International Conference on Data Mining Workshops (ICDMW)*, 709–716.
155. Poria, S., A. Gelbukh, E. Cambria, A. Hussain, and G.-B. Huang. 2014. EmoSenticSpace: A Novel Framework for Affective Common-sense Reasoning. *Proceedings of the Elsevier Knowledge-Based Systems* 69: 108–123.
156. Poria, S., A. Gelbukh, E. Cambria, P. Yang, A. Hussain, and T. Durrani. 2012. Merging SenticNet and WordNet-Affect Emotion Lists for Sentiment Analysis. *Proceedings of the IEEE International Conference on Signal Processing (ICSP)* 2: 1251–1255.
157. Poria, S., A. Gelbukh, A. Hussain, S. Bandyopadhyay, and N. Howard. 2013. Music Genre Classification: A Semi-Supervised Approach. In *Proceedings of the Springer Mexican Conference on Pattern Recognition*, 254–263.
158. Poria, S., N. Ofek, A. Gelbukh, A. Hussain, and L. Rokach. 2014. Dependency Tree-Based Rules for Concept-level Aspect-Based Sentiment Analysis. In *Proceedings of the Springer Semantic Web Evaluation Challenge*, 41–47.
159. Poria, S., H. Peng, A. Hussain, N. Howard, and E. Cambria. 2017. Ensemble Application of Convolutional Neural Networks and Multiple Kernel Learning for Multimodal Sentiment Analysis. *In Proceedings of the Elsevier Neurocomputing*.
160. Pye, D., N.J. Hollinghurst, T.J. Mills, and K.R. Wood. 1998. Audio-visual Segmentation for Content-Based Retrieval. In *Proceedings of the International Conference on Spoken Language Processing*.
161. Qiao, Z., P. Zhang, C. Zhou, Y. Cao, L. Guo, and Y. Zhang. 2014. *Event Recommendation in Event-Based Social Networks*.

162. Raad, E.J. and R. Chbeir. 2014. Foto2Events: From Photos to Event Discovery and Linking in Online Social Networks. In *Proceedings of the IEEE Big Data and Cloud Computing*, 508–515, .
163. Radsch, C.C.. 2013. The Revolutions will be Blogged: Cyberactivism and the 4th Estate in Egypt. Doctoral Disseration. American University.
164. Rae, A., B. Sigurbjörnssön, and R. van Zwol. 2010. Improving Tag Recommendation using Social Networks. In *Proceedings of the Adaptivity, Personalization and Fusion of Heterogeneous Information*, 92–99.
165. Rahmani, H., B. Piccart, D. Fierens, and H. Blockeel. 2010. Three Complementary Approaches to Context Aware Movie Recommendation. In *Proceedings of the ACM Workshop on Context-Aware Movie Recommendation*, 57–60.
166. Rattenbury, T., N. Good, and M. Naaman. 2007. Towards Automatic Extraction of Event and Place Semantics from Flickr Tags. In *Proceedings of the ACM Special Interest Group on Information Retrieval*.
167. Rawat, Y. and M. S. Kankanhalli. 2016. ConTagNet: Exploiting User Context for Image Tag Recommendation. In *Proceedings of the ACM International Conference on Multimedia*, 1102–1106.
168. Repp, S., A. Groß, and C. Meinel. 2008. Browsing within Lecture Videos Based on the Chain Index of Speech Transcription. *Proceedings of the IEEE Transactions on Learning Technologies* 1(3): 145–156.
169. Repp, S. and C. Meinel. 2006. Semantic Indexing for Recorded Educational Lecture Videos. In *Proceedings of the IEEE International Conference on Pervasive Computing and Communications Workshops*, 5.
170. Repp, S., J. Waitelonis, H. Sack, and C. Meinel. 2007. Segmentation and Annotation of Audiovisual Recordings Based on Automated Speech Recognition. In *Proceedings of the Springer Intelligent Data Engineering and Automated Learning*, 620–629.
171. Russell, J.A. 1980. A Circumplex Model of Affect. *Proceedings of the Journal of Personality and Social Psychology* 39: 1161–1178.
172. Sahidullah, M., and G. Saha. 2012. Design, Analysis and Experimental Evaluation of Block Based Transformation in MFCC Computation for Speaker Recognition. *Proceedings of the Speech Communication* 54: 543–565.
173. Salamon, J., J. Serra, and E. Gomez. 2013. Tonal Representations for Music Retrieval: From Version Identification to Query-by-Humming. In *Proceedings of the Springer International Journal of Multimedia Information Retrieval* 2(1): 45–58.
174. Schedl, M. and D. Schnitzer. 2014. Location-Aware Music Artist Recommendation. In *Proceedings of the Springer MultiMedia Modeling*, 205–213.
175. M. Schedl and F. Zhou. 2016. Fusing Web and Audio Predictors to Localize the Origin of Music Pieces for Geospatial Retrieval. In *Proceedings of the Springer European Conference on Information Retrieval*, 322–334.
176. Scherp, A., and V. Mezaris. 2014. Survey on Modeling and Indexing Events in Multimedia. *Proceedings of the Springer Multimedia Tools and Applications* 70(1): 7–23.
177. Scherp, A., V. Mezaris, B. Ionescu, and F. De Natale. 2014. HuEvent '14: Workshop on Human-Centered Event Understanding from Multimedia. In *Proceedings of the ACM International Conference on Multimedia*, 1253–1254.
178. Schmitz, P.. 2006. Inducing Ontology from Flickr Tags. In *Proceedings of the Collaborative Web Tagging Workshop at ACM World Wide Web Conference*, vol 50.
179. Schuller, B., C. Hage, D. Schuller, and G. Rigoll. 2010. Mister DJ, Cheer Me Up!: Musical and Textual Features for Automatic Mood Classification. *Proceedings of the Journal of New Music Research* 39(1): 13–34.
180. Shah, R.R., M. Hefeeda, R. Zimmermann, K. Harras, C.-H. Hsu, and Y. Yu. 2016. NEWSMAN: Uploading Videos over Adaptive Middleboxes to News Servers In Weak Network Infrastructures. In *Proceedings of the Springer International Conference on Multimedia Modeling*, 100–113.

181. Shah, R.R., A. Samanta, D. Gupta, Y. Yu, S. Tang, and R. Zimmermann. 2016. PROMPT: Personalized User Tag Recommendation for Social Media Photos Leveraging Multimodal Information. In *Proceedings of the ACM International Conference on Multimedia*, 486–492.
182. Shah, R.R., A.D. Shaikh, Y. Yu, W. Geng, R. Zimmermann, and G. Wu. 2015. EventBuilder: Real-time Multimedia Event Summarization by Visualizing Social Media. In *Proceedings of the ACM International Conference on Multimedia*, 185–188.
183. Shah, R.R., Y. Yu, A.D. Shaikh, S. Tang, and R. Zimmermann. 2014. ATLAS: Automatic Temporal Segmentation and Annotation of Lecture Videos Based on Modelling Transition Time. In *Proceedings of the ACM International Conference on Multimedia*, 209–212.
184. Shah, R.R., Y. Yu, A.D. Shaikh, and R. Zimmermann. 2015. TRACE: A Linguistic-Based Approach for Automatic Lecture Video Segmentation Leveraging Wikipedia Texts. In *Proceedings of the IEEE International Symposium on Multimedia*, 217–220.
185. Shah, R.R., Y. Yu, S. Tang, S. Satoh, A. Verma, and R. Zimmermann. 2016. Concept-Level Multimodal Ranking of Flickr Photo Tags via Recall Based Weighting. In *Proceedings of the MMCommon's Workshop at ACM International Conference on Multimedia*, 19–26.
186. Shah, R.R., Y. Yu, A. Verma, S. Tang, A.D. Shaikh, and R. Zimmermann. 2016. Leveraging Multimodal Information for Event Summarization and Concept-level Sentiment Analysis. In *Proceedings of the Elsevier Knowledge-Based Systems*, 102–109.
187. Shah, R.R., Y. Yu, and R. Zimmermann. 2014. ADVISOR: Personalized Video Soundtrack Recommendation by Late Fusion with Heuristic Rankings. In *Proceedings of the ACM International Conference on Multimedia*, 607–616.
188. Shah, R.R., Y. Yu, and R. Zimmermann. 2014. User Preference-Aware Music Video Generation Based on Modeling Scene Moods. In *Proceedings of the ACM International Conference on Multimedia Systems*, 156–159.
189. Shaikh, A.D., M. Jain, M. Rawat, R.R. Shah, and M. Kumar. 2013. Improving Accuracy of SMS Based FAQ Retrieval System. In *Proceedings of the Springer Multilingual Information Access in South Asian Languages*, 142–156.
190. Shaikh, A.D., R.R. Shah, and R. Shaikh. 2013. SMS Based FAQ Retrieval for Hindi, English and Malayalam. In *Proceedings of the ACM Forum on Information Retrieval Evaluation*, 9.
191. Shamma, D.A., R. Shaw, P.L. Shafton, and Y. Liu. 2007. Watch What I Watch: Using Community Activity to Understand Content. In *Proceedings of the ACM International Workshop on Multimedia Information Retrieval*, 275–284.
192. Shaw, B., J. Shea, S. Sinha, and A. Hogue. 2013. Learning to Rank for Spatiotemporal Search. In *Proceedings of the ACM International Conference on Web Search and Data Mining*, 717–726.
193. Sigurbjörnsson, B. and R. Van Zwol. 2008. Flickr Tag Recommendation Based on Collective Knowledge. In *Proceedings of the ACM World Wide Web Conference*, 327–336.
194. Snoek, C.G., M. Worring, and A.W. Smeulders. 2005. Early versus Late Fusion in Semantic Video Analysis. In *Proceedings of the ACM International Conference on Multimedia*, 399–402.
195. Snoek, C.G., M. Worring, J.C. Van Gemert, J.-M. Geusebroek, and A.W. Smeulders. 2006. The Challenge Problem for Automated Detection of 101 Semantic Concepts in Multimedia. In *Proceedings of the ACM International Conference on Multimedia*, 421–430.
196. Soleymani, M., J.J.M. Kierkels, G. Chanel, and T. Pun. 2009. A Bayesian Framework for Video Affective Representation. In *Proceedings of the IEEE International Conference on Affective Computing and Intelligent Interaction and Workshops*, 1–7.
197. Stober, S., and A. . Nürnberger. 2013. Adaptive Music Retrieval – A State of the Art. *Proceedings of the Springer Multimedia Tools and Applications* 65(3): 467–494.
198. Stoyanov, V., N. Gilbert, C. Cardie, and E. Riloff. 2009. Conundrums in Noun Phrase Coreference Resolution: Making Sense of the State-of-the-art. In *Proceedings of the ACL International Joint Conference on Natural Language Processing of the Asian Federation of Natural Language Processing*, 656–664.

199. Stupar, A. and S. Michel. 2011. Picasso: Automated Soundtrack Suggestion for Multimodal Data. In *Proceedings of the ACM Conference on Information and Knowledge Management*, 2589–2592.
200. Thayer, R.E. 1989. *The Biopsychology of Mood and Arousal*. New York: Oxford University Press.
201. Thomee, B., B. Elizalde, D.A. Shamma, K. Ni, G. Friedland, D. Poland, D. Borth, and L.-J. Li. 2016. YFCC100M: The New Data in Multimedia Research. *Proceedings of the Communications of the ACM* 59(2): 64–73.
202. Tirumala, A., F. Qin, J. Dugan, J. Ferguson, and K. Gibbs. 2005. Iperf: The TCP/UDP Bandwidth Measurement Tool. http://dast.nlanr.net/Projects/Iperf/
203. Torralba, A., R. Fergus, and W.T. Freeman. 2008. 80 Million Tiny Images: A Large Data set for Nonparametric Object and Scene Recognition. *Proceedings of the IEEE Transactions on Pattern Analysis and Machine Intelligence* 30(11): 1958–1970.
204. Toutanova, K., D. Klein, C.D. Manning, and Y. Singer. 2003. Feature-Rich Part-of-Speech Tagging with a Cyclic Dependency Network. In *Proceedings of the North American Chapter of the Association for Computational Linguistics on Human Language Technology*, 173–180.
205. Toutanova, K. and C.D. Manning. 2000. Enriching the Knowledge Sources used in a Maximum Entropy Part-of-Speech Tagger. In *Proceedings of the Annual Meeting of the Association for Computational Linguistics*, 63–70.
206. Utiyama, M. and H. Isahara. 2001. A Statistical Model for Domain-Independent Text Segmentation. In *Proceedings of the Annual Meeting on Association for Computational Linguistics*, 499–506.
207. Vishal, K., C. Jawahar, and V. Chari. 2015. Accurate Localization by Fusing Images and GPS Signals. In *Proceedings of the IEEE Computer Vision and Pattern Recognition Workshops*, 17–24.
208. Wang, C., F. Jing, L. Zhang, and H.-J. Zhang. 2008. Scalable Search-Based Image Annotation. *Proceedings of the Springer Multimedia Systems* 14(4): 205–220.
209. Wang, H.L., and L.F. Cheong. 2006. Affective Understanding in Film. *Proceedings of the IEEE Transactions on Circuits and Systems for Video Technology* 16(6): 689–704.
210. Wang, J., J. Zhou, H. Xu, T. Mei, X.-S. Hua, and S. Li. 2014. Image Tag Refinement by Regularized Latent Dirichlet Allocation. *Proceedings of the Elsevier Computer Vision and Image Understanding* 124: 61–70.
211. Wang, P., H. Wang, M. Liu, and W. Wang. 2010. An Algorithmic Approach to Event Summarization. In *Proceedings of the ACM Special Interest Group on Management of Data*, 183–194.
212. Wang, X., Y. Jia, R. Chen, and B. Zhou. 2015. Ranking User Tags in Micro-Blogging Website. In *Proceedings of the IEEE ICISCE*, 400–403.
213. Wang, X., L. Tang, H. Gao, and H. Liu. 2010. Discovering Overlapping Groups in Social Media. In *Proceedings of the IEEE International Conference on Data Mining*, 569–578.
214. Wang, Y. and M.S. Kankanhalli. 2015. Tweeting Cameras for Event Detection. In *Proceedings of the IW3C2 International Conference on World Wide Web*, 1231–1241.
215. Webster, A.A., C.T. Jones, M.H. Pinson, S.D. Voran, and S. Wolf. 1993. Objective Video Quality Assessment System Based on Human Perception. In *Proceedings of the IS&T/SPIE's Symposium on Electronic Imaging: Science and Technology*, 15–26. International Society for Optics and Photonics.
216. Wei, C.Y., N. Dimitrova, and S.-F. Chang. 2004. Color-Mood Analysis of Films Based on Syntactic and Psychological Models. In *Proceedings of the IEEE International Conference on Multimedia and Expo*, 831–834.
217. Whissel, C. 1989. The Dictionary of Affect in Language. In *Emotion: Theory, Research and Experience. Vol. 4. The Measurement of Emotions*, ed. R. Plutchik and H. Kellerman, 113–131. New York: Academic.

218. Wu, L., L. Yang, N. Yu, and X.-S. Hua. 2009. Learning to Tag. In *Proceedings of the ACM World Wide Web Conference*, 361–370.
219. Xiao, J., W. Zhou, X. Li, M. Wang, and Q. Tian. 2012. Image Tag Re-ranking by Coupled Probability Transition. In *Proceedings of the ACM International Conference on Multimedia*, 849–852.
220. Xie, D., B. Qian, Y. Peng, and T. Chen. 2009. A Model of Job Scheduling with Deadline for Video-on-Demand System. In *Proceedings of the IEEE International Conference on Web Information Systems and Mining*, 661–668.
221. Xu, M., L.-Y. Duan, C. Xu, M. Kankanhalli, and Q. Tian. 2003. Event Detection in Basketball Video using Multiple Modalities. *Proceedings of the IEEE Joint Conference of the Fourth International Conference on Information, Communications and Signal Processing, and Fourth Pacific Rim Conference on Multimedia* 3: 1526–1530.
222. Xu, M., N.C. Maddage, C. Xu, M. Kankanhalli, and Q. Tian. 2003. Creating Audio Keywords for Event Detection in Soccer Video. In *Proceedings of the IEEE International Conference on Multimedia and Expo*, 2:II–281.
223. Yamamoto, N., J. Ogata, and Y. Ariki. 2003. Topic Segmentation and Retrieval System for Lecture Videos Based on Spontaneous Speech Recognition. In *Proceedings of the INTERSPEECH*, 961–964.
224. Yang, H., M. Siebert, P. Luhne, H. Sack, and C. Meinel. 2011. Automatic Lecture Video Indexing using Video OCR Technology. In *Proceedings of the IEEE International Symposium on Multimedia*, 111–116.
225. Yang, Y.H., Y.C. Lin, Y.F. Su, and H.H. Chen. 2008. A Regression Approach to Music Emotion Recognition. *Proceedings of the IEEE Transactions on Audio, Speech, and Language Processing* 16(2): 448–457.
226. Ye, G., D. Liu, I.-H. Jhuo, and S.-F. Chang. 2012. Robust Late Fusion with Rank Minimization. In *Proceedings of the IEEE Conference on Computer Vision and Pattern Recognition*, 3021–3028.
227. Ye, Q., Q. Huang, W. Gao, and D. Zhao. 2005. Fast and Robust Text Detection in Images and Video Frames. *Proceedings of the Elsevier Image and Vision Computing* 23(6): 565–576.
228. Yin, Y., Z. Shen, L. Zhang, and R. Zimmermann. 2015. Spatial temporal Tag Mining for Automatic Geospatial Video Annotation. *Proceedings of the ACM Transactions on Multimedia Computing, Communications, and Applications* 11(2): 29.
229. Yoon, S. and V. Pavlovic. 2014. Sentiment Flow for Video Interestingness Prediction. In *Proceedings of the Workshop on HuEvent at the ACM International Conference on Multimedia*, 29–34.
230. Yu, Y., K. Joe, V. Oria, F. Moerchen, J.S. Downie, and L. Chen. 2009. Multiversion Music Search using Acoustic Feature Union and Exact Soft Mapping. *Proceedings of the World Scientific International Journal of Semantic Computing* 3(02): 209–234.
231. Yu, Y., Z. Shen, and R. Zimmermann. 2012. Automatic Music Soundtrack Generation for Outdoor Videos from Contextual Sensor Information. In *Proceedings of the ACM International Conference on Multimedia*, 1377–1378.
232. Zaharieva, M., M. Zeppelzauer, and C. Breiteneder. 2013. Automated Social Event Detection in Large Photo Collections. In *Proceedings of the ACM International Conference on Multimedia Retrieval*, 167–174.
233. Zhang, J., X. Liu, L. Zhuo, and C. Wang. 2015. Social Images Tag Ranking Based on Visual Words in Compressed Domain. *Proceedings of the Elsevier Neurocomputing* 153: 278–285.
234. Zhang, J., S. Wang, and Q. Huang. 2015. Location-Based Parallel Tag Completion for Geo-tagged Social Image Retrieval. In *Proceedings of the ACM International Conference on Multimedia Retrieval*, 355–362.
235. Zhang, M., J. Wong, W. Tavanapong, J. Oh, and P. de Groen. 2004. Media Uploading Systems with Hard Deadlines. In *Proceedings of the Citeseer International Conference on Internet and Multimedia Systems and Applications*, 305–310.

236. Zhang, M., J. Wong, W. Tavanapong, J. Oh, and P. de Groen. 2008. Deadline-constrained Media Uploading Systems. *Proceedings of the Springer Multimedia Tools and Applications* 38(1): 51–74.
237. Zhang, W., J. Lin, X. Chen, Q. Huang, and Y. Liu. 2006. Video Shot Detection using Hidden Markov Models with Complementary Features. *Proceedings of the IEEE International Conference on Innovative Computing, Information and Control* 3: 593–596.
238. Zheng, L., V. Noroozi, and P.S. Yu. 2017. Joint Deep Modeling of Users and Items using Reviews for Recommendation. In *Proceedings of the ACM International Conference on Web Search and Data Mining*, 425–434.
239. Zhou, X.S. and T.S. Huang. 2000. CBIR: from Low-level Features to High-level Semantics. In *Proceedings of the International Society for Optics and Photonics Electronic Imaging*, 426–431.
240. Zhuang, J. and S.C. Hoi. 2011. A Two-view Learning Approach for Image Tag Ranking. In *Proceedings of the ACM International Conference on Web Search and Data Mining*, 625–634.
241. Zimmermann, R. and Y. Yu. 2013. Social Interactions over Geographic-aware Multimedia Systems. In *Proceedings of the ACM International Conference on Multimedia*, 1115–1116.
242. Shah, R.R. 2016. Multimodal-based Multimedia Analysis, Retrieval, and Services in Support of Social Media Applications. In *Proceedings of the ACM International Conference on Multimedia*, 1425–1429.
243. Shah, R.R. 2016. Multimodal Analysis of User-Generated Content in Support of Social Media Applications. In *Proceedings of the ACM International Conference in Multimedia Retrieval*, 423–426.
244. Yin, Y., R.R. Shah, and R. Zimmermann. 2016. A General Feature-based Map Matching Framework with Trajectory Simplification. In *Proceedings of the ACM SIGSPATIAL International Workshop on GeoStreaming*, 7.

Chapter 7
Adaptive News Video Uploading

Abstract An interesting recent trend, enabled by the ubiquitous availability of mobile devices, is that regular citizens report events which news providers then disseminate, *e.g.*, CNN iReport. Often such news are captured in places with very weak network infrastructures and it is imperative that a citizen journalist can quickly and reliably upload videos in the face of slow, unstable, and intermittent Internet access. We envision that some middleboxes are deployed to collect these videos over energy-efficient short-range wireless networks. Multiple videos may need to be prioritized, and then optimally transcoded and scheduled. In this study we introduce an adaptive middlebox design, called NEWSMAN, to support citizen journalists. NEWSMAN jointly considers two aspects under varying network conditions: (i) choosing the optimal transcoding parameters, and (ii) determining the uploading schedule for news videos. We design, implement, and evaluate an efficient scheduling algorithm to maximize a user-specified objective function. We conduct a series of experiments using trace-driven simulations, which confirm that our approach is practical and performs well. For instance, NEWSMAN outperforms the existing algorithms (i) by 12 times in terms of system utility (*i.e.*, sum of utilities of all uploaded videos), and (ii) by four times in terms of the number of videos uploaded before their deadline.

Keywords Adaptive news videos uploading • Citizen journalism • Videos uploading • Video transcoding • Adaptive middleboxes • NEWSMAN

7.1 Introduction

Owing to technical advances in mobile devices and wireless communications, user-generated news videos have become popular since they can be easily captured using most modern smartphones and tablets in sufficiently high quality. Moreover, in the era of globalization, most news providers cover news from every part of the world, while on many occasions, reporters send news materials to editing rooms over the Internet. Therefore, in addition to traditional news reporting, the concept of citizen journalism, which allows people to play active roles in the process of collecting news reports, is also gaining much popularity. For instance, Cable News Network

R. Shah, R. Zimmermann, *Multimodal Analysis of User-Generated Multimedia Content*, Socio-Affective Computing 6, DOI 10.1007/978-3-319-61807-4_7

Table 7.1 Notations used in the adaptive news video uploading chapter

Symbols	Meanings
$\bar{B}$	Number of breaking news B_1 to $B_{\bar{B}}$
$\bar{N}$	Number of traditional (normal) news N_1 to $N_{\bar{N}}$
G_c	The number of news categories
j_i	i^{th} job (a video which is either breaking or normal news)
$\bar{A}$	Arrival times of jobs
$\bar{D}$	Deadlines of jobs
$\bar{M}$	Metadata consisting of users' reputations and video information such as bitrates and fps (frames per second)
$\bar{\mu}(j_i)$	Weight for boosting or ignoring the importance of any particular news type or category
$\bar{\xi}(j_i)$	Score for the video length of j_i
$\bar{\lambda}(j_i)$	Score for the news-location of j_i
$\bar{\gamma}(r)$	Score for the user reputation of a reporter r
$\bar{\sigma}$	Editor-specified minimum required video quality (in PSNR)
p_i	The transcoded video quality of j_i
$\bar{p}_i$	The original video quality of j_i
b_i	The transcoded bitrate of j_i
$\bar{b}_i$	The original bitrate of j_i
t_c	Current time
$\bar{\omega}(t_c)$	The available disk size at t_c
$\bar{s}_i$	The original file size of j_i
s_i	The transcoded file size of j_i
$\bar{\eta}(s_i)$	Time required to transcode j_i with file size s_i
$\bar{\beta}(t_1, t_2)$	Average throughput between time interval t_1 and t_2
$\bar{\delta}(j_i)$	The video length (in seconds) of j_i
$\bar{\tau}$	The time interval of running the scheduler in a middlebox
$\bar{u}(j_i)$	The news importance of j_i
$\bar{v}(j_i)$	The news decay rate of j_i
$\bar{\rho}(j_i)$	The news utility value of j_i
χ	The number of possible video qualities
$\mathcal{U}$	Total utility value for the NEWSMAN system
Q	The list of all jobs arrived till time t_c at the middlebox
$\bar{L}$	The list of all jobs scheduled at the middlebox

(CNN) allows citizens to report news using modern smartphones and tablets through its CNN iReport service. It is, however, quite challenging for reporters to timely upload news videos, especially from developing countries, where Internet access is slow or even intermittent. Hence, it is crucial to deploy *adaptive middleboxes*, which upload news videos respecting the varying network conditions. Such middleboxes will allow citizen reporters to quickly *drop* the news videos over energy-efficient short-range wireless networks, and continue their daily life. All notations used in this chapter are listed in Table 7.1.

Journalists can upload news videos to middleboxes or news providers, either by using cellular or WiFi networks if available. Since an energy-efficient short-range wireless network between mobile devices and middleboxes can be leveraged using optimized mobile applications, we focus on a scheduling algorithm tuned for varying network conditions which can adaptively schedule the uploads of videos. Middleboxes can be placed in cloud servers or strategic places in towns such as city centers, coffee shops, train and bus stations, *etc.*, so that when reporters frequent these places then the short-range wireless communication can be leveraged for uploading videos. One can envision that an efficient smartphone application can further improve such communication among different reporters based on collaborative models. Shops at these places may host such middleboxes incentivized by the following reasons: (i) advertisement companies can sponsor the cost of resources (*e.g.*, several companies already sponsor Internet connectivity at airports), (ii) news providers can sponsor resources since they will receive news on time with less investment, (iii) more customers may be attracted to visit these shops, and (iv) a collaborative model of information sharing based on crowdsourcing is gaining popularity. Moreover, middleboxes can be used to decide whether reporters can directly upload videos to news providers based on current network conditions.

In designing the adaptive middlebox, we consider two categories of news videos, first, *breaking news* and, second, *traditional news*. Usually, the breaking news videos have stricter deadlines than those of the traditional news videos. There is significant competition among news organizations to be the first to report breaking news. Hence, ubiquitous availability of mobile devices and the concept of citizen journalism help with fast reporting of news videos, using the mobile applications and the web sites of news providers. However, many times, the uploading of news videos is delayed due to reporters' slow Internet access and the big sizes of news videos. In pilot experiments among news reporters in early 2015, we noticed low throughput and non-trivial network interruptions in some of our test cases, as summarized in Table 7.2. Reporters tested uploading from a few locations in India, Pakistan, Argentina, and the USA, mostly through cellular networks. For example, when news reporters uploaded their videos over the Internet to an editing room in New York City for a leading news provider, they suffered from as many as seven interrupts per upload. Without our proposed adaptive middleboxes, news reporters may be frustrated and eventually give up, because of long uploading times. This necessitates carefully designed adaptive middleboxes which run a scheduling algorithm to determine an uploading schedule for news videos considering factors such as optimal bitrates, videos deadlines, and network conditions.

In this study, we propose NEWSMAN, which maximizes the system utility by optimizing the number and quality of the videos uploaded before their deadlines from users to news editors under varying network conditions. We place middleboxes between reporters and news editors, to decouple the local upload from the long-haul transmission to the editing room, in order to optimize both network *segments*, which have diverse characteristics. To optimize the system performance, we design an efficient scheduling algorithm in the middlebox to derive the uploading schedule and to transcode news videos (if required, to meet

Table 7.2 Real-world results of news uploading

Location	India	Pakistan	Argentina	USA
Throughput	500 ~ 600 Kbps	300 ~ 500 Kbps	200 ~ 300 Kbps	20 ~ 23 Mbps
File sizes	100 ~ 200 MB	50 ~ 100 MB	500 ~ 600 MB	100 ~ 200 MB
#Interruptions	6	3	7	0

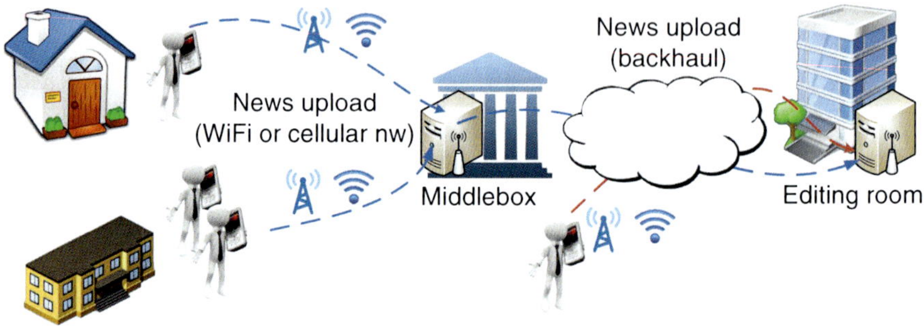

Fig. 7.1 Architecture of the proposed NEWSMAN system

their deadlines) adaptively following a practical video quality model. The NEWSMAN scheduling process is described as follows: (i) reporters directly upload news videos to the news organizations if the Internet connectivity is good, otherwise (ii) reporters upload news videos to the middlebox, and (iii) the scheduler in the middlebox determines an uploading schedule and optimal bitrates for transcoding. Since multimodal information of user-generated content is useful in several applications [189, 190, 242, 243] such as video understanding [183, 184, 187, 188] and event and tag understanding [181, 182, 185, 186], we use it to optimally schedule the uploading of videos. Figure 7.1 presents the architecture of the NEWSMAN system [180].

The key contribution of this study is an efficient scheduling algorithm to upload news videos to a cloud server such that: (i) the system utility is maximized, (ii) the number of news videos uploaded before their deadlines is maximized, and (iii) news videos are delivered in the best possible video qualities under varying network conditions. We conducted extensive trace-driven simulations using real datasets of 130 online news videos. The results from the simulations show the merits of NEWSMAN as it outperforms the current algorithms: (i) by 1200% in terms of system utility and (ii) by 400% in terms of the number of videos uploaded before their deadlines. Furthermore, NEWSMAN achieves low average delay of the uploaded news videos.

The chapter is organized as follows. In Sect. 2, we describe the NEWSMAN system. Sect. 3 discuss problem formulation to maximize the system utility. The evaluation results are presented in Sect. 4. Finally, we conclude the chapter with a summary in Sect. 5.

7.2 Adaptive News Video Uploading

We refer to the uploading of a news video as a *job* in this study. NEWSMAN schedules jobs such that videos are uploaded before their deadlines in the highest possible qualities with optimally selected coding parameters for video transcoding.

7.2.1 NEWSMAN Scheduling Algorithm

Figure 7.2 shows the architecture of the scheduler. Reporters upload jobs to a middlebox. For every job arriving at the middlebox, the scheduler performs the following actions when the scheduling interval expires: (i) it computes the job's importance, (ii) it sorts all jobs based on news importance, and (iii) it estimates the job's uploading schedule and the optimal bitrate for transcoding. The scheduling algorithm is described in details in Sect. 3. As Fig. 7.2 shows, we consider c video qualities for a job j_i and select the optimal bitrate for transcoding of j_i to meet its deadline, under current network conditions.

7.2.2 Rate–Distortion (R–D) Model

Traditional digital video transmission and storage systems either fully upload a news video to a news editor or not at all due to fixed spatio-temporal format of the video signal. The key idea for transcoding videos with optimal bitrates is to compress videos for transmission to adaptively transfer video content before their deadlines, under varying network conditions. More motion in adjacent frames indicates higher TI (temporal perceptual information) values and scenes with minimal spatial detail result in low SI (spatial perceptual information). For instance,

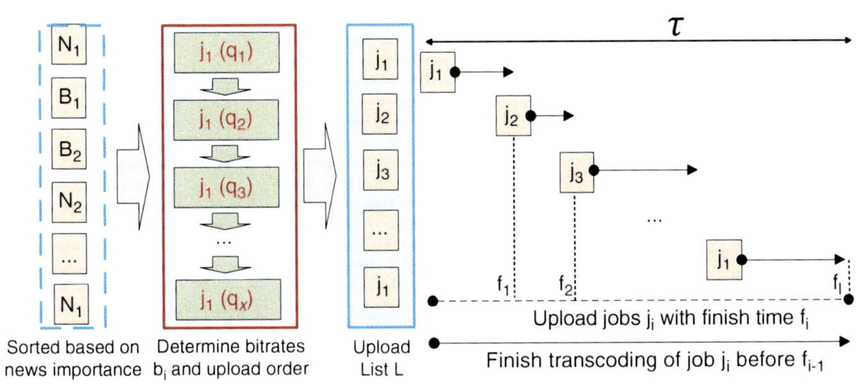

Fig. 7.2 Scheduler architecture in a middlebox

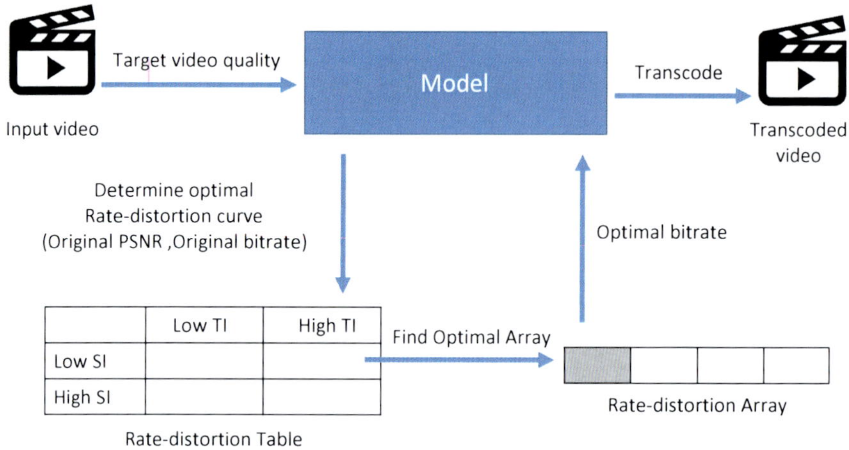

Fig. 7.3 Quality model to determine optimal bitrate

a scene from a football game contains a large amount of motion (*i.e.*, high TI) as well as spatial detail (*i.e.*, high SI). Since two different scenes with the same TI/SI values produce similar perceived quality [215], news videos can be classified in G_c news categories. Therefore, news videos can be categorized into different categories such as sport videos, interviews, *etc*., based on their TI/SI values. Although, news editors may be willing to sacrifice some video quality to meet deadlines, but question arises, how much quality is renounced for how much savings in video size (or transmission time) while uploading? We determine the suitable coding bitrates (hence, transcoded video size) adaptively for an editor-specified video quality (say in PSNR, peak signal-to-noise ratio) for previews and full videos, using R–D curves, which we construct for the four video clusters (four news categories) based on TI and SI values of news videos (see Fig. 7.3). Three video segments of length 5 s each are randomly selected from a video to compute the average TI and SI values of the video. After determining the average TI and SI values, a suitable R–D curve can be selected to compute optimal bitrate for a given editor-specified video quality.

7.3 Problem Formulation

7.3.1 Formulation

The news importance $\bar{u}$ of a job j_i is defined as $\bar{u}(j_i) = \overline{\mu}(j_i) \cdot \big(w_1\bar{\xi}(j_i) + w_2\bar{\lambda}(j_i) + w_3\overline{\gamma}(r)\big)$, where the multiplier $\overline{\mu}(j_i)$ is a weight for boosting or ignoring the importance of any particular news type or category. *E.g.*, in our experiments the value of $\overline{\mu}(j_i)$ is 1 if job j_i is traditional news and 2 if job j_i is breaking news. By considering news categories such as sports a news provider can boost videos during a sports events such as the FIFA world cup. Moreover, the news decay function $\bar{v}$ is defined as:

$$\bar{v}(f_i) = \begin{cases} 1, & \text{if } f_i \leq d_i \\ e^{-\alpha(f_i - d_i)}, & \text{otherwise, where } d_i \text{ and } f_i \text{ are the deadline and finish time of job } j_i, \text{ respectively. } \alpha \text{ is an exponential decay constant.} \end{cases}$$

The utility score of a news video j_i depends on the following factors: (i) the importance of j_i, (ii) how quickly the importance of j_i decays, and (iii) the delivered video quality of j_i. Thus, we define the news utility r for job j_i as $\rho(j_i) = \bar{u}(j_i)\bar{v}(f_i)p_i$. With the above notations and functions, we state the problem formulation as:

$$\max \sum_{i=1}^{\bar{B}+\bar{N}} \overline{\rho}(j_i) \tag{7.1a}$$

$$s.t.\ \overline{\sigma} \leq p_i \leq \overline{p_i}\ \forall 1 \leq i \leq \bar{B} + \bar{N} \tag{7.1b}$$

$$(f_i - f_{i-1})\beta(f_{i-1}, f_i) \geq b_i \overline{\delta}(j_i) \tag{7.1c}$$

$$\sum_{\forall j_k \in K} \overline{\eta}(s_k) < f_i, \text{where } K = \{j_k | j_k \text{ is scheduled before } j_i\} \tag{7.1d}$$

$$\sum_{i=1}^{\bar{B}+\bar{N}} S_i \leq \overline{\omega}(t_c) - \sum_{i=1}^{\bar{B}+\bar{N}} \overline{S_i} \tag{7.1e}$$

$$f_i \leq f_k, \forall 1 \leq i \leq k \leq \bar{B} + \bar{N} \tag{7.1f}$$

$$0 \leq f_i, \forall 1 \leq i \leq \bar{B} + \bar{N} \tag{7.1g}$$

$$0 \leq b_i \leq \overline{b_i}, \forall 1 \leq i \leq \bar{B} + \bar{N} \tag{7.1h}$$

$$j_i \in \{B_1, \ldots, B_{\bar{B}}, N_1, \ldots, N_{\bar{N}}\} \tag{7.1i}$$

The objective function in Eq. (7.1a) maximizes the sum of news utility (*i.e.*, the product of importance, decay value and video quality) for all jobs. Eq. (7.1b) makes sure that the video quality of the transcoded video is atleast the minimum video quality $\overline{\sigma}$. Eq. (7.1c) ensures bandwidth constraints for NEWSMAN. Eq. (7.1d) enforces that the transcoding of a video completes before its uploading starts and Eq. (7.1e) ensures disk constraints of a middlebox. Eq. (7.1f) ensures that the scheduler uploads jobs in the order scheduled by NEWSMAN. Eqs. (7.1g) and (7.1h) define the ranges of the decision variables. Finally, Eq. (7.1i) indicates that all jobs are either breaking news or traditional news.

Lemma Let $j_{i_{i=1}}^{n}$ be a set of n jobs in a middlebox at time t_c, and $d_{i_{i=1}}^{n}$ their respective deadlines for uploading. The scheduler is executed when either the scheduling interval $\overline{\tau}$ expires or when all jobs in the middlebox have been uploaded before $\overline{\tau}$ expires. Thus, the average throughput $\overline{\beta}(t_c, t_c + \overline{\tau})$ (or $\overline{\beta}$ in short) during the scheduling interval is distributed among several jobs selected for parallel uploading,[1] and as a consequence, the sequential upload of jobs has higher utility than parallel uploading.

Proof Sketch Let k jobs $j_{i_{i=1}}^{k}$ with transcoded sizes $s_{i_{i=1}}^{k}$ be selected in parallel uploading. Let k_t of them $\left(j_{i_{i=1}}^{k_t}\right)$ require transcoding. Thus, it takes some time for their transcoding $\left(i.e., \overline{\eta}_p(s_i)_{i=1}^{k_t} \geq 0\right)$ before the actual uploading starts. Hence, uploading throughput is wasted during the transcoding of these jobs in parallel uploading. During sequential uploading, NEWSMAN ensures that transcoding of a job is finished (if required) before the uploading of the job is started. Thus, it results in a net transcoding time of zero $\left(i.e., \overline{\eta}_s(s_i)_{i=1}^{k_t} = 0\right)$ in sequential uploading, and it fully utilizes the uploading throughput $\overline{\beta}$. Let t_u be the time (excluding transcoding time) to upload jobs $j_{i_{i=1}}^{n}$. Thus, t_u is equal for both sequential and parallel uploading since the same uploading throughput is divided among parallel jobs. Let t_p $\left(i.e., t_u + \overline{\eta}_p\right)$ and $t_s\left(i.e., t_u + \overline{\eta}_s\right)$ be the uploading time for all jobs $\left(j_{i_{i=1}}^{k_t}\right)$ when the jobs are uploaded in parallel or sequential manner, respectively. Hence, the actual time required to upload in a parallel manner ($i.e.$, t_p) is greater than the time required to upload in a sequential manner ($i.e.$, t_s). Moreover, the uploading of important jobs is delayed in parallel uploading since throughput is divided among several other selected jobs ($\overline{\beta}/k$ for each job). Therefore, the sequential uploading of jobs is better than the parallel uploading.

7.3.2 Upload Scheduling Algorithm

We design an efficient scheduling algorithm to solve the above formulation. Algorithm 7.1 shows the main procedure of scheduling a list of jobs at a middlebox. If it is not possible to upload any job within its deadline, NEWSMAN uploads the transcoded news videos to meet the deadline. Algorithm 7.2 shows the procedure of calculating the encoding parameters for transcoding under current network conditions and $\overline{\sigma}$. Algorithm 7.2 is invoked on line 18 of Algorithm 7.1 whenever necessary.

---

[1]Some videos may require transcoding first before uploading to meet deadlines in the NEWSMAN system

Algorithm 7.1 A procedure for scheduling news videos

procedure SCHEDULEANDUPLOADNEWSREPORTS
  INPUT: $\overline{B}+\overline{N}$ jobs with given arrival times A and deadlines D
  OUTPUT: A uploading schedule for jobs at a middlebox
  Initialization: $t = t_c$, $\mathscr{U} = 0$ ▷ t_c is current time and $\mathscr{U}$ is total utility.
  Q = getAllJobs(t_c) ▷ Q is the list of all jobs arrived till time t_c.
  compNewsImporatnce(Q) ▷ Compute news importance for all jobs.
  compEstmFinishTime(Q, χ) ▷ Compute estimated finish time for χ video qualities.
  sortJobs(Q) ▷ Sort all jobs in Q based on their news importance scores.
  for each job j_i in Q **do** ▷ Read j_i with the highest importance first.
    t_d = getDeadline(j_i, A) ▷ Get the deadline for j_i.
    t_h = getEstmFinishTime(j_i, t) ▷ Estimated time to upload j_i in original quality.
    if ($t_h \leq t_d$) **then** ▷ Indicates j_i can be uploaded in original video before deadline.
      $\overline{\rho}(j_i) = \overline{u}(j_i)\, \overline{v}(t_h)\, p_i$ ▷ Utility based on its importance, decay, and quality.
      $\mathscr{U} += \overline{\rho}(j_i)$ ▷ Total utility for jobs scheduled so far.
      addJobToUploadList(j_i, $\overline{L}$) ▷ Add j_i to $\overline{L}$ in original quality.
      $t += t_h$ ▷ Update total scheduled time so that more jobs can be added to $\overline{L}$.
    else ▷ Try to add j_i to $\overline{L}$ in lower video quality, if possible.
      ϕ = addJobInLowerQuality(j_i, t_d, t, $\overline{L}$, $\mathscr{U}$) ▷ See Algorithm 10.
    end if ▷ ϕ is $TRUE$ if j_i added to $\overline{L}$ after lowering qualities of some videos.
    if (ϕ == $FALSE$) **then** ▷ ϕ is $FALSE$ if j_i couldn't added to $\overline{L}$.
      addJobToScheduleLater(j_i, $Q\prime$) ▷ Jobs in $Q\prime$ could not scheduled.
    end if
  end for ▷ Output the uploading order and b_i (decision variables in formulation) of all j_i.
  transcodeAndUpload(L) ▷ Set f_i (decision variable in formulation) for j_i after upload.
end procedure

The NEWSMAN scheduler considers χ possible video qualities (hence, smaller video size and shorter upload time are possible) for a job. NEWSMAN considers $\overline{\sigma}$ as a threshold and divides a region between $\overline{\sigma}$ (minimum required video quality) and $\bar{p}_i$ (original video quality) among c discrete qualities (say, $qi_{i=1}^{\chi}$, with $q_1 = \overline{\sigma}$ and $q_\chi = \bar{p}_i$). The scheduler keeps checking lower, but acceptable, video qualities starting with the least important job first, to accommodate j in $\overline{L}$ such that: (i) the total estimated system utility increases after adding j, and (ii) all jobs in $\overline{L}$ still meet their deadlines (maybe with lower video qualities), if they are estimated to meet deadlines earlier. However, if the scheduler is not able to add j in the uploading list, then this job is added to a missed-deadline list whose deadline can be modified later by news-editors based on news importance. Once the scheduling of all jobs is done, NEWSMAN starts uploading news videos from the middlebox to the editing room and transcodes (in parallel with uploading) the rest of the news videos (if required) in the uploading list $\bar{L}$.

Algorithm 7.2 is invoked when it is not possible to add a job with the original video quality to $\bar{L}$. This procedure keeps checking jobs at lower video qualities until all jobs in the list are added to $\bar{L}$ with estimated uploading times within their deadlines. The *isJobAccomodatedWihinDeadline()* method on line 13 of Algorithm 7.2 ensures that: (i) the selected video quality q_k is lower than the current video quality q_c (*i.e.*, q_kq_c) since some jobs are already set to lower video qualities in earlier steps, (ii) the utility value is increased after adding the job (*i.e.*, $\bar{\mathscr{U}} \geq \mathscr{U}$), (iii) all jobs in L is completed (estimated) within their deadlines, and (iv) a job with higher importance comes first in $\bar{L}$.

7.4 Evaluation

7.4.1 Real-Life Datasets

We collected 130 online news video sequences from Al Jazeera, CNN, and BBC YouTube channels during mid-February 2015. The shortest and longest duration of videos are 0.33 and 26 min, and the smallest and biggest news video sizes are 4 and 340 MB, respectively. We also collected network traces from different PCs across the globe, such as (Delhi and Hyderabad) India, and (Nanjing) China, which emulate middleboxes in our system. More specifically, we use IPERF [202] to collect throughput from the PCs to an Amazon EC2 (Amazon Elastic Compute Cloud) server in Singapore (see Table 7.3). The news and network datasets are used to drive our simulator.

7.4.2 Piecewise Linear R–D Model

It is important to determine the category (or TI/SI values) of a news video, so that we can select appropriate R–D models for these categories. A scene with little motion and limited spatial detail (such as a head and shoulders shot of a newscaster) may be compressed to 384 kbits/sec and decompressed with relatively little distortion. Another scene (such as from a soccer game) which contains a large amount of motion as well as spatial detail will appear quite distorted at the same bit rate [215]. Therefore, it is important to consider different R–D models for all categories. Empirical piecewise linear R–D models can be constructed for individual TI/SI pairs (see Fig. 7.4). We encode online news videos with diverse content complexities and empirically analyze their R–D characteristics. We consider four categories

Table 7.3 Statistics of network traces

Location	Dates	Avg. throughput
Delhi	2015-03-12 to 2015-03-14	409 Kbps
Hyderabad	2015-03-14 to 2015-03-18	297 Kbps
Nanjing	2015-03-23 to 2015-03-27	1138 Kbps

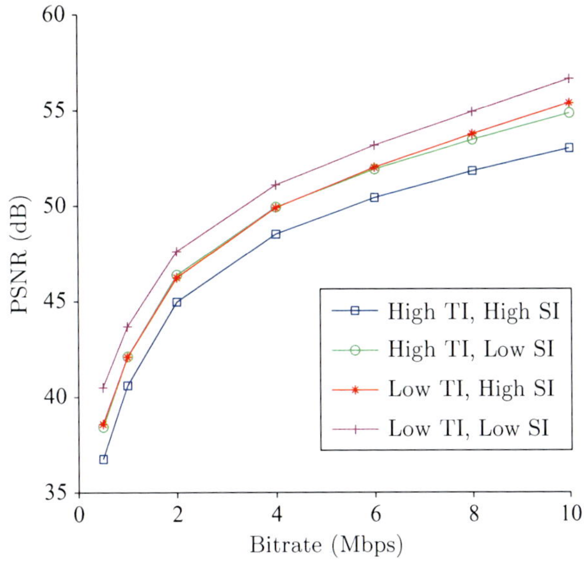

Fig. 7.4 R–D curves for news categories

(*i.e.*, $G_c = 4$) in our experiments; corresponding to *high TI/high SI*, *high TI/low SI*, *low TI/high SI*, and *low TI/low SI*. We adaptively determine the suitable coding bitrates for an editor-specified video quality for videos, using these piecewise linear R–D models.

7.4.3 Simulator Implementation and Scenarios

We implemented a trace–driven simulator for NEWSMAN using Java. Our focus is on the proposed scheduling algorithm under varying network conditions. The scheduler runs once every scheduling interval $\overline{\tau}$ (say, 5 min) in our simulator. The scheduler reads randomly generated new jobs following the Poisson process [140] as inputs. We consider 0.1, 0.5, 1, 5, and 10 per min as mean job arrival rate and randomly mark a job as breaking news or traditional news in our experiments. In the computation of news importance for videos, we randomly generate a real number in [0,1] for user reputations and location importance in simulations. We set deadlines for news videos randomly in the following time intervals: (i) [1, 2] hours for breaking news, and (ii) [2, 3] hours for traditional news. We implemented two baseline algorithms: (i) earlier deadline first (EDF), and (ii) first in first out (FIFO) scheduling algorithms. For fair comparisons, we run the simulations for 24 hours and repeat each simulation scenario 20 times. If not otherwise specified, we use the first–day network trace to drive the simulator. We use the same set of jobs (with the same arrival times, deadlines, news types, user reputations, location importance, *etc.*) for three algorithms, in a simulation iteration. We report the average performance with 95% confidence intervals whenever applicable.

Algorithm 7.2 A procedure for determining transcoding parameters for jobs in L, if j is added to the uploading list

procedure ADDJOBINLOWERQUALITY
    Initialization: $\bar{t} = t$, $\overline{\mathscr{U}} = \mathscr{U}$ ▷ Initialize updated utility and scheduled time after adding j.
    addJobToUploadList(j, $\bar{L}$) ▷ Add j to $\bar{L}$ in original quality and reduce quality later.
    n = getNumberOfJobs($\bar{L}$) ▷ Check all existing jobs in $\bar{L}$ to determine scheduling list.
    for $(m = n;\ m >= 1;\ m--)$ **do**
        for each q_k of j_m **do** ▷ q_k is k^{th} video quality of j_m.
            t_e = getEstmFinishTimeLowQlty(j_m, t, q_c) ▷ Time to upload j_m in current quality.
            $\bar{\rho}_k = \bar{u}(j_m)\ \bar{v}(t_k)\ q_k$ ▷ Compute the utility of j_m for k^{th} quality.
            $\bar{\rho}_c = \bar{u}(j_m)\ \bar{v}(t_e)\ q_c$ ▷ Compute the utility of j_m for current quality.
            $\bar{t}$ = getUpdatedScheduledTime(j_m, t_k, t_e) ▷ Get the updated scheduled time.
            $\overline{\mathscr{U}}$ = getUpdatedUtility(j_m, $\bar{\rho}_c$, $\bar{\rho}_k$, $\mathscr{U}$, $\overline{\mathscr{U}}$) ▷ Updated total utility.
            ϕ = isJobAccomodatedWihinDeadline(m, t_d, $\bar{t}$, t, $\bar{L}$, $\mathscr{U}$, $\overline{\mathscr{U}}$) ▷ All jobs in $\bar{L}$ (including j) can be scheduled within deadline after lower qualities of jobs (from j_m to j_n).
            if $(\phi == TRUE)$ **then** ▷ Set the video quality of j_m to q_k.
                updateTranscodingParam(m, $\bar{L}$, q_k) ▷ Set quality of jobs (j_{m+1} to j_n) to $\bar{\sigma}$.
                $\mathscr{U} = \overline{\mathscr{U}}$ ▷ Update total utility value after changing qualities of some jobs.
                $t = \bar{t}$ ▷ Update the total uploading time after scheduling all jobs in $\bar{L}$.
                return *TRUE*
            end if
        end for
    end for
    removeJobFromUploadList(j, $\bar{L}$) ▷ j is removed from $\bar{L}$ since it could not be scheduled.
    return *FALSE* ▷ j could not be added to $\bar{L}$ despite lowering qualities of jobs.
end procedure

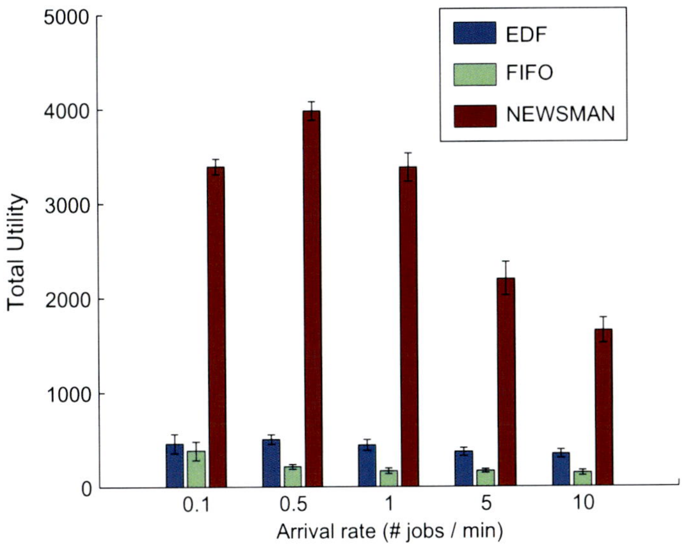

Fig. 7.5 System utility

7.4.4 *Results*

Figures 7.5, 7.6, 7.7 and 7.8 show results after running the simulator for 24 h using network traces from Delhi, India. Figures 7.9 and 7.10 show results after running the simulator for 24 h using network traces from different locations. Similarly, Figs. 7.11 and 7.12 show results after running the simulator for 24 h using network traces on different dates.

NEWSMAN delivers the most news videos in time, and achieves the highest system utility. Figures 7.5, 7.9 and 7.11 show that NEWSMAN performs up to 1200% better than baseline algorithms in terms of system utility. Figures 7.6 and 7.7 show that our system outperforms baselines (i) by up to 400% in terms of number of videos uploaded before their deadlines, and (ii) by up to 150% in terms of total number of uploaded videos. That is, NEWSMAN significantly outperforms the baselines either when news editors set hard deadlines (4$\times$ improvement) or soft deadlines (1.5$\times$ improvement).

NEWSMAN achieves low average lateness. Despite delivering the most news videos in time, and achieving the highest system utility for Delhi, NEWSMAN achieves fairly low average lateness (see Figs. 7.8, 7.10 and 7.12).

NEWSMAN performs well under all network infrastructures. Figure 7.9 shows that NEWSMAN outperforms baselines under all network conditions such as low average throughput in India, and higher average throughput in China (see Table 7.3). In the future, we would like to leverage map matching techniques to determine the importance of videos, hence, uploading order [244].

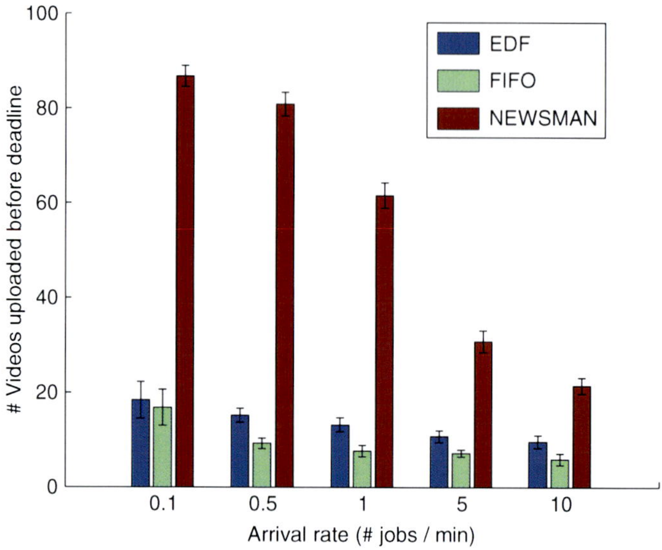

Fig. 7.6 Number of videos uploaded before deadline

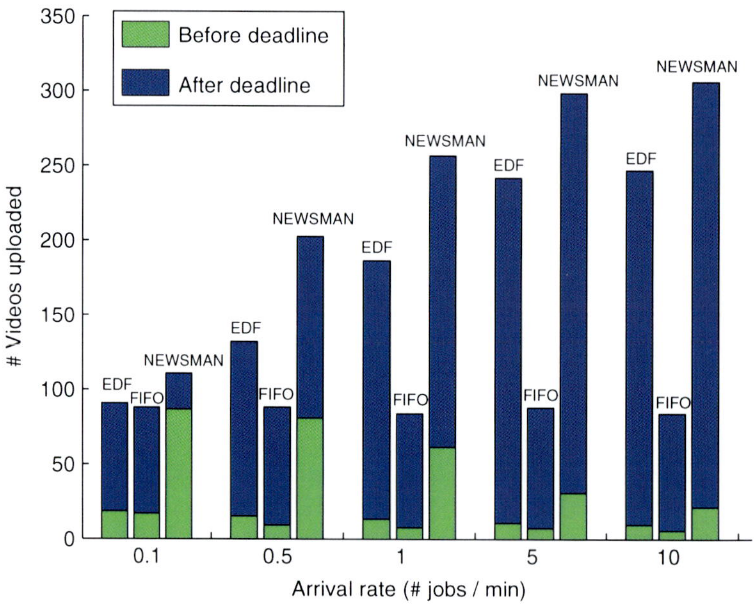

Fig. 7.7 Total number of uploaded videos

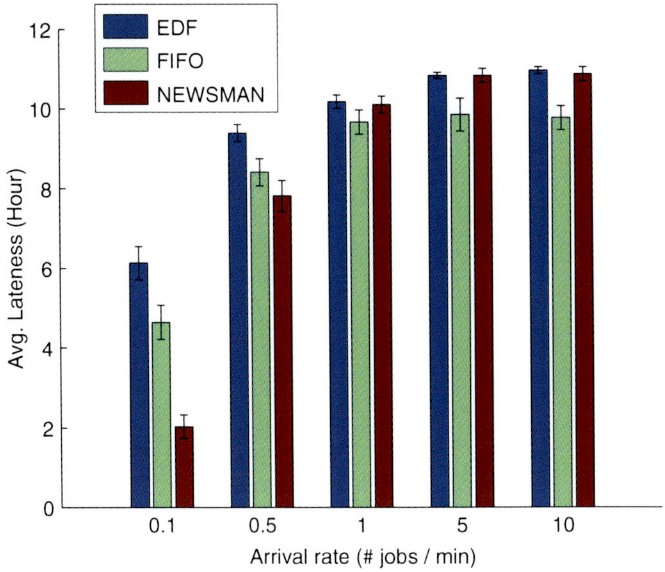

Fig. 7.8 Average lateness in uploading a job

Fig. 7.9 System utility from different locations

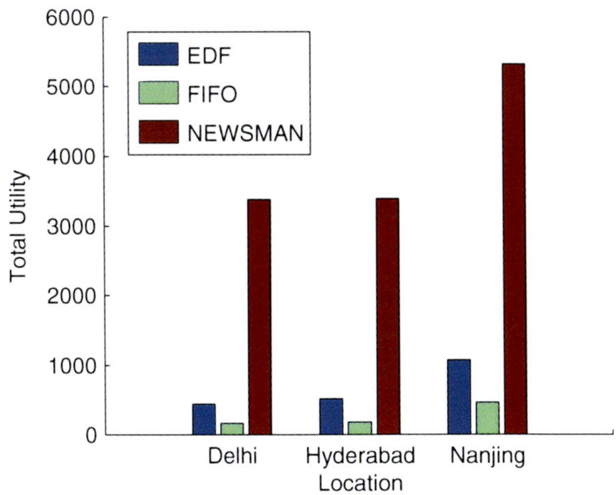

7.5 Summary

We present an innovative design for efficient uploading of news videos with deadlines under weak network infrastructures. In our proposed news reporting system called NEWSMAN, we use middleboxes with a novel scheduling and transcoding selection algorithm for uploading news videos under varying network conditions. The system intelligently schedules news videos based on their characteristics and underlying

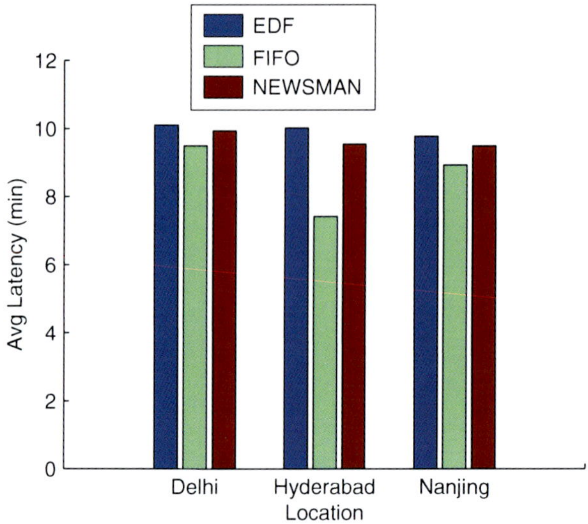

Fig. 7.10 System utility on different dates

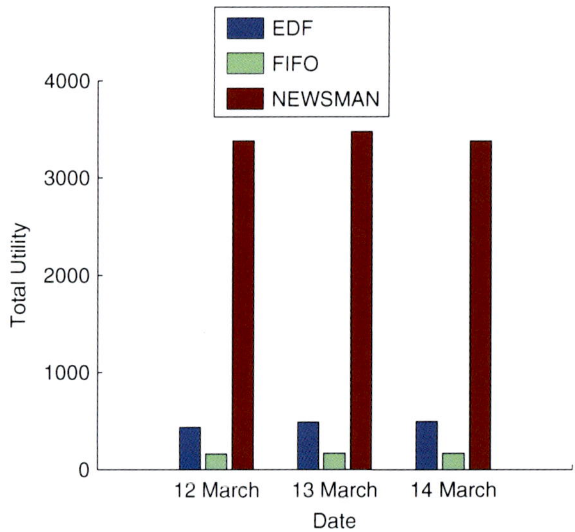

Fig. 7.11 Total number of uploaded videos

network conditions such that: (i) it maximizes the system utility, (ii) it uploads news videos in the best possible qualities, and (iii) it achieves low average lateness of the uploaded videos. We formulated this scheduling problem into a mathematical optimization problem. Furthermore, we developed a trace-driven simulator to conduct a series of extensive experiments using real datasets and network traces collected between a Singapore EC2 server and different PCs in Asia. The simulation results indicate that our proposed scheduling algorithm improves system performance. We are planning to deploy NEWSMAN in developing countries to demonstrate its practicality and efficiency in practice.

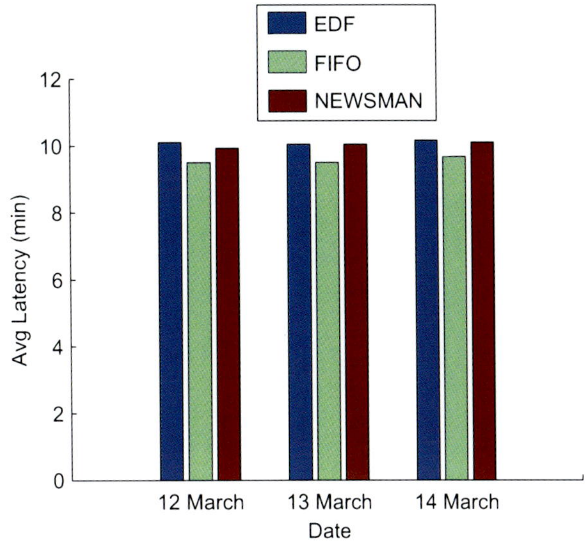

Fig. 7.12 Average lateness on different dates

References

1. Apple Denies Steve Jobs Heart Attack Report: "It Is Not True". http://www.businessinsider.com/2008/10/apple-s-steve-jobs-rushed-to-er-after-heart-attack-says-cnn-citizen-journalist/. October 2008. Online: Last Accessed Sept 2015.
2. SVM^{hmm}: Sequence Tagging with Structural Support Vector Machines. https://www.cs.cornell.edu/people/tj/svm light/svm hmm.html. August 2008. Online: Last Accessed May 2016.
3. NPTEL. 2009, December. http://www.nptel.ac.in. Online; Accessed Apr 2015.
4. iReport at 5: Nearly 900,000 contributors worldwide. http://www.niemanlab.org/2011/08/ireport-at-5-nearly-900000-contributors-worldwide/. August 2011. Online: Last Accessed Sept 2015.
5. Meet the million: 999,999 iReporters + you! http://www.ireport.cnn.com/blogs/ireport-blog/2012/01/23/meet-the-million-999999-ireporters-you. January 2012. Online: Last Accessed Sept 2015.
6. 5 Surprising Stats about User-generated Content. 2014, April. http://www.smartblogs.com/social-media/2014/04/11/6.-surprising-stats-about-user-generated-content/. Online: Last Accessed Sept 2015.
7. The Citizen Journalist: How Ordinary People are Taking Control of the News. 2015, June. http://www.digitaltrends.com/features/the-citizen-journalist-how-ordinary-people-are-taking-control-of-the-news/. Online: Last Accessed Sept 2015.
8. Wikipedia API. 2015, April. http://tinyurl.com/WikiAPI-AI. API: Last Accessed Apr 2015.
9. Apache Lucene. 2016, June. https://lucene.apache.org/core/. Java API: Last Accessed June 2016.
10. By the Numbers: 14 Interesting Flickr Stats. 2016, May. http://www.expandedramblings.com/index.php/flickr-stats/. Online: Last Accessed May 2016.
11. By the Numbers: 180+ Interesting Instagram Statistics (June 2016). 2016, June. http://www.expandedramblings.com/index.php/important-instagram-stats/. Online: Last Accessed July 2016.
12. Coursera. 2016, May. https://www.coursera.org/. Online: Last Accessed May 2016.

13. FourSquare API. 2016, June. https://developer.foursquare.com/. Last Accessed June 2016.
14. Google Cloud Vision API. 2016, December. https://cloud.google.com/vision/. Online: Last Accessed Dec 2016.
15. Google Forms. 2016, May. https://docs.google.com/forms/. Online: Last Accessed May 2016.
16. MIT Open Course Ware. 2016, May. http://www.ocw.mit.edu/. Online: Last Accessed May 2016.
17. Porter Stemmer. 2016, May. https://tartarus.org/martin/PorterStemmer/. Online: Last Accessed May 2016.
18. SenticNet. 2016, May. http://www.sentic.net/computing/. Online: Last Accessed May 2016.
19. Sentics. 2016, May. https://en.wiktionary.org/wiki/sentics. Online: Last Accessed May 2016.
20. VideoLectures.Net. 2016, May. http://www.videolectures.net/. Online: Last Accessed May, 2016.
21. YouTube Statistics. 2016, July. http://www.youtube.com/yt/press/statistics.html. Online: Last Accessed July, 2016.
22. Abba, H.A., S.N.M. Shah, N.B. Zakaria, and A.J. Pal. 2012. Deadline based performance evaluation of job scheduling algorithms. In *Proceedings of the IEEE International Conference on Cyber-Enabled Distributed Computing and Knowledge Discovery*, 106–110.
23. Achanta, R.S., W.-Q. Yan, and M.S. Kankanhalli. 2006. Modeling Intent for Home Video Repurposing. *Proceedings of the IEEE MultiMedia* 45(1): 46–55.
24. Adcock, J., M. Cooper, A. Girgensohn, and L. Wilcox. 2005. Interactive Video Search Using Multilevel Indexing. In *Proceedings of the Springer Image and Video Retrieval*, 205–214.
25. Agarwal, B., S. Poria, N. Mittal, A. Gelbukh, and A. Hussain. 2015. Concept-level Sentiment Analysis with Dependency-based Semantic Parsing: A Novel Approach. In *Proceedings of the Springer Cognitive Computation*, 1–13.
26. Aizawa, K., D. Tancharoen, S. Kawasaki, and T. Yamasaki. 2004. Efficient Retrieval of Life Log based on Context and Content. In *Proceedings of the ACM Workshop on Continuous Archival and Retrieval of Personal Experiences*, 22–31.
27. Altun, Y., I. Tsochantaridis, and T. Hofmann. 2003. Hidden Markov Support Vector Machines. In *Proceedings of the International Conference on Machine Learning*, 3–10.
28. Anderson, A., K. Ranghunathan, and A. Vogel. 2008. Tagez: Flickr Tag Recommendation. In *Proceedings of the Association for the Advancement of Artificial Intelligence*.
29. Atrey, P.K., A. El Saddik, and M.S. Kankanhalli. 2011. Effective Multimedia Surveillance using a Human-centric Approach. *Proceedings of the Springer Multimedia Tools and Applications* 51(2): 697–721.
30. Barnard, K., P. Duygulu, D. Forsyth, N. De Freitas, D.M. Blei, and M.I. Jordan. 2003. Matching Words and Pictures. *Proceedings of the Journal of Machine Learning Research* 3: 1107–1135.
31. Basu, S., Y. Yu, V.K. Singh, and R. Zimmermann. 2016. Videopedia: Lecture Video Recommendation for Educational Blogs Using Topic Modeling. In *Proceedings of the Springer International Conference on Multimedia Modeling*, 238–250.
32. Basu, S., Y. Yu, and R. Zimmermann. 2016. Fuzzy Clustering of Lecture Videos Based on Topic Modeling. In *Proceedings of the IEEE International Workshop on Content-Based Multimedia Indexing*, 1–6.
33. Basu, S., R. Zimmermann, K.L. OHalloran, S. Tan, and K. Marissa. 2015. Performance Evaluation of Students Using Multimodal Learning Systems. In *Proceedings of the Springer International Conference on Multimedia Modeling*, 135–147.
34. Beeferman, D., A. Berger, and J. Lafferty. 1999. Statistical Models for Text Segmentation. *Proceedings of the Springer Machine Learning* 34(1–3): 177–210.
35. Bernd, J., D. Borth, C. Carrano, J. Choi, B. Elizalde, G. Friedland, L. Gottlieb, K. Ni, R. Pearce, D. Poland, et al. 2015. Kickstarting the Commons: The YFCC100M and the YLI Corpora. In *Proceedings of the ACM Workshop on Community-Organized Multimodal Mining: Opportunities for Novel Solutions*, 1–6.

36. Bhatt, C.A., and M.S. Kankanhalli. 2011. Multimedia Data Mining: State of the Art and Challenges. *Proceedings of the Multimedia Tools and Applications* 51(1): 35–76.
37. Bhatt, C.A., A. Popescu-Belis, M. Habibi, S. Ingram, S. Masneri, F. McInnes, N. Pappas, and O. Schreer. 2013. Multi-factor Segmentation for Topic Visualization and Recommendation: the MUST-VIS System. In *Proceedings of the ACM International Conference on Multimedia*, 365–368.
38. Bhattacharjee, S., W.C. Cheng, C.-F. Chou, L. Golubchik, and S. Khuller. 2000. BISTRO: A Framework for Building Scalable Wide-Area Upload Applications. *Proceedings of the ACM SIGMETRICS Performance Evaluation Review* 28(2): 29–35.
39. Cambria, E., J. Fu, F. Bisio, and S. Poria. 2015. AffectiveSpace 2: Enabling Affective Intuition for Concept-level Sentiment Analysis. In *Proceedings of the AAAI Conference on Artificial Intelligence*, 508–514.
40. Cambria, E., A. Livingstone, and A. Hussain. 2012. The Hourglass of Emotions. In *Proceedings of the Springer Cognitive Behavioural Systems*, 144–157.
41. Cambria, E., D. Olsher, and D. Rajagopal. 2014. SenticNet 3: A Common and Common-sense Knowledge Base for Cognition-driven Sentiment Analysis. In *Proceedings of the AAAI Conference on Artificial Intelligence*, 1515–1521.
42. Cambria, E., S. Poria, R. Bajpai, and B. Schuller. 2016. SenticNet 4: A Semantic Resource for Sentiment Analysis based on Conceptual Primitives. In *Proceedings of the International Conference on Computational Linguistics (COLING)*, 2666–2677.
43. Cambria, E., S. Poria, F. Bisio, R. Bajpai, and I. Chaturvedi. 2015. The CLSA Model: A Novel Framework for Concept-Level Sentiment Analysis. In *Proceedings of the Springer Computational Linguistics and Intelligent Text Processing*, 3–22.
44. Cambria, E., S. Poria, A. Gelbukh, and K. Kwok. 2014. Sentic API: A Common-sense based API for Concept-level Sentiment Analysis. *CEUR Workshop Proceedings* 144: 19–24.
45. Cao, J., Z. Huang, and Y. Yang. 2015. Spatial-aware Multimodal Location Estimation for Social Images. In *Proceedings of the ACM Conference on Multimedia Conference*, 119–128.
46. Chakraborty, I., H. Cheng, and O. Javed. 2014. Entity Centric Feature Pooling for Complex Event Detection. In *Proceedings of the Workshop on HuEvent at the ACM International Conference on Multimedia*, 1–5.
47. Che, X., H. Yang, and C. Meinel. 2013. Lecture Video Segmentation by Automatically Analyzing the Synchronized Slides. In *Proceedings of the ACM International Conference on Multimedia*, 345–348.
48. Chen, B., J. Wang, Q. Huang, and T. Mei. 2012. Personalized Video Recommendation through Tripartite Graph Propagation. In *Proceedings of the ACM International Conference on Multimedia*, 1133–1136.
49. Chen, S., L. Tong, and T. He. 2011. Optimal Deadline Scheduling with Commitment. In *Proceedings of the IEEE Annual Allerton Conference on Communication, Control, and Computing*, 111–118.
50. Chen, W.-B., C. Zhang, and S. Gao. 2012. Segmentation Tree based Multiple Object Image Retrieval. In *Proceedings of the IEEE International Symposium on Multimedia*, 214–221.
51. Chen, Y., and W.J. Heng. 2003. Automatic Synchronization of Speech Transcript and Slides in Presentation. *Proceedings of the IEEE International Symposium on Circuits and Systems* 2: 568–571.
52. Cohen, J. 1960. A Coefficient of Agreement for Nominal Scales. *Proceedings of the Durham Educational and Psychological Measurement* 20(1): 37–46.
53. Cristani, M., A. Pesarin, C. Drioli, V. Murino, A. Rodà,M. Grapulin, and N. Sebe. 2010. Toward an Automatically Generated Soundtrack from Low-level Cross-modal Correlations for Automotive Scenarios. In *Proceedings of the ACM International Conference on Multimedia*, 551–560.
54. Dang-Nguyen, D.-T., L. Piras, G. Giacinto, G. Boato, and F.G. De Natale. 2015. A Hybrid Approach for Retrieving Diverse Social Images of Landmarks. In *Proceedings of the IEEE International Conference on Multimedia and Expo (ICME)*, 1–6.

55. Fabro, M. Del, A. Sobe, and L. Böszörmenyi. 2012. Summarization of Real-life Events Based on Community-contributed Content. In *Proceedings of the International Conferences on Advances in Multimedia*, 119–126.
56. Du, L., W.L. Buntine, and M. Johnson. 2013. Topic Segmentation with a Structured Topic Model. In *Proceedings of the North American Chapter of the Association for Computational Linguistics: Human Language Technologies*, 190–200.
57. Fan, Q., K. Barnard, A. Amir, A. Efrat, and M. Lin. 2006. Matching Slides to Presentation Videos using SIFT and Scene Background Matching. In *Proceedings of the ACM International Conference on Multimedia*, 239–248.
58. Filatova, E. and V. Hatzivassiloglou. 2004. Event-Based Extractive Summarization. In *Proceedings of the ACL Workshop on Summarization*, 104–111.
59. Firan, C.S., M. Georgescu, W. Nejdl, and R. Paiu. 2010. Bringing Order to Your Photos: Event-driven Classification of Flickr Images Based on Social Knowledge. In *Proceedings of the ACM International Conference on Information and Knowledge Management*, 189–198.
60. Gao, S., C. Zhang, and W.-B. Chen. 2012. An Improvement of Color Image Segmentation through Projective Clustering. In *Proceedings of the IEEE International Conference on Information Reuse and Integration*, 152–158.
61. Garg, N. and I. Weber. 2008. Personalized, Interactive Tag Recommendation for Flickr. In *Proceedings of the ACM Conference on Recommender Systems*, 67–74.
62. Ghias, A., J. Logan, D. Chamberlin, and B.C. Smith. 1995. Query by Humming: Musical Information Retrieval in an Audio Database. In *Proceedings of the ACM International Conference on Multimedia*, 231–236.
63. Golder, S.A., and B.A. Huberman. 2006. Usage Patterns of Collaborative Tagging Systems. *Proceedings of the Journal of Information Science* 32(2): 198–208.
64. Gozali, J.P., M.-Y. Kan, and H. Sundaram. 2012. Hidden Markov Model for Event Photo Stream Segmentation. In *Proceedings of the IEEE International Conference on Multimedia and Expo Workshops*, 25–30.
65. Guo, Y., L. Zhang, Y. Hu, X. He, and J. Gao. 2016. Ms-celeb-1m: Challenge of recognizing one million celebrities in the real world. *Proceedings of the Society for Imaging Science and Technology Electronic Imaging* 2016(11): 1–6.
66. Hanjalic, A., and L.-Q. Xu. 2005. Affective Video Content Representation and Modeling. *Proceedings of the IEEE Transactions on Multimedia* 7(1): 143–154.
67. Haubold, A. and J.R. Kender. 2005. Augmented Segmentation and Visualization for Presentation Videos. In *Proceedings of the ACM International Conference on Multimedia*, 51–60.
68. Healey, J.A., and R.W. Picard. 2005. Detecting Stress during Real-world Driving Tasks using Physiological Sensors. *Proceedings of the IEEE Transactions on Intelligent Transportation Systems* 6(2): 156–166.
69. Hefeeda, M., and C.-H. Hsu. 2010. On Burst Transmission Scheduling in Mobile TV Broadcast Networks. *Proceedings of the IEEE/ACM Transactions on Networking* 18(2): 610–623.
70. Hevner, K. 1936. Experimental Studies of the Elements of Expression in Music. *Proceedings of the American Journal of Psychology* 48: 246–268.
71. Hochbaum, D.S. 1996. Approximating Covering and Packing Problems: Set Cover, Vertex Cover, Independent Set, and related Problems. In *Proceedings of the PWS Approximation algorithms for NP-hard problems*, 94–143.
72. Hong, R., J. Tang, H.-K. Tan, S. Yan, C. Ngo, and T.-S. Chua. 2009. Event Driven Summarization for Web Videos. In *Proceedings of the ACM SIGMM Workshop on Social Media*, 43–48.
73. P. ITU-T Recommendation. 1999. Subjective Video Quality Assessment Methods for Multimedia Applications.
74. Jiang, L., A.G. Hauptmann, and G. Xiang. 2012. Leveraging High-level and Low-level Features for Multimedia Event Detection. In *Proceedings of the ACM International Conference on Multimedia*, 449–458.

75. Joachims, T., T. Finley, and C.-N. Yu. 2009. Cutting-plane Training of Structural SVMs. *Proceedings of the Machine Learning Journal* 77(1): 27–59.
76. Johnson, J., L. Ballan, and L. Fei-Fei. 2015. Love Thy Neighbors: Image Annotation by Exploiting Image Metadata. In *Proceedings of the IEEE International Conference on Computer Vision*, 4624–4632.
77. Johnson, J., A. Karpathy, and L. Fei-Fei. 2015. Densecap: Fully Convolutional Localization Networks for Dense Captioning. *In Proceedings of the arXiv preprint arXiv:1511.07571*.
78. Jokhio, F., A. Ashraf, S. Lafond, I. Porres, and J. Lilius. 2013. Prediction-Based dynamic resource allocation for video transcoding in cloud computing. In *Proceedings of the IEEE International Conference on Parallel, Distributed and Network-Based Processing*, 254–261.
79. Kaminskas, M., I. Fernández-Tobías, F. Ricci, and I. Cantador. 2014. Knowledge-Based Identification of Music Suited for Places of Interest. *Proceedings of the Springer Information Technology & Tourism* 14(1): 73–95.
80. Kaminskas, M. and F. Ricci. 2011. Location-adapted Music Recommendation using Tags. In *Proceedings of the Springer User Modeling, Adaption and Personalization*, 183–194.
81. Kan, M.-Y. 2001. Combining Visual Layout and Lexical Cohesion Features for Text Segmentation. *In Proceedings of the Citeseer*.
82. Kan, M.-Y. 2003. *Automatic Text Summarization as Applied to Information Retrieval*. PhD thesis, Columbia University.
83. Kan, M.-Y., J.L. Klavans, and K.R. McKeown. 1998. Linear Segmentation and Segment Significance. *In Proceedings of the arXiv preprint cs/9809020*.
84. Kan, M.-Y., K.R. McKeown, and J.L. Klavans. 2001. Applying Natural Language Generation to Indicative Summarization. *Proceedings of the ACL European Workshop on Natural Language Generation* 8: 1–9.
85. Kang, H.B. 2003. Affective Content Detection using HMMs. In *Proceedings of the ACM International Conference on Multimedia*, 259–262.
86. Kang, Y.-L., J.-H. Lim, M.S. Kankanhalli, C.-S. Xu, and Q. Tian. 2004. Goal Detection in Soccer Video using Audio/Visual Keywords. *Proceedings of the IEEE International Conference on Image Processing* 3: 1629–1632.
87. Kang, Y.-L., J.-H. Lim, Q. Tian, and M.S. Kankanhalli. 2003. Soccer Video Event Detection with Visual Keywords. *Proceedings of the Joint Conference of International Conference on Information, Communications and Signal Processing, and Pacific Rim Conference on Multimedia* 3: 1796–1800.
88. Kankanhalli, M.S., and T.-S. Chua. 2000. Video Modeling using Strata-Based Annotation. *Proceedings of the IEEE MultiMedia* 7(1): 68–74.
89. Kennedy, L., M. Naaman, S. Ahern, R. Nair, and T. Rattenbury. 2007. How Flickr Helps us Make Sense of the World: Context and Content in Community-Contributed Media Collections. In *Proceedings of the ACM International Conference on Multimedia*, 631–640.
90. Kennedy, L.S., S.-F. Chang, and I.V. Kozintsev. 2006. To Search or to Label?: Predicting the Performance of Search-Based Automatic Image Classifiers. In *Proceedings of the ACM International Workshop on Multimedia Information Retrieval*, 249–258.
91. Kim, Y.E., E.M. Schmidt, R. Migneco, B.G. Morton, P. Richardson, J. Scott, J.A. Speck, and D. Turnbull. 2010. Music Emotion Recognition: A State of the Art Review. In *Proceedings of the International Society for Music Information Retrieval*, 255–266.
92. Klavans, J.L., K.R. McKeown, M.-Y. Kan, and S. Lee. 1998. Resources for Evaluation of Summarization Techniques. *In Proceedings of the arXiv preprint cs/9810014*.
93. Ko, Y. 2012. A Study of Term Weighting Schemes using Class Information for Text Classification. In *Proceedings of the ACM Special Interest Group on Information Retrieval*, 1029–1030.
94. Kort, B., R. Reilly, and R.W. Picard. 2001. An Affective Model of Interplay between Emotions and Learning: Reengineering Educational Pedagogy-Building a Learning Companion. *Proceedings of the IEEE International Conference on Advanced Learning Technologies* 1: 43–47.

95. Kucuktunc, O., U. Gudukbay, and O. Ulusoy. 2010. Fuzzy Color Histogram-Based Video Segmentation. *Proceedings of the Computer Vision and Image Understanding* 114(1): 125–134.
96. Kuo, F.-F., M.-F. Chiang, M.-K. Shan, and S.-Y. Lee. 2005. Emotion-Based Music Recommendation by Association Discovery from Film Music. In *Proceedings of the ACM International Conference on Multimedia*, 507–510.
97. Lacy, S., T. Atwater, X. Qin, and A. Powers. 1988. Cost and Competition in the Adoption of Satellite News Gathering Technology. *Proceedings of the Taylor & Francis Journal of Media Economics* 1(1): 51–59.
98. Lambert, P., W. De Neve, P. De Neve, I. Moerman, P. Demeester, and R. Van de Walle. 2006. Rate-distortion performance of H. 264/AVC compared to state-of-the-art video codecs. *Proceedings of the IEEE Transactions on Circuits and Systems for Video Technology* 16(1): 134–140.
99. Laurier, C., M. Sordo, J. Serrà, and P. Herrera. 2009. Music Mood Representations from Social Tags. In *Proceedings of the International Society for Music Information Retrieval*, 381–386.
100. Li, C.T. and M.K. Shan. 2007. Emotion-Based Impressionism Slideshow with Automatic Music Accompaniment. In *Proceedings of the ACM International Conference on Multimedia*, 839–842.
101. Li, J., and J.Z. Wang. 2008. Real-Time Computerized Annotation of Pictures. *Proceedings of the IEEE Transactions on Pattern Analysis and Machine Intelligence* 30(6): 985–1002.
102. Li, X., C.G. Snoek, and M. Worring. 2009. Learning Social Tag Relevance by Neighbor Voting. *Proceedings of the IEEE Transactions on Multimedia* 11(7): 1310–1322.
103. Li, X., T. Uricchio, L. Ballan, M. Bertini, C.G. Snoek, and A.D. Bimbo. 2016. Socializing the Semantic Gap: A Comparative Survey on Image Tag Assignment, Refinement, and Retrieval. *Proceedings of the ACM Computing Surveys (CSUR)* 49(1): 14.
104. Li, Z., Y. Huang, G. Liu, F. Wang, Z.-L. Zhang, and Y. Dai. 2012. Cloud Transcoder: Bridging the Format and Resolution Gap between Internet Videos and Mobile Devices. In *Proceedings of the ACM International Workshop on Network and Operating System Support for Digital Audio and Video*, 33–38.
105. Liang, C., Y. Guo, and Y. Liu. 2008. Is Random Scheduling Sufficient in P2P Video Streaming? In *Proceedings of the IEEE International Conference on Distributed Computing Systems*, 53–60. IEEE.
106. Lim, J.-H., Q. Tian, and P. Mulhem. 2003. Home Photo Content Modeling for Personalized Event-Based Retrieval. *Proceedings of the IEEE MultiMedia* 4: 28–37.
107. Lin, M., M. Chau, J. Cao, and J.F. Nunamaker Jr. 2005. Automated Video Segmentation for Lecture Videos: A Linguistics-Based Approach. *Proceedings of the IGI Global International Journal of Technology and Human Interaction* 1(2): 27–45.
108. Liu, C.L., and J.W. Layland. 1973. Scheduling Algorithms for Multiprogramming in a Hard-real-time Environment. *Proceedings of the ACM Journal of the ACM* 20(1): 46–61.
109. Liu, D., X.-S. Hua, L. Yang, M. Wang, and H.-J. Zhang. 2009. Tag Ranking. In *Proceedings of the ACM World Wide Web Conference*, 351–360.
110. Liu, T., C. Rosenberg, and H.A. Rowley. 2007. Clustering Billions of Images with Large Scale Nearest Neighbor Search. In *Proceedings of the IEEE Winter Conference on Applications of Computer Vision*, 28–28.
111. Liu, X. and B. Huet. 2013. Event Representation and Visualization from Social Media. In *Proceedings of the Springer Pacific-Rim Conference on Multimedia*, 740–749.
112. Liu, Y., D. Zhang, G. Lu, and W.-Y. Ma. 2007. A Survey of Content-Based Image Retrieval with High-level Semantics. *Proceedings of the Elsevier Pattern Recognition* 40(1): 262–282.
113. Livingston, S., and D.A.V. Belle. 2005. The Effects of Satellite Technology on Newsgathering from Remote Locations. *Proceedings of the Taylor & Francis Political Communication* 22(1): 45–62.

114. Long, R., H. Wang, Y. Chen, O. Jin, and Y. Yu. 2011. Towards Effective Event Detection, Tracking and Summarization on Microblog Data. In *Proceedings of the Springer Web-Age Information Management*, 652–663.
115. L. Lu, H. You, and H. Zhang. 2001. A New Approach to Query by Humming in Music Retrieval. In *Proceedings of the IEEE International Conference on Multimedia and Expo*, 22–25.
116. Lu, Y., H. To, A. Alfarrarjeh, S.H. Kim, Y. Yin, R. Zimmermann, and C. Shahabi. 2016. GeoUGV: User-generated Mobile Video Dataset with Fine Granularity Spatial Metadata. In *Proceedings of the ACM International Conference on Multimedia Systems*, 43.
117. Mao, J., W. Xu, Y. Yang, J. Wang, Z. Huang, and A. Yuille. 2014. Deep Captioning with Multimodal Recurrent Neural Networks (M-RNN). *In Proceedings of the arXiv preprint arXiv:1412.6632*.
118. Matusiak, K.K. 2006. Towards User-Centered Indexing in Digital Image Collections. *Proceedings of the OCLC Systems & Services: International Digital Library Perspectives* 22(4): 283–298.
119. McDuff, D., R. El Kaliouby, E. Kodra, and R. Picard. 2013. Measuring Voter's Candidate Preference Based on Affective Responses to Election Debates. In *Proceedings of the IEEE Humaine Association Conference on Affective Computing and Intelligent Interaction*, 369–374.
120. McKeown, K.R., J.L. Klavans, and M.-Y. Kan. Method and System for Topical Segmentation, Segment Significance and Segment Function, 29 2002. US Patent 6,473,730.
121. Mezaris, V., A. Scherp, R. Jain, M. Kankanhalli, H. Zhou, J. Zhang, L. Wang, and Z. Zhang. 2011. Modeling and Rrepresenting Events in Multimedia. In *Proceedings of the ACM International Conference on Multimedia*, 613–614.
122. Mezaris, V., A. Scherp, R. Jain, and M.S. Kankanhalli. 2014. Real-life Events in Multimedia: Detection, Representation, Retrieval, and Applications. *Proceedings of the Springer Multimedia Tools and Applications* 70(1): 1–6.
123. Miller, G., and C. Fellbaum. 1998. *Wordnet: An Electronic Lexical Database*. Cambridge, MA: MIT Press.
124. Miller, G.A. 1995. WordNet: A Lexical Database for English. *Proceedings of the Communications of the ACM* 38(11): 39–41.
125. Moxley, E., J. Kleban, J. Xu, and B. Manjunath. 2009. Not All Tags are Created Equal: Learning Flickr Tag Semantics for Global Annotation. In *Proceedings of the IEEE International Conference on Multimedia and Expo*, 1452–1455.
126. Mulhem, P., M.S. Kankanhalli, J. Yi, and H. Hassan. 2003. Pivot Vector Space Approach for Audio-Video Mixing. *Proceedings of the IEEE MultiMedia* 2: 28–40.
127. Naaman, M. 2012. Social Multimedia: Highlighting Opportunities for Search and Mining of Multimedia Data in Social Media Applications. *Proceedings of the Springer Multimedia Tools and Applications* 56(1): 9–34.
128. Natarajan, P., P.K. Atrey, and M. Kankanhalli. 2015. Multi-Camera Coordination and Control in Surveillance Systems: A Survey. *Proceedings of the ACM Transactions on Multimedia Computing, Communications, and Applications* 11(4): 57.
129. Nayak, M.G. 2004. *Music Synthesis for Home Videos*. PhD thesis.
130. Neo, S.-Y., J. Zhao, M.-Y. Kan, and T.-S. Chua. 2006. Video Retrieval using High Level Features: Exploiting Query Matching and Confidence-Based Weighting. In *Proceedings of the Springer International Conference on Image and Video Retrieval*, 143–152.
131. Ngo, C.-W., F. Wang, and T.-C. Pong. 2003. Structuring Lecture Videos for Distance Learning Applications. In *Proceedings of the IEEE International Symposium on Multimedia Software Engineering*, 215–222.
132. Nguyen, V.-A., J. Boyd-Graber, and P. Resnik. 2012. SITS: A Hierarchical Nonparametric Model using Speaker Identity for Topic Segmentation in Multiparty Conversations. In *Proceedings of the Annual Meeting of the Association for Computational Linguistics*, 78–87.

133. Nwana, A.O. and T. Chen. 2016. Who Ordered This?: Exploiting Implicit User Tag Order Preferences for Personalized Image Tagging. *In Proceedings of the arXiv preprint arXiv:1601.06439*.
134. Papagiannopoulou, C. and V. Mezaris. 2014. Concept-Based Image Clustering and Summarization of Event-related Image Collections. In *Proceedings of the Workshop on HuEvent at the ACM International Conference on Multimedia*, 23–28.
135. Park, M.H., J.H. Hong, and S.B. Cho. 2007. Location-Based Recommendation System using Bayesian User's Preference Model in Mobile Devices. In *Proceedings of the Springer Ubiquitous Intelligence and Computing*, 1130–1139.
136. Petkos, G., S. Papadopoulos, V. Mezaris, R. Troncy, P. Cimiano, T. Reuter, and Y. Kompatsiaris. 2014. Social Event Detection at MediaEval: a Three-Year Retrospect of Tasks and Results. In *Proceedings of the Workshop on Social Events in Web Multimedia at ACM International Conference on Multimedia Retrieval*.
137. Pevzner, L., and M.A. Hearst. 2002. A Critique and Improvement of an Evaluation Metric for Text Segmentation. *Proceedings of the Computational Linguistics* 28(1): 19–36.
138. Picard, R.W., and J. Klein. 2002. Computers that Recognise and Respond to User Emotion: Theoretical and Practical Implications. *Proceedings of the Interacting with Computers* 14(2): 141–169.
139. Picard, R.W., E. Vyzas, and J. Healey. 2001. Toward machine emotional intelligence: Analysis of affective physiological state. *Proceedings of the IEEE Transactions on Pattern Analysis and Machine Intelligence* 23(10): 1175–1191.
140. Poisson, S.D. and C.H. Schnuse. 1841. Recherches Sur La Pprobabilité Des Jugements En Mmatieré Criminelle Et En Matieré Civile. Meyer.
141. Poria, S., E. Cambria, R. Bajpai, and A. Hussain. 2017. A Review of Affective Computing: From Unimodal Analysis to Multimodal Fusion. *Proceedings of the Elsevier Information Fusion* 37: 98–125.
142. Poria, S., E. Cambria, and A. Gelbukh. 2016. Aspect Extraction for Opinion Mining with a Deep Convolutional Neural Network. *Proceedings of the Elsevier Knowledge-Based Systems* 108: 42–49.
143. Poria, S., E. Cambria, A. Gelbukh, F. Bisio, and A. Hussain. 2015. Sentiment Data Flow Analysis by Means of Dynamic Linguistic Patterns. *Proceedings of the IEEE Computational Intelligence Magazine* 10(4): 26–36.
144. Poria, S., E. Cambria, and A.F. Gelbukh. 2015. Deep Convolutional Neural Network Textual Features and Multiple Kernel Learning for Utterance-level Multimodal Sentiment Analysis. In *Proceedings of the EMNLP*, 2539–2544.
145. Poria, S., E. Cambria, D. Hazarika, N. Mazumder, A. Zadeh, and L.-P. Morency. 2017. Context-Dependent Sentiment Analysis in User-Generated Videos. *In Proceedings of the Association for Computational Linguistics*.
146. Poria, S., E. Cambria, D. Hazarika, and P. Vij. 2016. A Deeper Look into Sarcastic Tweets using Deep Convolutional Neural Networks. In *Proceedings of the International Conference on Computational Linguistics (COLING)*.
147. Poria, S., E. Cambria, N. Howard, G.-B. Huang, and A. Hussain. 2016. Fusing Audio Visual and Textual Clues for Sentiment Analysis from Multimodal Content. *Proceedings of the Elsevier Neurocomputing* 174: 50–59.
148. Poria, S., E. Cambria, N. Howard, and A. Hussain. 2015. Enhanced SenticNet with Affective Labels for Concept-Based Opinion Mining: Extended Abstract. In *Proceedings of the International Joint Conference on Artificial Intelligence*.
149. Poria, S., E. Cambria, A. Hussain, and G.-B. Huang. 2015. Towards an Intelligent Framework for Multimodal Affective Data Analysis. *Proceedings of the Elsevier Neural Networks* 63: 104–116.
150. Poria, S., E. Cambria, L.-W. Ku, C. Gui, and A. Gelbukh. 2014. A Rule-Based Approach to Aspect Extraction from Product Reviews. In *Proceedings of the Second Workshop on Natural Language Processing for Social Media (SocialNLP)*, 28–37.

151. Poria, S., I. Chaturvedi, E. Cambria, and F. Bisio. 2016. Sentic LDA: Improving on LDA with Semantic Similarity for Aspect-Based Sentiment Analysis. In *Proceedings of the IEEE International Joint Conference on Neural Networks (IJCNN)*, 4465–4473.
152. Poria, S., I. Chaturvedi, E. Cambria, and A. Hussain. 2016. Convolutional MKL Based Multimodal Emotion Recognition and Sentiment Analysis. In *Proceedings of the IEEE International Conference on Data Mining (ICDM)*, 439–448.
153. Poria, S., A. Gelbukh, B. Agarwal, E. Cambria, and N. Howard. 2014. Sentic Demo: A Hybrid Concept-level Aspect-Based Sentiment Analysis Toolkit. In *Proceedings of the ESWC*.
154. Poria, S., A. Gelbukh, E. Cambria, D. Das, and S. Bandyopadhyay. 2012. Enriching SenticNet Polarity Scores Through Semi-Supervised Fuzzy Clustering. In *Proceedings of the IEEE International Conference on Data Mining Workshops (ICDMW)*, 709–716.
155. Poria, S., A. Gelbukh, E. Cambria, A. Hussain, and G.-B. Huang. 2014. EmoSenticSpace: A Novel Framework for Affective Common-sense Reasoning. *Proceedings of the Elsevier Knowledge-Based Systems* 69: 108–123.
156. Poria, S., A. Gelbukh, E. Cambria, P. Yang, A. Hussain, and T. Durrani. 2012. Merging SenticNet and WordNet-Affect Emotion Lists for Sentiment Analysis. *Proceedings of the IEEE International Conference on Signal Processing (ICSP)* 2: 1251–1255.
157. Poria, S., A. Gelbukh, A. Hussain, S. Bandyopadhyay, and N. Howard. 2013. Music Genre Classification: A Semi-Supervised Approach. In *Proceedings of the Springer Mexican Conference on Pattern Recognition*, 254–263.
158. Poria, S., N. Ofek, A. Gelbukh, A. Hussain, and L. Rokach. 2014. Dependency Tree-Based Rules for Concept-level Aspect-Based Sentiment Analysis. In *Proceedings of the Springer Semantic Web Evaluation Challenge*, 41–47.
159. Poria, S., H. Peng, A. Hussain, N. Howard, and E. Cambria. 2017. Ensemble Application of Convolutional Neural Networks and Multiple Kernel Learning for Multimodal Sentiment Analysis. *In Proceedings of the Elsevier Neurocomputing*.
160. Pye, D., N.J. Hollinghurst, T.J. Mills, and K.R. Wood. 1998. Audio-visual Segmentation for Content-Based Retrieval. In *Proceedings of the International Conference on Spoken Language Processing*.
161. Qiao, Z., P. Zhang, C. Zhou, Y. Cao, L. Guo, and Y. Zhang. 2014. *Event Recommendation in Event-Based Social Networks*.
162. Raad, E.J. and R. Chbeir. 2014. Foto2Events: From Photos to Event Discovery and Linking in Online Social Networks. In *Proceedings of the IEEE Big Data and Cloud Computing*, 508–515, .
163. Radsch, C.C. 2013. The Revolutions will be Blogged: Cyberactivism and the 4th Estate in Egypt. Doctoral Disseration. American University.
164. Rae, A., B. Sigurbjörnssön, and R. van Zwol. 2010. Improving Tag Recommendation using Social Networks. In *Proceedings of the Adaptivity, Personalization and Fusion of Heterogeneous Information*, 92–99.
165. Rahmani, H., B. Piccart, D. Fierens, and H. Blockeel. 2010. Three Complementary Approaches to Context Aware Movie Recommendation. In *Proceedings of the ACM Workshop on Context-Aware Movie Recommendation*, 57–60.
166. Rattenbury, T., N. Good, and M. Naaman. 2007. Towards Automatic Extraction of Event and Place Semantics from Flickr Tags. In *Proceedings of the ACM Special Interest Group on Information Retrieval*.
167. Rawat, Y. and M. S. Kankanhalli. 2016. ConTagNet: Exploiting User Context for Image Tag Recommendation. In *Proceedings of the ACM International Conference on Multimedia*, 1102–1106.
168. Repp, S., A. Groß, and C. Meinel. 2008. Browsing within Lecture Videos Based on the Chain Index of Speech Transcription. *Proceedings of the IEEE Transactions on Learning Technologies* 1(3): 145–156.

169. Repp, S. and C. Meinel. 2006. Semantic Indexing for Recorded Educational Lecture Videos. In *Proceedings of the IEEE International Conference on Pervasive Computing and Communications Workshops*, 5.
170. Repp, S., J. Waitelonis, H. Sack, and C. Meinel. 2007. Segmentation and Annotation of Audiovisual Recordings Based on Automated Speech Recognition. In *Proceedings of the Springer Intelligent Data Engineering and Automated Learning*, 620–629.
171. Russell, J.A. 1980. A Circumplex Model of Affect. *Proceedings of the Journal of Personality and Social Psychology* 39: 1161–1178.
172. Sahidullah, M., and G. Saha. 2012. Design, Analysis and Experimental Evaluation of Block Based Transformation in MFCC Computation for Speaker Recognition. *Proceedings of the Speech Communication* 54: 543–565.
173. Salamon, J., J. Serra, and E. Gomez. 2013. Tonal Representations for Music Retrieval: From Version Identification to Query-by-Humming. In *Proceedings of the Springer International Journal of Multimedia Information Retrieval* 2(1): 45–58.
174. Schedl, M. and D. Schnitzer. 2014. Location-Aware Music Artist Recommendation. In *Proceedings of the Springer MultiMedia Modeling*, 205–213.
175. M. Schedl and F. Zhou. 2016. Fusing Web and Audio Predictors to Localize the Origin of Music Pieces for Geospatial Retrieval. In *Proceedings of the Springer European Conference on Information Retrieval*, 322–334.
176. Scherp, A., and V. Mezaris. 2014. Survey on Modeling and Indexing Events in Multimedia. *Proceedings of the Springer Multimedia Tools and Applications* 70(1): 7–23.
177. Scherp, A., V. Mezaris, B. Ionescu, and F. De Natale. 2014. HuEvent '14: Workshop on Human-Centered Event Understanding from Multimedia. In *Proceedings of the ACM International Conference on Multimedia*, 1253–1254.
178. Schmitz, P. 2006. Inducing Ontology from Flickr Tags. In *Proceedings of the Collaborative Web Tagging Workshop at ACM World Wide Web Conference*, vol 50.
179. Schuller, B., C. Hage, D. Schuller, and G. Rigoll. 2010. Mister DJ, Cheer Me Up!: Musical and Textual Features for Automatic Mood Classification. *Proceedings of the Journal of New Music Research* 39(1): 13–34.
180. Shah, R.R., M. Hefeeda, R. Zimmermann, K. Harras, C.-H. Hsu, and Y. Yu. 2016. NEWSMAN: Uploading Videos over Adaptive Middleboxes to News Servers In Weak Network Infrastructures. In *Proceedings of the Springer International Conference on Multimedia Modeling*, 100–113.
181. Shah, R.R., A. Samanta, D. Gupta, Y. Yu, S. Tang, and R. Zimmermann. 2016. PROMPT: Personalized User Tag Recommendation for Social Media Photos Leveraging Multimodal Information. In *Proceedings of the ACM International Conference on Multimedia*, 486–492.
182. Shah, R.R., A.D. Shaikh, Y. Yu, W. Geng, R. Zimmermann, and G. Wu. 2015. EventBuilder: Real-time Multimedia Event Summarization by Visualizing Social Media. In *Proceedings of the ACM International Conference on Multimedia*, 185–188.
183. Shah, R.R., Y. Yu, A.D. Shaikh, S. Tang, and R. Zimmermann. 2014. ATLAS: Automatic Temporal Segmentation and Annotation of Lecture Videos Based on Modelling Transition Time. In *Proceedings of the ACM International Conference on Multimedia*, 209–212.
184. Shah, R.R., Y. Yu, A.D. Shaikh, and R. Zimmermann. 2015. TRACE: A Linguistic-Based Approach for Automatic Lecture Video Segmentation Leveraging Wikipedia Texts. In *Proceedings of the IEEE International Symposium on Multimedia*, 217–220.
185. Shah, R.R., Y. Yu, S. Tang, S. Satoh, A. Verma, and R. Zimmermann. 2016. Concept-Level Multimodal Ranking of Flickr Photo Tags via Recall Based Weighting. In *Proceedings of the MMCommon's Workshop at ACM International Conference on Multimedia*, 19–26.
186. Shah, R.R., Y. Yu, A. Verma, S. Tang, A.D. Shaikh, and R. Zimmermann. 2016. Leveraging Multimodal Information for Event Summarization and Concept-level Sentiment Analysis. In *Proceedings of the Elsevier Knowledge-Based Systems*, 102–109.

187. Shah, R.R., Y. Yu, and R. Zimmermann. 2014. ADVISOR: Personalized Video Soundtrack Recommendation by Late Fusion with Heuristic Rankings. In *Proceedings of the ACM International Conference on Multimedia*, 607–616.
188. Shah, R.R., Y. Yu, and R. Zimmermann. 2014. User Preference-Aware Music Video Generation Based on Modeling Scene Moods. In *Proceedings of the ACM International Conference on Multimedia Systems*, 156–159.
189. Shaikh, A.D., M. Jain, M. Rawat, R.R. Shah, and M. Kumar. 2013. Improving Accuracy of SMS Based FAQ Retrieval System. In *Proceedings of the Springer Multilingual Information Access in South Asian Languages*, 142–156.
190. Shaikh, A.D., R.R. Shah, and R. Shaikh. 2013. SMS Based FAQ Retrieval for Hindi, English and Malayalam. In *Proceedings of the ACM Forum on Information Retrieval Evaluation*, 9.
191. Shamma, D.A., R. Shaw, P.L. Shafton, and Y. Liu. 2007. Watch What I Watch: Using Community Activity to Understand Content. In *Proceedings of the ACM International Workshop on Multimedia Information Retrieval*, 275–284.
192. Shaw, B., J. Shea, S. Sinha, and A. Hogue. 2013. Learning to Rank for Spatiotemporal Search. In *Proceedings of the ACM International Conference on Web Search and Data Mining*, 717–726.
193. Sigurbjörnsson, B. and R. Van Zwol. 2008. Flickr Tag Recommendation Based on Collective Knowledge. In *Proceedings of the ACM World Wide Web Conference*, 327–336.
194. Snoek, C.G., M. Worring, and A.W. Smeulders. 2005. Early versus Late Fusion in Semantic Video Analysis. In *Proceedings of the ACM International Conference on Multimedia*, 399–402.
195. Snoek, C.G., M. Worring, J.C. Van Gemert, J.-M. Geusebroek, and A.W. Smeulders. 2006. The Challenge Problem for Automated Detection of 101 Semantic Concepts in Multimedia. In *Proceedings of the ACM International Conference on Multimedia*, 421–430.
196. Soleymani, M., J.J.M. Kierkels, G. Chanel, and T. Pun. 2009. A Bayesian Framework for Video Affective Representation. In *Proceedings of the IEEE International Conference on Affective Computing and Intelligent Interaction and Workshops*, 1–7.
197. Stober, S., and A. . Nürnberger. 2013. Adaptive Music Retrieval – A State of the Art. *Proceedings of the Springer Multimedia Tools and Applications* 65(3): 467–494.
198. Stoyanov, V., N. Gilbert, C. Cardie, and E. Riloff. 2009. Conundrums in Noun Phrase Coreference Resolution: Making Sense of the State-of-the-art. In *Proceedings of the ACL International Joint Conference on Natural Language Processing of the Asian Federation of Natural Language Processing*, 656–664.
199. Stupar, A. and S. Michel. 2011. Picasso: Automated Soundtrack Suggestion for Multimodal Data. In *Proceedings of the ACM Conference on Information and Knowledge Management*, 2589–2592.
200. Thayer, R.E. 1989. *The Biopsychology of Mood and Arousal*. New York: Oxford University Press.
201. Thomee, B., B. Elizalde, D.A. Shamma, K. Ni, G. Friedland, D. Poland, D. Borth, and L.-J. Li. 2016. YFCC100M: The New Data in Multimedia Research. *Proceedings of the Communications of the ACM* 59(2): 64–73.
202. Tirumala, A., F. Qin, J. Dugan, J. Ferguson, and K. Gibbs. 2005. Iperf: The TCP/UDP Bandwidth Measurement Tool. http://dast.nlanr.net/Projects/Iperf/
203. Torralba, A., R. Fergus, and W.T. Freeman. 2008. 80 Million Tiny Images: A Large Data set for Nonparametric Object and Scene Recognition. *Proceedings of the IEEE Transactions on Pattern Analysis and Machine Intelligence* 30(11): 1958–1970.
204. Toutanova, K., D. Klein, C.D. Manning, and Y. Singer. 2003. Feature-Rich Part-of-Speech Tagging with a Cyclic Dependency Network. In *Proceedings of the North American Chapter of the Association for Computational Linguistics on Human Language Technology*, 173–180.

205. Toutanova, K. and C.D. Manning. 2000. Enriching the Knowledge Sources used in a Maximum Entropy Part-of-Speech Tagger. In *Proceedings of the Annual Meeting of the Association for Computational Linguistics*, 63–70.
206. Utiyama, M. and H. Isahara. 2001. A Statistical Model for Domain-Independent Text Segmentation. In *Proceedings of the Annual Meeting on Association for Computational Linguistics*, 499–506.
207. Vishal, K., C. Jawahar, and V. Chari. 2015. Accurate Localization by Fusing Images and GPS Signals. In *Proceedings of the IEEE Computer Vision and Pattern Recognition Workshops*, 17–24.
208. Wang, C., F. Jing, L. Zhang, and H.-J. Zhang. 2008. Scalable Search-Based Image Annotation. *Proceedings of the Springer Multimedia Systems* 14(4): 205–220.
209. Wang, H.L., and L.F. Cheong. 2006. Affective Understanding in Film. *Proceedings of the IEEE Transactions on Circuits and Systems for Video Technology* 16(6): 689–704.
210. Wang, J., J. Zhou, H. Xu, T. Mei, X.-S. Hua, and S. Li. 2014. Image Tag Refinement by Regularized Latent Dirichlet Allocation. *Proceedings of the Elsevier Computer Vision and Image Understanding* 124: 61–70.
211. Wang, P., H. Wang, M. Liu, and W. Wang. 2010. An Algorithmic Approach to Event Summarization. In *Proceedings of the ACM Special Interest Group on Management of Data*, 183–194.
212. Wang, X., Y. Jia, R. Chen, and B. Zhou. 2015. Ranking User Tags in Micro-Blogging Website. In *Proceedings of the IEEE ICISCE*, 400–403.
213. Wang, X., L. Tang, H. Gao, and H. Liu. 2010. Discovering Overlapping Groups in Social Media. In *Proceedings of the IEEE International Conference on Data Mining*, 569–578.
214. Wang, Y. and M.S. Kankanhalli. 2015. Tweeting Cameras for Event Detection. In *Proceedings of the IW3C2 International Conference on World Wide Web*, 1231–1241.
215. Webster, A.A., C.T. Jones, M.H. Pinson, S.D. Voran, and S. Wolf. 1993. Objective Video Quality Assessment System Based on Human Perception. In *Proceedings of the IS&T/SPIE's Symposium on Electronic Imaging: Science and Technology*, 15–26. International Society for Optics and Photonics.
216. Wei, C.Y., N. Dimitrova, and S.-F. Chang. 2004. Color-Mood Analysis of Films Based on Syntactic and Psychological Models. In *Proceedings of the IEEE International Conference on Multimedia and Expo*, 831–834.
217. Whissel, C. 1989. The Dictionary of Affect in Language. In *Emotion: Theory, Research and Experience. Vol. 4. The Measurement of Emotions*, ed. R. Plutchik and H. Kellerman, 113–131. New York: Academic.
218. Wu, L., L. Yang, N. Yu, and X.-S. Hua. 2009. Learning to Tag. In *Proceedings of the ACM World Wide Web Conference*, 361–370.
219. Xiao, J., W. Zhou, X. Li, M. Wang, and Q. Tian. 2012. Image Tag Re-ranking by Coupled Probability Transition. In *Proceedings of the ACM International Conference on Multimedia*, 849–852.
220. Xie, D., B. Qian, Y. Peng, and T. Chen. 2009. A Model of Job Scheduling with Deadline for Video-on-Demand System. In *Proceedings of the IEEE International Conference on Web Information Systems and Mining*, 661–668.
221. Xu, M., L.-Y. Duan, C. Xu, M. Kankanhalli, and Q. Tian. 2003. Event Detection in Basketball Video using Multiple Modalities. *Proceedings of the IEEE Joint Conference of the Fourth International Conference on Information, Communications and Signal Processing, and Fourth Pacific Rim Conference on Multimedia* 3: 1526–1530.
222. Xu, M., N.C. Maddage, C. Xu, M. Kankanhalli, and Q. Tian. 2003. Creating Audio Keywords for Event Detection in Soccer Video. In *Proceedings of the IEEE International Conference on Multimedia and Expo*, 2:II–281.
223. Yamamoto, N., J. Ogata, and Y. Ariki. 2003. Topic Segmentation and Retrieval System for Lecture Videos Based on Spontaneous Speech Recognition. In *Proceedings of the INTERSPEECH*, 961–964.

224. Yang, H., M. Siebert, P. Luhne, H. Sack, and C. Meinel. 2011. Automatic Lecture Video Indexing using Video OCR Technology. In *Proceedings of the IEEE International Symposium on Multimedia*, 111–116.
225. Yang, Y.H., Y.C. Lin, Y.F. Su, and H.H. Chen. 2008. A Regression Approach to Music Emotion Recognition. *Proceedings of the IEEE Transactions on Audio, Speech, and Language Processing* 16(2): 448–457.
226. Ye, G., D. Liu, I.-H. Jhuo, and S.-F. Chang. 2012. Robust Late Fusion with Rank Minimization. In *Proceedings of the IEEE Conference on Computer Vision and Pattern Recognition*, 3021–3028.
227. Ye, Q., Q. Huang, W. Gao, and D. Zhao. 2005. Fast and Robust Text Detection in Images and Video Frames. *Proceedings of the Elsevier Image and Vision Computing* 23(6): 565–576.
228. Yin, Y., Z. Shen, L. Zhang, and R. Zimmermann. 2015. Spatial temporal Tag Mining for Automatic Geospatial Video Annotation. *Proceedings of the ACM Transactions on Multimedia Computing, Communications, and Applications* 11(2): 29.
229. Yoon, S. and V. Pavlovic. 2014. Sentiment Flow for Video Interestingness Prediction. In *Proceedings of the Workshop on HuEvent at the ACM International Conference on Multimedia*, 29–34.
230. Yu, Y., K. Joe, V. Oria, F. Moerchen, J.S. Downie, and L. Chen. 2009. Multiversion Music Search using Acoustic Feature Union and Exact Soft Mapping. *Proceedings of the World Scientific International Journal of Semantic Computing* 3(02): 209–234.
231. Yu, Y., Z. Shen, and R. Zimmermann. 2012. Automatic Music Soundtrack Generation for Outdoor Videos from Contextual Sensor Information. In *Proceedings of the ACM International Conference on Multimedia*, 1377–1378.
232. Zaharieva, M., M. Zeppelzauer, and C. Breiteneder. 2013. Automated Social Event Detection in Large Photo Collections. In *Proceedings of the ACM International Conference on Multimedia Retrieval*, 167–174.
233. Zhang, J., X. Liu, L. Zhuo, and C. Wang. 2015. Social Images Tag Ranking Based on Visual Words in Compressed Domain. *Proceedings of the Elsevier Neurocomputing* 153: 278–285.
234. Zhang, J., S. Wang, and Q. Huang. 2015. Location-Based Parallel Tag Completion for Geo-tagged Social Image Retrieval. In *Proceedings of the ACM International Conference on Multimedia Retrieval*, 355–362.
235. Zhang, M., J. Wong, W. Tavanapong, J. Oh, and P. de Groen. 2004. Media Uploading Systems with Hard Deadlines. In *Proceedings of the Citeseer International Conference on Internet and Multimedia Systems and Applications*, 305–310.
236. Zhang, M., J. Wong, W. Tavanapong, J. Oh, and P. de Groen. 2008. Deadline-constrained Media Uploading Systems. *Proceedings of the Springer Multimedia Tools and Applications* 38(1): 51–74.
237. Zhang, W., J. Lin, X. Chen, Q. Huang, and Y. Liu. 2006. Video Shot Detection using Hidden Markov Models with Complementary Features. *Proceedings of the IEEE International Conference on Innovative Computing, Information and Control* 3: 593–596.
238. Zheng, L., V. Noroozi, and P.S. Yu. 2017. Joint Deep Modeling of Users and Items using Reviews for Recommendation. In *Proceedings of the ACM International Conference on Web Search and Data Mining*, 425–434.
239. Zhou, X.S. and T.S. Huang. 2000. CBIR: from Low-level Features to High-level Semantics. In *Proceedings of the International Society for Optics and Photonics Electronic Imaging*, 426–431.
240. Zhuang, J. and S.C. Hoi. 2011. A Two-view Learning Approach for Image Tag Ranking. In *Proceedings of the ACM International Conference on Web Search and Data Mining*, 625–634.
241. Zimmermann, R. and Y. Yu. 2013. Social Interactions over Geographic-aware Multimedia Systems. In *Proceedings of the ACM International Conference on Multimedia*, 1115–1116.

242. Shah, R.R. 2016. Multimodal-based Multimedia Analysis, Retrieval, and Services in Support of Social Media Applications. In *Proceedings of the ACM International Conference on Multimedia*, 1425–1429.
243. Shah, R.R. 2016. Multimodal Analysis of User-Generated Content in Support of Social Media Applications. In *Proceedings of the ACM International Conference in Multimedia Retrieval*, 423–426.
244. Yin, Y., R.R. Shah, and R. Zimmermann. 2016. A General Feature-based Map Matching Framework with Trajectory Simplification. In *Proceedings of the ACM SIGSPATIAL International Workshop on GeoStreaming*, 7.

Chapter 8
Conclusion and Future Work

Abstract This book studied several significant multimedia analytics problems and presented their solutions leveraging multimodal information. The multimodal information of user-generated multimedia content (UGC) is very useful in an effective search, retrieval, and recommendation services on social media. Specifically, we determine semantics and sentics information from UGC, and leverage them in building improved systems for several significant multimedia analytics problems. We collected and created the significant amount of user-generated multimedia content in our study. To benefit from the multimodal information, we extract knowledge structures from different modalities and exploit them in our solutions for several significant multimedia-based applications. We presented our solution on event understanding from UGIs, tag ranking and recommendation for UGIs, soundtrack recommendation for UGVs, lecture videos segmentation, and news videos uploading in the area with weak network infrastructures leveraging multimodal information. Here we summarize our contributions and future work for several significant multimedia analytics problems.

Keywords Multimodal analysis • User-generated multimedia content • SmartTutor • Google cloud vision API • Multimedia analytics problems • E-learning agent

8.1 Event Understanding

For event understanding, we presented two real-time multimedia summarization systems: (i) EventBuilder and (ii) EventSensor. They perform semantics and sentics analysis on UGIs at social media platforms such as Flickr, respectively. Our systems present multimedia summaries by enabling users to generate summaries by selecting an event name, a timestamp, and a mood tag. Our systems produce multimedia summaries in real-time and facilitate an effective way to get an overview of an event based on input semantics and sentics queries. EventBuilder [182] performs an offline event detection and next produces real-time multimedia summaries for a given event by solving an optimization problem. EventSensor [186] enables users to obtain sentics-based multimedia summaries such as the slideshow of UGIs with matching soundtracks. If users select a mood tag

R. Shah, R. Zimmermann, *Multimodal Analysis of User-Generated Multimedia Content*, Socio-Affective Computing 6, DOI 10.1007/978-3-319-61807-4_8

as an input then soundtracks corresponding to the input mood tag are selected. If users choose an event as input then soundtracks corresponding to the most frequent mood tags of UGIs in the representative set for the event are attached to the slideshow. Experimental results on the YFCC100M dataset confirm that our systems outperform their baselines. Specifically, EventBuilder outperforms its baseline by 11.41% in terms of event detection (see Table 3.7). Moreover, EventBuilder outperforms its baseline for text summaries of events by (i) 19.36% in terms of informative rating, (ii) 27.70% in terms of experience rating, and (ii) 21.58% in terms of acceptance rating (see Table 3.11 and Fig. 3.9). Our EventSensor system investigated the fusion of multimodal information (*i.e.*, user tags, title, description, and visual concepts) to determine sentics details of UGIs. Experimental results indicate that features based on user tags are salient and the most useful in determining sentics details of UGIs (see Fig. 3.10).

In our future work, we plan to add two new characteristics to the EventSensor system: (i) introducing diversity in multimedia summaries by leveraging visual concepts of UGIs and (ii) enabling users to obtain multimedia summaries for a given event and mood tag. Since relevance and diversity are the two main characteristics of a good multimedia summary [54], we would like to consider them in our produced summaries. However, the selection of the representative set R in EventBuilder lacks diversity because R is constructed based on relevance scores of UGIs only. Thus, we plan to address the diversity criterion in our enhanced systems by performing the clustering of UGIs during pre-processing. Clusters are formed based on visual concepts derived from the visual content of UGIs and helpful in producing diverse multimedia summaries. For instance, clustering based on visual concepts helps in producing a multimedia summary with visually dissimilar photos (*i.e.*, from different clusters). Next, to enable users to obtain multimedia summaries for any input event, we plan to compute the semantics similarity between the input event and all known events, clusters, and mood tags. We can compute the semantics similarity of an input event with 1756 visual concepts and known events using Apache Lucene and WordNet. In our current work [186], we do not evaluate how good our produced summary as compared to other possible summaries. Earlier work [84, 92] suggest the task of creating indicative summaries that help a user decide whether to read a particular document is a difficult task. Thus, in future work, we can examine different summaries to produce a summary that is easy to understand. Furthermore, in our current work [186], we selected photos in random order to generate a slideshow from UGIs of a given event or mood and attach only one soundtrack for full slideshow. However, this selection can be improved further by a method based on Hidden Markov Model for event photo stream segmentation [64].

Due to advancements in computing power and deep neural networks (DNN), it is now feasible to quickly recognize a huge number of concepts in UGIs and UGVs. Thus, DNN-based new image representations are considered to be very useful in image and video retrieval. For instance, Google Cloud Vision API [14] can quickly classify photos into thousands of categories. Such categories provide much semantics information for UGIs and UGVs. These semantics categories can be further used to construct high-level features and train learning models to solve several significant multimedia analytics problems such as surveillance, users' preferences, privacies, and e–commerce. Moreover, the amount of UGC on a web (specifically on social media websites) has increased rapidly due to advancements in smartphones, digital cameras, and wireless technologies. Furthermore, UGC in social media platforms is not just multimedia content but a lot of contextual information such as spatial and temporal information, annotations, and other sensor data are also associated with it. Thus, categories determined by Google Cloud Vision API from UGIs and UGVs can be fused with other available contextual information and existing knowledge bases for multimodal indexing and storage of multimedia data. We can determine the fusion weights for different modalities based on DNN techniques.

In the near future, first, we would like to leverage knowledge structures from heterogeneous signals to address several significant problems related to perception, cognition, and interaction. Advancements in deep neural networks help us in analyzing affective information from UGC. For instance, in addition to determining the thousands of categories from photos, Google Cloud Vision API can also analyze emotional facial attributes of people in photos such as joy, sorrow, and anger. Thus, such information will help in developing methods and techniques to make UGIs and UGVs available, searchable and accessible in the context of user needs. We would like to bridge the gap between knowledge representation and interactive exploration of user-generated multimedia content leveraging domain knowledge in addition to content analysis. Simultaneously, we would like to explore links among unconnected multimedia data available on the web. Specifically, we would like to explore hypergraph structures for multimedia documents and create explicit and meaningful links between them based on not only content-based proximity but also exploiting domain knowledge and other multimodal information to provide focused relations between documents. We would also like to leverage social media network characteristics to create useful links among multimedia content. Social media network information is very useful in providing the personalized solutions for different multimedia analytics problems based on the friendship network of a user. Finally, we would like to bridge the gap between knowledge representation and interactive exploration of multimedia content by applying the notion of knowledge representation and management, data mining, social media network analysis, and visualization. We can employ this solution in a number of application domains such as tourism, journalism, distance learning, and surveillance.

8.2 Tag Recommendation and Ranking

Subsequently, we further focus on the semantics understanding of UGIs by computing their tag relevance scores. Based on tag relevance scores, we presented our solutions for tag recommendation and ranking. User tags are very useful in an effective multimedia search, retrieval, and recommendation. They are also very useful in semantics and sentics based multimedia summarization [182, 186]. In our tag relevance computation work, first, we presented our tag recommendation system, called, PROMPT, that predicts user tags for UGIs in the following four steps: (i) it determines a group of users who have similar tagging behavior as the user of a given photo, (ii) it computes relevance scores of tags in candidate sets determined from tag co-occurrence and neighbor voting, (iii) it fuses tags and their relevance scores of candidate sets determined from different modalities after normalizing scores between 0 to 1, and (iv) it predicts the top five tags with the highest relevance scores from the merged candidate tag lists. We construct feature vectors for users based on their past annotated UGIs using the bag-of-the-words model and compute similarities among them using the cosine similarity metric. Since it is very difficult to predict user tags from a virtually endless pool of tags, we consider the 1540 most frequent tags used in the YFCC100M dataset for tag prediction. Our PROMPT [181] system recommends user tags with 76% accuracy, 26% precision, and 20% recall for five predicted tags on the test set with 46,700 photos from Flickr (see Figs. 4.8, 4.9, and 4.10). Thus, there is an improvement of 11.34%, 17.84%, and 17.5% in terms of accuracy, precision, and recall evaluation metrics, respectively, in the performance of the PROMPT system as compared to the best performing state-of-the-art for tag recommendation (*i.e.*, an approach based on random walk, see Sect. 4.2.1). In our next tag relevance computation work, we presented a tag ranking system.

We presented a tag ranking system, called, CRAFT [185], that ranks tags of UGIs based on three proposed novel high-level features. We construct such high-level features using the bag-of-the-words model based on concepts derived from different modalities. We determine semantically similar neighbors of UGIs leveraging concepts derived in the earlier step. We compute tag relevance for UGIs for different modalities based on vote counts accumulated from semantically similar neighbors. Finally, we compute the final tag relevance for UGIs by performing a late fusion based on weights determined by the recall of modalities. The NDCG score of tags ranked by our CRAFT system is 0.886264, *i.e.*, there is an improvement of 22.24% in the NDCG score for the original order of tags (the baseline). Moreover, there is an improvement of 5.23% and 9.28% in the tag ranking performance (in terms of NDCG scores) of the CRAFT system than the following two most popular state-of-the-arts, respectively: (i) a probabilistic random walk approach (PRW) [109] and (ii) a neighbor voting approach (NVLV) [102] (see Fig. 4.13 and Sect. 4.3.2 for details). Furthermore, our proposed recall-based late fusion technique for tag ranking results in 9.23% improvement in terms of the NDCG score than the early fusion technique (see Fig. 4.12). Results from our CRAFT system is consistent with different numbers of neighbors (see Fig. 4.14). Recently, Li et al. [103] presented a comparative survey on tag assignment, refinement, and retrieval. This indicates that deep neural network models are getting much attention to solve these problems. Thus, in our future work for tag recommendation and ranking, we would like to leverage deep neural network techniques to compute tag relevance.

Since finding photo neighbors is a very important component in our tag recommendation and ranking systems, we would like to determine photo neighbors leveraging deep neural network (DNN) techniques. Such techniques are able to learn DNN-based new representations that contribute performance improvement in neighbors computing. Specifically, in the future, we would like to determine neighbors of UGIs leveraging photo metadata nonparametrically, then use a deep neural network to blend visual information from the photo and its neighbors [76]. Since spatial information is also an important component in our techniques to compute tag relevance computation, we would like to improve this component through the work by Shaw et al. [192] further. They investigated the problem of mapping a noisy estimate of a user's current location to a semantically meaningful point of interest, such as a home, park, restaurant, or store. They suggested that despite the poor accuracy of GPS on current mobile devices and the relatively high density of places in urban areas, it is possible to predict a user's location with considerable precision by explicitly modeling both places and users by combining a variety of signals about a user's current context. Furthermore, in the future, we plan to leverage the field-of-view (FoV) model [116, 228] to accurately determine tags based on the location of the user and objects in UGIs. FoV model is very important since often objects in UGIs and UGVs are located a bit far from the camera location (*e.g.*, a user captures the photo of a bridge from a skyscraper located a few hundred meters away from the bridge). In future, we would also like to leverage social media network characteristics in order to accurately learn users' preferences and tag graph.

In our future work, we would also like to work on tag recommendation and ranking for 360 videos (*i.e.*, 360-degree videos). Recently, 360 videos are getting much popular since they not only include scenes from one direction but cover omnidirectional scenes. Thus, a frame of such a video consists several scenes and regions that can be described by different annotations (*e.g.*, tags and captions). Therefore, a very interesting problem that we would like to work in future is to recommend tags and captions at frame, segment, and video level for 360 videos. Moreover, the availability of several sensor information (*e.g.*, GPS, compass, light sensors, and motion sensors) in devices that can capture 360 videos (*e.g.*, Samsung 360 cameras), opens several interesting research problems. For instance, a multimedia summary for 360 videos can be created leveraging both content and contextual information. Recommending and ranking of tags for such 360 videos are also very useful in determining summaries of the 360 videos by ranking the important regions, frames, and segments. We can also leverage the Google Cloud Vision API to solve such problems because it can quickly classify regions in photos into thousands of known-categories. Semantics information derived from the classified categories provide an overview of 360 videos. Thus, in our future work, we can leverage deep neural network technologies to determine semantics and sentics details of 360 videos.

8.3 Soundtrack Recommendation for UGVs

We further focus on the sentics understanding of UGVs by determining scene moods from videos and recommending matching soundtracks. We presented the ADVISOR system [187, 188] for user preference-aware video soundtrack generation. Our work represents one of the first attempts for user preference-aware video soundtrack generation. User-generated heterogeneous data augments video soundtrack recommendations for individual users by leveraging user activity logs from multiple modalities by using semantics concepts. The ADVISOR system exploits content and contextual information to automatically generates a matching soundtrack for a UGV in four steps. In particular, first, it recognizes scene moods in the UGV using a learning model based on the late fusion of geo- and visual features. Second, a list of songs is recommended based on the predicted scene moods using our proposed novel heuristic method. Third, this list of songs is re-ranked based on the user's listening history. Finally, a music video is generated automatically by selecting the most appropriate song using a learning model based on the late fusion of visual and concatenated audio features using our Android application. ADVISOR investigated the emotion prediction accuracies of several learning models (see Table 5.3) based on geo- and visual features of outdoor UGVs. We found that the proposed model M_{GVC} based on the late fusion of learning models M_G and M_F (proposed baselines) that are built from geo- and visual features, respectively, performed the best. Particularly, M_{GVC} performs 30.83%, 13.93%, and 14.26% better than M_F, M_G, and M_{Cat}, respectively. M_{Cat} is the model build by concatenating geo- and visual features for training. Moreover, the emotion prediction accuracy (70.0%) of the generated soundtrack UGVs from D_{GeoVid} by the ADVISOR system is comparable to the emotion prediction accuracy (68.8%) of soundtrack videos from $D_{Hollywood}$ of the Hollywood movies. Since the deep neural network is getting much popularity these days due to affordable computing resources, we would like to compare the emotion prediction accuracies of our ADVISOR system with the system based on the deep neural network. Similar to our above mentioned problems, we would like to derive affective details from videos using Google Cloud Vision API. Particularly, in our future work, we would like to build a personalized and location-aware soundtrack recommendation system for outdoor UGVs based on emotion predicted by a learning model based on a deep neural network.

In the future, we would like to address the automated mining of salient sequences of actions in the UGVs for an effective sentics analysis. Moreover, since our ADVISOR system predicts scene moods for sensor-rich UGVs leveraging spatial information, it is important to use accurate location information for sentics analysis. Thus, similar to our extended future work on tag recommendation and ranking, we plan to leverage both the field-of-view (FoV) model [116, 228] and the model based on the combination of a variety of signals about a user's current context [192] to accurately determine geo categories for every segments of UGVs. Furthermore, we would like to collect additional semantics and sentics details of UGVs using Google Cloud Vision API that can quickly classify photos (keyframes of video segments) into thousands of semantics categories and identify emotional facial attributes of people in photos such as joy, sorrow, and anger. We would like to associate semantics and sentics details further, as methods described in our earlier work [186]. We can associate semantics and sentics details using existing knowledge bases such as SenticNet, EmoSenticNet, and WordNet. Subsequently, we want to use a deep neural network to blend visual information from keyframes of video segments and (semantics and sentics) information derived using Google Cloud Vision API. For recommending a location-aware matching music for a UGV, we can leverage knowledge-based identification of music suited for places of interest [79]. Since location information is useful in map matching for trajectory simplification [244], we can also explore this direction to analyze the trajectory of UGVs for a better soundtrack recommendation.

Currently, the ADVISOR system determines scene moods for a UGV based on visual content and geo information only. It has totally ignored the audio content of the UGV in the prediction of scene moods. Since sound is an important aspect of video, we would like to extract useful knowledge structures from the audio of a UGV for improving the emotion prediction accuracy of our ADVISOR system further. For instance, despite a UGV has ambient background noise, some of its segments may have meaningful information, *e.g.*, a crowd cheering when your kid hits a baseball, or a baby laughing. We can use a deep neural network (DNN) to identify such high-level action categories from audio signals since DNN technologies have yielded immense success in computer vision, natural language processing (NLP), and speech processing. We can also model the variation in audio energy to identify salient video segments, and then identify scene moods for the UGV. We would also like to investigate the correlation between audio signals and geo categories to determine scene moods of UGVs further. Inspired by the earlier work [175] that localize the origin of music pieces for geospatial retrieval by fusing web and audio predictors, we would like to use a deep neural network to blend geo-, visual, and audio content to predict scene moods for different videos segments of UGVs. Next, we would like to use the deep neural network to learn the association between scenes (visual content) and music (audio signals) from professional video soundtracks from Hollywood movies or official music albums. This model will help us to determine a matching soundtrack for different segments of UGVs automatically. We would also like to investigate how can we efficiently combine the recommended soundtrack with the existing audio of a UGV, *i.e.*, we would like to determine weights for the recommended soundtrack and the audio for different video segments of the UGV.

Since our work on recommending soundtrack for UGVs are among the first attempts for this problem, we would like to evaluate the generated soundtrack video to some strong baselines. Minimally, we can use the following baselines to compare against automatically generated version: (i) randomly select any soundtrack from the music dataset and attach it to a UGV, (ii) randomly select any soundtrack from the music dataset, which has the same scene mood as predicted by emotion prediction model, and then attach it to the UGV, and (iii) the soundtrack is selected by a human evaluator and attach it to the UGV. Among the above baselines settings, we can also consider the personalization factor to compare our system with strong baselines.

8.4 Lecture Video Segmentation

We further focused on the semantics understanding of UGVs by determining segment boundaries for lecture videos. We presented two solutions for automatic lecture video segmentation. First, we proposed the ATLAS system which provides a novel way to automatically determine the temporal segmentation of lecture videos leveraging visual content and transcripts of UGVs. Next, we proposed the TRACE system which provides a novel approach to automatically resolve segment boundaries of lecture videos by leveraging Wikipedia texts. To the best of our knowledge, our work is the first attempt to compute segment boundaries using crowdsourced knowledge base such as Wikipedia. The proposed ATLAS system [183] works in two steps. In the first step, it determines the temporal segmentation by fusing transition cues computed from the visual content and speech transcripts of lecture videos. In the next step, it annotates titles corresponding to determined temporal transitions. In the proposed TRACE system [184], first, we compute segment boundaries from visual content and SRT (speech transcript), as described in state-of-the-arts [107, 183]. Next, we compute segment boundaries by leveraging Wikipedia texts. We further investigated their (Wikipedia segment boundaries) fusion with segment boundaries determined from the visual content and SRT of lecture videos, as described in state-of-the-arts [107, 183]. Experimental results confirm that the ATLAS and TRACE systems can effectively segment lecture videos to facilitate the accessibility and traceability within their content although video qualities are not sufficiently high. Specifically, the segment boundaries derived from the Wikipedia knowledge base outperforms state-of-the-arts in terms of precision, *i.e.*, 25.54% and 29.78% better than approaches when only visual content [183] and speech transcript [107] are used in segment boundaries detection from lecture videos, respectively (see Table 6.5). Moreover, the segment boundaries derived from the Wikipedia knowledge base outperforms state-of-the-arts in terms of F1 score, *i.e.*, 48.04% and 12.53% better than approaches when only visual content [183] and speech transcript [107] are used in segment boundaries detection from lecture videos, respectively. Finally, the fusion of segment boundaries derived from visual content, speech transcript, and Wikipedia knowledge base results in the highest recall score.

In the future, we plan to use the statistical approach proposed by Beeferman et al. [34] to automatically partitioning text (speech transcript) into coherent segments. Their proposed models use two classes of features: (i) topicality features and (ii) cue-word features. The former features use adaptive language models in a novel way to detect broad changes of topic. The later features detect occurrences of specific words such that (i) they may be domain-specific and (ii) they tend to be used near segment boundaries. Furthermore, Beeferman et al. [34] proposed a new probabilistically motivated error metric, called the P_k evaluation metric, for the assessment of segmentation approaches. Furthermore, we would like to use a new evaluation metric, called WindowDiff, for text segmentation proposed by Pevzner and Hearst [137] that addresses the problems in the P_k evaluation metric. For text (speech transcript), the unit can be a group of words or sentences. We may also quantize the lecture video in time or segment into chunks based on pauses (low energy) in audio signals. Thus, similar to earlier work [56, 132], we would like to employ P_k and WindowDiff instead of as precision, recall, and F1 score to evaluate our lecture video segmentation system in the future.

Once we get the correct segment boundaries from lecture videos, we assist e–learning through automatically determining topics for different lecture video segments. Thus, in the future, we would also like to focus on the topic modeling for different segments in lecture videos. Basu et al. [31, 32] used topic modeling to map videos (*e.g.*, YouTube and VideoLectures.Net) and blogs (Wikipedia and Edublogs) in the common semantic space of topics. These work perform topic modeling based on text processing only. Thus, we would like to improve topic modeling using a deep neural network to blend information from visual content, audio signals, speech transcript, and information from other available knowledge bases further. Next, we would like to evaluate the performance of students using multimodal learning system [33]. We plan to introduce a browsing tool for use and evaluation by students; that is based on segment boundaries derived from our proposed systems and topics determined through topic modeling techniques mentioned above. Considering the immense success of deep neural network technologies in computer vision, natural language processing (NLP), and speech processing, we would like exploit DNN-based new representations to contribute performance improvement in lecture videos segmentation.

In our long-term future work, we would like to build an intelligent tutor, called SmartTutor, that can provide a lecture to a student based on the student's need. Figure 8.1 shows the motivation for our SmartTutor system. Similar to a real tutor who can understand expressions of a student for different topics and teach accordingly (say, we exploit emotional facial attributes using Google Cloud Vision API to determine affective states), SmartTutor can adaptively change its teaching content, style, speed, and medium of instructions to facilitate students. That is, our SmartTutor system can adjust itself based on a student's needs and comfortable zone. Specifically, first, it can automatically analyze and collect a huge amount of multimedia data for any given topic, considering the student's interests, affective state, and learning history. Next, it will prepare a unified teaching material from multimedia data collected from multiple sources. Finally, SmartTutor adaptively controls its teaching speed, language, style, and content based on continuous signals collected from the student such as facial expressions, eye gaze tracking, and other signals. Figure 8.2 shows the system framework of our SmartTutor system. It has two main components: (i) knowledge base and (ii) controller. The knowledge base component keeps track of all dataset, ontologies, and other available data. The controller component process all data and signals to adaptively decide teaching content and strategies. The controller follows closed-loop learning. Thus, it actively learns the teaching strategies and provides a personalized teaching experience. Moreover, SmartTutor could be very much useful for the persons with disabilities since it is based on analyzing signals from heterogeneous sources and act accordingly.

Fig. 8.1 Motivation for SmartTutor

8.5 Adaptive News Video Uploading

Subsequently, we focus on automatic news videos uploading to support users (say, citizen journalists) in the areas under weak network infrastructures. Since news is very sensitive to time and need to be broadcasted before it is too late, it is required to timely uploaded to news servers. To address this problem, we present an innovative design, called, NEWSMAN, for efficient uploading of news videos with deadlines under weak network infrastructures. We use middleboxes with a novel scheduling and transcoding selection algorithm for uploading news videos under varying network conditions. NEWSMAN schedules news videos based on their characteristics and underlying network conditions. It solves an optimization problem to maximize the system utility, upload news videos in the best possible qualities, and achieve low average lateness of the uploaded videos. We conduct a series of experiments using trace-driven simulations, which confirm that our approach is practical and performs well. Experimental results confirm that NEWSMAN outperforms the existing algorithms: (i) by 12 times in terms of system utility (*i.e.*, sum of utilities of all uploaded videos), and (ii) by 4 times in terms of the number of videos uploaded before their deadline. Since our current work is mainly based on trace-driven simulations, we would like to perform extend our experiments with real-world scenarios in future. Since not all news of same importance, not all citizen journalists of same reputations, and not all locations are of same importance at different times, we would like to efficiently determine these factors adaptively to prioritize news videos for uploading. Given the immense success of deep neural network technologies in computer vision, natural language processing (NLP), and speech processing, we would like exploit DNN-based new representations to determine the importance of news videos and optimal bitrate for transcodeing news videos.

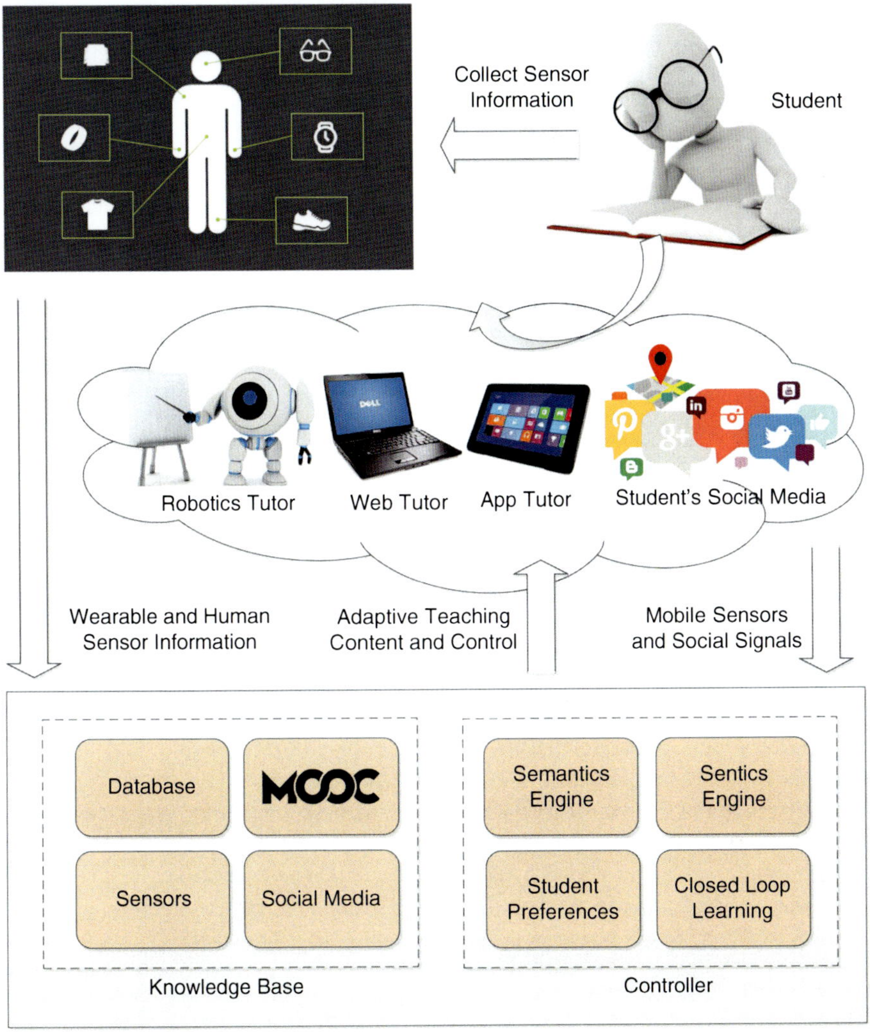

Fig. 8.2 System framework of SmartTutor

8.6 SMS and MMS-Based Search and Retrieval System

In addition to five multimedia analytics problems discussed in this book, we would also like to focus on the multimedia search and retrieval system over Short Message Service (SMS) and Multimedia Messaging Service (MMS). The increased penetration of Internet makes multimedia information available at any place and time from any devices connected to the Internet. However, there is still a significant number of mobile users those do not have access to the Internet due to its high price, network infrastructures, and others reasons, especially in developing countries. Thus, it requires an efficient information retrieval technique to retrieve relevant information from a huge amount of information spread over the Internet. SMS has become very popular as the one of the easiest, fastest, and cheapest way of communication due to the ubiquitous availability of mobile devices. Shaikh et al. [189, 190] presented a system for SMS-based FAQ (Frequently Asked Questions) retrieval by performing a match between SMS queries and FAQ database. In the future, we can extend this concept to build an MMS-based news retrieval system. In developed countries, MMS is also getting much popularity due to advancements in smartphones and network infrastructures. Shah et al. [180] presented a news video uploading system, called NEWSMAN. Thus, we can also think of extending this work to MMS-based news uploading and retrieval system. In fact, these concepts can be applied to any e–commerce system to advertise or enquire price, review, and other information in real-time. Recent studies confirm that deep neural network (DNN) technologies also yield success in recommendation system problems. Thus, in the future, we would like to focus on DNN-based news recommendation system.

8.7 Multimodal Sentiment Analysis of UGC

Affective computing is an interdisciplinary research area that bring together researchers from various fields such as AI, NLP, and cognitive and social sciences. Due to availability of huge amount of contextual information together with UGC, affective computing research has increasingly evolved from conventional unimodal analysis to more complex forms of multimodal analysis [141, 144, 147]. Multimodality is defined by the presence of more than one modality or channel, *e.g.*, visual, audio, text, gestures, eye gage, other contextual information. Multimodal information is claimed to be very useful in semantics and sentics understanding of user-generated multimedia content [242, 243]. Due to the increasing popularity and success of Deep Neural Network (DNN) technologies, Poria et al. [152] proposed a convolutional multiple kernel learning (MKL) based approach for multimodal emotion recognition and sentiment analysis. Recently, Poria et al. [159] ensemble application of convolutional neural networks and MKL for multimodal sentiment analysis. They proposed a multimodal affective data analysis framework to extract user opinion and emotions from multimodal information leveraging multiple kernel learning. In an another work, Poria et al. [146] took a deeper look into sarcastic tweets using deep convolutional neural networks. However, this research area is still in an early stage and needs further exploration for multimodal sentiment analysis of UGC. We can exploit user's contextual and personalized information using deep convolutional neural networks in the multimodal sentiment analysis of UGC. We can leverage APIs such as Google Cloud Vision API which determines affective and semantics categories from content.

Aspect-based opinion mining is one of the fundamental challenges within sentiment analysis. Exploiting common-sense knowledge and sentence dependency trees from product reviews to detect both explicit and implicit aspects based on a rule-based approach is a possible direction [150]. Poria et al. [151] further proposed

Sentic LDA that exploits common-sense reasoning to shift LDA clustering from a syntactic to a semantic level. Sentic LDA leverages on the semantics associated with words and multi-word expressions to improve clustering rather than looking at word co-occurrence frequencies. Next, they exploited a deep convolutional neural network to extract aspects for opinion mining [142]. Recently, Poria et al. [145] explored context-dependent sentiment analysis in user-generated videos. This inspires us to focus on the multimodal sentiment analysis of UGC leveraging deep neural network technologies.

8.8 DNN-Based Event Detection and Recommendation

With the advent of smartphones and auto-uploaders, user-generated content (*e.g.*, tweets, photos, and videos) uploads on social media have become more numerous and asynchronous. Thus, it is difficult and time taking for users to manually search (detect) interesting events. It requires for social media companies to automatically detect events and subsequently recommend them to their users. An automatic event detection is also very useful in an efficient search and retrieval of UGC. Furthermore, since the number of users and events on event-based social networks (EBSN) is increasing rapidly, it is not feasible for users to manually find the personalized events of their interest. We would like to further explore events on EBSN such as Meetup for different multimedia analytics projects such as recommending events, groups, and friends to users [161]. We would like to use Deep Neural Network (DNN) technologies due to their immense success to address interesting problems on recommendation systems.

In recent years, advances in DNN technologies have yielded immense success in computer vision, natural language processing (NLP), and speech processing. Especially, DNN has enabled significant performance boost in many visual tasks including image and video semantic classification, object detection, face matching and retrieval, text detection and recognition in natural scenes, image and video captioning, text classification, speech classification, item recommendation, and others. Recent studies confirm that DNN-based new representations contribute performance improvement in recommendation problems [238]. In the future, we would like to explore the new directions and the technologies of DNN-based event detection and recommendation. Moreover, in the future, we would like to present a general DNN-based framework for any recommendation problems (*i.e.*, not only limited to event recommendation).

References

1. Apple Denies Steve Jobs Heart Attack Report: "It Is Not True". http://www.businessinsider.com/2008/10/apple-s-steve-jobs-rushed-to-er-after-heart-attack-says-cnn-citizen-journalist/. October 2008. Online: Last Accessed Sept 2015.
2. SVM^{hmm}: Sequence Tagging with Structural Support Vector Machines. https://www.cs.cornell.edu/people/tj/svm light/svm hmm.html. August 2008. Online: Last Accessed May 2016.

3. NPTEL. 2009, December. http://www.nptel.ac.in. Online; Accessed Apr 2015.
4. iReport at 5: Nearly 900,000 contributors worldwide. http://www.niemanlab.org/2011/08/ireport-at-5-nearly-900000-contributors-worldwide/. August 2011. Online: Last Accessed Sept 2015.
5. Meet the million: 999,999 iReporters + you! http://www.ireport.cnn.com/blogs/ireport-blog/2012/01/23/meet-the-million-999999-ireporters-you. January 2012. Online: Last Accessed Sept 2015.
6. 5 Surprising Stats about User-generated Content. 2014, April. http://www.smartblogs.com/social-media/2014/04/11/6.-surprising-stats-about-user-generated-content/. Online: Last Accessed Sept 2015.
7. The Citizen Journalist: How Ordinary People are Taking Control of the News. 2015, June. http://www.digitaltrends.com/features/the-citizen-journalist-how-ordinary-people-are-taking-control-of-the-news/. Online: Last Accessed Sept 2015.
8. Wikipedia API. 2015, April. http://tinyurl.com/WikiAPI-AI. API: Last Accessed Apr 2015.
9. Apache Lucene. 2016, June. https://lucene.apache.org/core/. Java API: Last Accessed June 2016.
10. By the Numbers: 14 Interesting Flickr Stats. 2016, May. http://www.expandedramblings.com/index.php/flickr-stats/. Online: Last Accessed May 2016.
11. By the Numbers: 180+ Interesting Instagram Statistics (June 2016). 2016, June. http://www.expandedramblings.com/index.php/important-instagram-stats/. Online: Last Accessed July 2016.
12. Coursera. 2016, May. https://www.coursera.org/. Online: Last Accessed May 2016.
13. FourSquare API. 2016, June. https://developer.foursquare.com/. Last Accessed June 2016.
14. Google Cloud Vision API. 2016, December. https://cloud.google.com/vision/. Online: Last Accessed Dec 2016.
15. Google Forms. 2016, May. https://docs.google.com/forms/. Online: Last Accessed May 2016.
16. MIT Open Course Ware. 2016, May. http://www.ocw.mit.edu/. Online: Last Accessed May 2016.
17. Porter Stemmer. 2016, May. https://tartarus.org/martin/PorterStemmer/. Online: Last Accessed May 2016.
18. SenticNet. 2016, May. http://www.sentic.net/computing/. Online: Last Accessed May 2016.
19. Sentics. 2016, May. https://en.wiktionary.org/wiki/sentics. Online: Last Accessed May 2016.
20. VideoLectures.Net. 2016, May. http://www.videolectures.net/. Online: Last Accessed May, 2016.
21. YouTube Statistics. 2016, July. http://www.youtube.com/yt/press/statistics.html. Online: Last Accessed July, 2016.
22. Abba, H.A., S.N.M. Shah, N.B. Zakaria, and A.J. Pal. 2012. Deadline based performance evaluation of job scheduling algorithms. In *Proceedings of the IEEE International Conference on Cyber-Enabled Distributed Computing and Knowledge Discovery*, 106–110.
23. Achanta, R.S., W.-Q. Yan, and M.S. Kankanhalli. 2006. Modeling Intent for Home Video Repurposing. *Proceedings of the IEEE MultiMedia* 45(1): 46–55.
24. Adcock, J., M. Cooper, A. Girgensohn, and L. Wilcox. 2005. Interactive Video Search Using Multilevel Indexing. In *Proceedings of the Springer Image and Video Retrieval*, 205–214.
25. Agarwal, B., S. Poria, N. Mittal, A. Gelbukh, and A. Hussain. 2015. Concept-level Sentiment Analysis with Dependency-based Semantic Parsing: A Novel Approach. In *Proceedings of the Springer Cognitive Computation*, 1–13.
26. Aizawa, K., D. Tancharoen, S. Kawasaki, and T. Yamasaki. 2004. Efficient Retrieval of Life Log based on Context and Content. In *Proceedings of the ACM Workshop on Continuous Archival and Retrieval of Personal Experiences*, 22–31.
27. Altun, Y., I. Tsochantaridis, and T. Hofmann. 2003. Hidden Markov Support Vector Machines. In *Proceedings of the International Conference on Machine Learning*, 3–10.
28. Anderson, A., K. Ranghunathan, and A. Vogel. 2008. Tagez: Flickr Tag Recommendation. In *Proceedings of the Association for the Advancement of Artificial Intelligence*.

29. Atrey, P.K., A. El Saddik, and M.S. Kankanhalli. 2011. Effective Multimedia Surveillance using a Human-centric Approach. *Proceedings of the Springer Multimedia Tools and Applications* 51(2): 697–721.
30. Barnard, K., P. Duygulu, D. Forsyth, N. De Freitas, D.M. Blei, and M.I. Jordan. 2003. Matching Words and Pictures. *Proceedings of the Journal of Machine Learning Research* 3: 1107–1135.
31. Basu, S., Y. Yu, V.K. Singh, and R. Zimmermann. 2016. Videopedia: Lecture Video Recommendation for Educational Blogs Using Topic Modeling. In *Proceedings of the Springer International Conference on Multimedia Modeling*, 238–250.
32. Basu, S., Y. Yu, and R. Zimmermann. 2016. Fuzzy Clustering of Lecture Videos Based on Topic Modeling. In *Proceedings of the IEEE International Workshop on Content-Based Multimedia Indexing*, 1–6.
33. Basu, S., R. Zimmermann, K.L. OHalloran, S. Tan, and K. Marissa. 2015. Performance Evaluation of Students Using Multimodal Learning Systems. In *Proceedings of the Springer International Conference on Multimedia Modeling*, 135–147.
34. Beeferman, D., A. Berger, and J. Lafferty. 1999. Statistical Models for Text Segmentation. *Proceedings of the Springer Machine Learning* 34(1–3): 177–210.
35. Bernd, J., D. Borth, C. Carrano, J. Choi, B. Elizalde, G. Friedland, L. Gottlieb, K. Ni, R. Pearce, D. Poland, et al. 2015. Kickstarting the Commons: The YFCC100M and the YLI Corpora. In *Proceedings of the ACM Workshop on Community-Organized Multimodal Mining: Opportunities for Novel Solutions*, 1–6.
36. Bhatt, C.A., and M.S. Kankanhalli. 2011. Multimedia Data Mining: State of the Art and Challenges. *Proceedings of the Multimedia Tools and Applications* 51(1): 35–76.
37. Bhatt, C.A., A. Popescu-Belis, M. Habibi, S. Ingram, S. Masneri, F. McInnes, N. Pappas, and O. Schreer. 2013. Multi-factor Segmentation for Topic Visualization and Recommendation: the MUST-VIS System. In *Proceedings of the ACM International Conference on Multimedia*, 365–368.
38. Bhattacharjee, S., W.C. Cheng, C.-F. Chou, L. Golubchik, and S. Khuller. 2000. BISTRO: A Framework for Building Scalable Wide-Area Upload Applications. *Proceedings of the ACM SIGMETRICS Performance Evaluation Review* 28(2): 29–35.
39. Cambria, E., J. Fu, F. Bisio, and S. Poria. 2015. AffectiveSpace 2: Enabling Affective Intuition for Concept-level Sentiment Analysis. In *Proceedings of the AAAI Conference on Artificial Intelligence*, 508–514.
40. Cambria, E., A. Livingstone, and A. Hussain. 2012. The Hourglass of Emotions. In *Proceedings of the Springer Cognitive Behavioural Systems*, 144–157.
41. Cambria, E., D. Olsher, and D. Rajagopal. 2014. SenticNet 3: A Common and Common-sense Knowledge Base for Cognition-driven Sentiment Analysis. In *Proceedings of the AAAI Conference on Artificial Intelligence*, 1515–1521.
42. Cambria, E., S. Poria, R. Bajpai, and B. Schuller. 2016. SenticNet 4: A Semantic Resource for Sentiment Analysis based on Conceptual Primitives. In *Proceedings of the International Conference on Computational Linguistics (COLING)*, 2666–2677.
43. Cambria, E., S. Poria, F. Bisio, R. Bajpai, and I. Chaturvedi. 2015. The CLSA Model: A Novel Framework for Concept-Level Sentiment Analysis. In *Proceedings of the Springer Computational Linguistics and Intelligent Text Processing*, 3–22.
44. Cambria, E., S. Poria, A. Gelbukh, and K. Kwok. 2014. Sentic API: A Common-sense based API for Concept-level Sentiment Analysis. *CEUR Workshop Proceedings* 144: 19–24.
45. Cao, J., Z. Huang, and Y. Yang. 2015. Spatial-aware Multimodal Location Estimation for Social Images. In *Proceedings of the ACM Conference on Multimedia Conference*, 119–128.
46. Chakraborty, I., H. Cheng, and O. Javed. 2014. Entity Centric Feature Pooling for Complex Event Detection. In *Proceedings of the Workshop on HuEvent at the ACM International Conference on Multimedia*, 1–5.
47. Che, X., H. Yang, and C. Meinel. 2013. Lecture Video Segmentation by Automatically Analyzing the Synchronized Slides. In *Proceedings of the ACM International Conference on Multimedia*, 345–348.

48. Chen, B., J. Wang, Q. Huang, and T. Mei. 2012. Personalized Video Recommendation through Tripartite Graph Propagation. In *Proceedings of the ACM International Conference on Multimedia*, 1133–1136.
49. Chen, S., L. Tong, and T. He. 2011. Optimal Deadline Scheduling with Commitment. In *Proceedings of the IEEE Annual Allerton Conference on Communication, Control, and Computing*, 111–118.
50. Chen, W.-B., C. Zhang, and S. Gao. 2012. Segmentation Tree based Multiple Object Image Retrieval. In *Proceedings of the IEEE International Symposium on Multimedia*, 214–221.
51. Chen, Y., and W.J. Heng. 2003. Automatic Synchronization of Speech Transcript and Slides in Presentation. *Proceedings of the IEEE International Symposium on Circuits and Systems* 2: 568–571.
52. Cohen, J. 1960. A Coefficient of Agreement for Nominal Scales. *Proceedings of the Durham Educational and Psychological Measurement* 20(1): 37–46.
53. Cristani, M., A. Pesarin, C. Drioli, V. Murino, A. Rodà,M. Grapulin, and N. Sebe. 2010. Toward an Automatically Generated Soundtrack from Low-level Cross-modal Correlations for Automotive Scenarios. In *Proceedings of the ACM International Conference on Multimedia*, 551–560.
54. Dang-Nguyen, D.-T., L. Piras, G. Giacinto, G. Boato, and F.G. De Natale. 2015. A Hybrid Approach for Retrieving Diverse Social Images of Landmarks. In *Proceedings of the IEEE International Conference on Multimedia and Expo (ICME)*, 1–6.
55. Fabro, M. Del, A. Sobe, and L. Böszörmenyi. 2012. Summarization of Real-life Events Based on Community-contributed Content. In *Proceedings of the International Conferences on Advances in Multimedia*, 119–126.
56. Du, L., W.L. Buntine, and M. Johnson. 2013. Topic Segmentation with a Structured Topic Model. In *Proceedings of the North American Chapter of the Association for Computational Linguistics: Human Language Technologies*, 190–200.
57. Fan, Q., K. Barnard, A. Amir, A. Efrat, and M. Lin. 2006. Matching Slides to Presentation Videos using SIFT and Scene Background Matching. In *Proceedings of the ACM International Conference on Multimedia*, 239–248.
58. Filatova, E. and V. Hatzivassiloglou. 2004. Event-Based Extractive Summarization. In *Proceedings of the ACL Workshop on Summarization*, 104–111.
59. Firan, C.S., M. Georgescu, W. Nejdl, and R. Paiu. 2010. Bringing Order to Your Photos: Event-driven Classification of Flickr Images Based on Social Knowledge. In *Proceedings of the ACM International Conference on Information and Knowledge Management*, 189–198.
60. Gao, S., C. Zhang, and W.-B. Chen. 2012. An Improvement of Color Image Segmentation through Projective Clustering. In *Proceedings of the IEEE International Conference on Information Reuse and Integration*, 152–158.
61. Garg, N. and I. Weber. 2008. Personalized, Interactive Tag Recommendation for Flickr. In *Proceedings of the ACM Conference on Recommender Systems*, 67–74.
62. Ghias, A., J. Logan, D. Chamberlin, and B.C. Smith. 1995. Query by Humming: Musical Information Retrieval in an Audio Database. In *Proceedings of the ACM International Conference on Multimedia*, 231–236.
63. Golder, S.A., and B.A. Huberman. 2006. Usage Patterns of Collaborative Tagging Systems. *Proceedings of the Journal of Information Science* 32(2): 198–208.
64. Gozali, J.P., M.-Y. Kan, and H. Sundaram. 2012. Hidden Markov Model for Event Photo Stream Segmentation. In *Proceedings of the IEEE International Conference on Multimedia and Expo Workshops*, 25–30.
65. Guo, Y., L. Zhang, Y. Hu, X. He, and J. Gao. 2016. Ms-celeb-1m: Challenge of recognizing one million celebrities in the real world. *Proceedings of the Society for Imaging Science and Technology Electronic Imaging* 2016(11): 1–6.
66. Hanjalic, A., and L.-Q. Xu. 2005. Affective Video Content Representation and Modeling. *Proceedings of the IEEE Transactions on Multimedia* 7(1): 143–154.

67. Haubold, A. and J.R. Kender. 2005. Augmented Segmentation and Visualization for Presentation Videos. In *Proceedings of the ACM International Conference on Multimedia*, 51–60.
68. Healey, J.A., and R.W. Picard. 2005. Detecting Stress during Real-world Driving Tasks using Physiological Sensors. *Proceedings of the IEEE Transactions on Intelligent Transportation Systems* 6(2): 156–166.
69. Hefeeda, M., and C.-H. Hsu. 2010. On Burst Transmission Scheduling in Mobile TV Broadcast Networks. *Proceedings of the IEEE/ACM Transactions on Networking* 18(2): 610–623.
70. Hevner, K. 1936. Experimental Studies of the Elements of Expression in Music. *Proceedings of the American Journal of Psychology* 48: 246–268.
71. Hochbaum, D.S.. 1996. Approximating Covering and Packing Problems: Set Cover, Vertex Cover, Independent Set, and related Problems. In *Proceedings of the PWS Approximation algorithms for NP-hard problems*, 94–143.
72. Hong, R., J. Tang, H.-K. Tan, S. Yan, C. Ngo, and T.-S. Chua. 2009. Event Driven Summarization for Web Videos. In *Proceedings of the ACM SIGMM Workshop on Social Media*, 43–48.
73. P. ITU-T Recommendation. 1999. Subjective Video Quality Assessment Methods for Multimedia Applications.
74. Jiang, L., A.G. Hauptmann, and G. Xiang. 2012. Leveraging High-level and Low-level Features for Multimedia Event Detection. In *Proceedings of the ACM International Conference on Multimedia*, 449–458.
75. Joachims, T., T. Finley, and C.-N. Yu. 2009. Cutting-plane Training of Structural SVMs. *Proceedings of the Machine Learning Journal* 77(1): 27–59.
76. Johnson, J., L. Ballan, and L. Fei-Fei. 2015. Love Thy Neighbors: Image Annotation by Exploiting Image Metadata. In *Proceedings of the IEEE International Conference on Computer Vision*, 4624–4632.
77. Johnson, J., A. Karpathy, and L. Fei-Fei. 2015. Densecap: Fully Convolutional Localization Networks for Dense Captioning. *In Proceedings of the arXiv preprint arXiv:1511.07571*.
78. Jokhio, F., A. Ashraf, S. Lafond, I. Porres, and J. Lilius. 2013. Prediction-Based dynamic resource allocation for video transcoding in cloud computing. In *Proceedings of the IEEE International Conference on Parallel, Distributed and Network-Based Processing*, 254–261.
79. Kaminskas, M., I. Fernández-Tobías, F. Ricci, and I. Cantador. 2014. Knowledge-Based Identification of Music Suited for Places of Interest. *Proceedings of the Springer Information Technology & Tourism* 14(1): 73–95.
80. Kaminskas, M. and F. Ricci. 2011. Location-adapted Music Recommendation using Tags. In *Proceedings of the Springer User Modeling, Adaption and Personalization*, 183–194.
81. Kan, M.-Y.. 2001. Combining Visual Layout and Lexical Cohesion Features for Text Segmentation. *In Proceedings of the Citeseer*.
82. Kan, M.-Y. 2003. *Automatic Text Summarization as Applied to Information Retrieval*. PhD thesis, Columbia University.
83. Kan, M.-Y., J.L. Klavans, and K.R. McKeown.1998. Linear Segmentation and Segment Significance. *In Proceedings of the arXiv preprint cs/9809020*.
84. Kan, M.-Y., K.R. McKeown, and J.L. Klavans. 2001. Applying Natural Language Generation to Indicative Summarization. *Proceedings of the ACL European Workshop on Natural Language Generation* 8: 1–9.
85. Kang, H.B.. 2003. Affective Content Detection using HMMs. In *Proceedings of the ACM International Conference on Multimedia*, 259–262.
86. Kang, Y.-L., J.-H. Lim, M.S. Kankanhalli, C.-S. Xu, and Q. Tian. 2004. Goal Detection in Soccer Video using Audio/Visual Keywords. *Proceedings of the IEEE International Conference on Image Processing* 3: 1629–1632.
87. Kang, Y.-L., J.-H. Lim, Q. Tian, and M.S. Kankanhalli. 2003. Soccer Video Event Detection with Visual Keywords. *Proceedings of the Joint Conference of International Conference on*

Information, Communications and Signal Processing, and Pacific Rim Conference on Multimedia 3: 1796–1800.
88. Kankanhalli, M.S., and T.-S. Chua. 2000. Video Modeling using Strata-Based Annotation. *Proceedings of the IEEE MultiMedia* 7(1): 68–74.
89. Kennedy, L., M. Naaman, S. Ahern, R. Nair, and T. Rattenbury. 2007. How Flickr Helps us Make Sense of the World: Context and Content in Community-Contributed Media Collections. In *Proceedings of the ACM International Conference on Multimedia*, 631–640.
90. Kennedy, L.S., S.-F. Chang, and I.V. Kozintsev. 2006. To Search or to Label?: Predicting the Performance of Search-Based Automatic Image Classifiers. In *Proceedings of the ACM International Workshop on Multimedia Information Retrieval*, 249–258.
91. Kim, Y.E., E.M. Schmidt, R. Migneco, B.G. Morton, P. Richardson, J. Scott, J.A. Speck, and D. Turnbull. 2010. Music Emotion Recognition: A State of the Art Review. In *Proceedings of the International Society for Music Information Retrieval*, 255–266.
92. Klavans, J.L., K.R. McKeown, M.-Y. Kan, and S. Lee. 1998. Resources for Evaluation of Summarization Techniques. *In Proceedings of the arXiv preprint cs/9810014*.
93. Ko, Y.. 2012. A Study of Term Weighting Schemes using Class Information for Text Classification. In *Proceedings of the ACM Special Interest Group on Information Retrieval*, 1029–1030.
94. Kort, B., R. Reilly, and R.W. Picard. 2001. An Affective Model of Interplay between Emotions and Learning: Reengineering Educational Pedagogy-Building a Learning Companion. *Proceedings of the IEEE International Conference on Advanced Learning Technologies* 1: 43–47.
95. Kucuktunc, O., U. Gudukbay, and O. Ulusoy. 2010. Fuzzy Color Histogram-Based Video Segmentation. *Proceedings of the Computer Vision and Image Understanding* 114(1): 125–134.
96. Kuo, F.-F., M.-F. Chiang, M.-K. Shan, and S.-Y. Lee. 2005. Emotion-Based Music Recommendation by Association Discovery from Film Music. In *Proceedings of the ACM International Conference on Multimedia*, 507–510.
97. Lacy, S., T. Atwater, X. Qin, and A. Powers. 1988. Cost and Competition in the Adoption of Satellite News Gathering Technology. *Proceedings of the Taylor & Francis Journal of Media Economics* 1(1): 51–59.
98. Lambert, P., W. De Neve, P. De Neve, I. Moerman, P. Demeester, and R. Van de Walle. 2006. Rate-distortion performance of H. 264/AVC compared to state-of-the-art video codecs. *Proceedings of the IEEE Transactions on Circuits and Systems for Video Technology* 16(1): 134–140.
99. Laurier, C., M. Sordo, J. Serrà, and P. Herrera. 2009. Music Mood Representations from Social Tags. In *Proceedings of the International Society for Music Information Retrieval*, 381–386.
100. Li, C.T. and M.K. Shan. 2007. Emotion-Based Impressionism Slideshow with Automatic Music Accompaniment. In *Proceedings of the ACM International Conference on Multimedia*, 839–842.
101. Li, J., and J.Z. Wang. 2008. Real-Time Computerized Annotation of Pictures. *Proceedings of the IEEE Transactions on Pattern Analysis and Machine Intelligence* 30(6): 985–1002.
102. Li, X., C.G. Snoek, and M. Worring. 2009. Learning Social Tag Relevance by Neighbor Voting. *Proceedings of the IEEE Transactions on Multimedia* 11(7): 1310–1322.
103. Li, X., T. Uricchio, L. Ballan, M. Bertini, C.G. Snoek, and A.D. Bimbo. 2016. Socializing the Semantic Gap: A Comparative Survey on Image Tag Assignment, Refinement, and Retrieval. *Proceedings of the ACM Computing Surveys (CSUR)* 49(1): 14.
104. Li, Z., Y. Huang, G. Liu, F. Wang, Z.-L. Zhang, and Y. Dai. 2012. Cloud Transcoder: Bridging the Format and Resolution Gap between Internet Videos and Mobile Devices. In *Proceedings of the ACM International Workshop on Network and Operating System Support for Digital Audio and Video*, 33–38.

105. Liang, C., Y. Guo, and Y. Liu. 2008. Is Random Scheduling Sufficient in P2P Video Streaming? In *Proceedings of the IEEE International Conference on Distributed Computing Systems*, 53–60. IEEE.
106. Lim, J.-H., Q. Tian, and P. Mulhem. 2003. Home Photo Content Modeling for Personalized Event-Based Retrieval. *Proceedings of the IEEE MultiMedia* 4: 28–37.
107. Lin, M., M. Chau, J. Cao, and J.F. Nunamaker Jr. 2005. Automated Video Segmentation for Lecture Videos: A Linguistics-Based Approach. *Proceedings of the IGI Global International Journal of Technology and Human Interaction* 1(2): 27–45.
108. Liu, C.L., and J.W. Layland. 1973. Scheduling Algorithms for Multiprogramming in a Hard-real-time Environment. *Proceedings of the ACM Journal of the ACM* 20(1): 46–61.
109. Liu, D., X.-S. Hua, L. Yang, M. Wang, and H.-J. Zhang. 2009. Tag Ranking. In *Proceedings of the ACM World Wide Web Conference*, 351–360.
110. Liu, T., C. Rosenberg, and H.A. Rowley. 2007. Clustering Billions of Images with Large Scale Nearest Neighbor Search. In *Proceedings of the IEEE Winter Conference on Applications of Computer Vision*, 28–28.
111. Liu, X. and B. Huet. 2013. Event Representation and Visualization from Social Media. In *Proceedings of the Springer Pacific-Rim Conference on Multimedia*, 740–749.
112. Liu, Y., D. Zhang, G. Lu, and W.-Y. Ma. 2007. A Survey of Content-Based Image Retrieval with High-level Semantics. *Proceedings of the Elsevier Pattern Recognition* 40(1): 262–282.
113. Livingston, S., and D.A.V. Belle. 2005. The Effects of Satellite Technology on Newsgathering from Remote Locations. *Proceedings of the Taylor & Francis Political Communication* 22(1): 45–62.
114. Long, R., H. Wang, Y. Chen, O. Jin, and Y. Yu. 2011. Towards Effective Event Detection, Tracking and Summarization on Microblog Data. In *Proceedings of the Springer Web-Age Information Management*, 652–663.
115. L. Lu, H. You, and H. Zhang. 2001. A New Approach to Query by Humming in Music Retrieval. In *Proceedings of the IEEE International Conference on Multimedia and Expo*, 22–25.
116. Lu, Y., H. To, A. Alfarrarjeh, S.H. Kim, Y. Yin, R. Zimmermann, and C. Shahabi. 2016. GeoUGV: User-generated Mobile Video Dataset with Fine Granularity Spatial Metadata. In *Proceedings of the ACM International Conference on Multimedia Systems*, 43.
117. Mao, J., W. Xu, Y. Yang, J. Wang, Z. Huang, and A. Yuille. 2014. Deep Captioning with Multimodal Recurrent Neural Networks (M-RNN). *In Proceedings of the arXiv preprint arXiv:1412.6632*.
118. Matusiak, K.K. 2006. Towards User-Centered Indexing in Digital Image Collections. *Proceedings of the OCLC Systems & Services: International Digital Library Perspectives* 22(4): 283–298.
119. McDuff, D., R. El Kaliouby, E. Kodra, and R. Picard. 2013. Measuring Voter's Candidate Preference Based on Affective Responses to Election Debates. In *Proceedings of the IEEE Humaine Association Conference on Affective Computing and Intelligent Interaction*, 369–374.
120. McKeown, K.R., J.L. Klavans, and M.-Y. Kan. Method and System for Topical Segmentation, Segment Significance and Segment Function, 29 2002. US Patent 6,473,730.
121. Mezaris, V., A. Scherp, R. Jain, M. Kankanhalli, H. Zhou, J. Zhang, L. Wang, and Z. Zhang. 2011. Modeling and Rrepresenting Events in Multimedia. In *Proceedings of the ACM International Conference on Multimedia*, 613–614.
122. Mezaris, V., A. Scherp, R. Jain, and M.S. Kankanhalli. 2014. Real-life Events in Multimedia: Detection, Representation, Retrieval, and Applications. *Proceedings of the Springer Multimedia Tools and Applications* 70(1): 1–6.
123. Miller, G., and C. Fellbaum. 1998. *Wordnet: An Electronic Lexical Database*. Cambridge, MA: MIT Press.
124. Miller, G.A. 1995. WordNet: A Lexical Database for English. *Proceedings of the Communications of the ACM* 38(11): 39–41.

125. Moxley, E., J. Kleban, J. Xu, and B. Manjunath. 2009. Not All Tags are Created Equal: Learning Flickr Tag Semantics for Global Annotation. In *Proceedings of the IEEE International Conference on Multimedia and Expo*, 1452–1455.
126. Mulhem, P., M.S. Kankanhalli, J. Yi, and H. Hassan. 2003. Pivot Vector Space Approach for Audio-Video Mixing. *Proceedings of the IEEE MultiMedia* 2: 28–40.
127. Naaman, M. 2012. Social Multimedia: Highlighting Opportunities for Search and Mining of Multimedia Data in Social Media Applications. *Proceedings of the Springer Multimedia Tools and Applications* 56(1): 9–34.
128. Natarajan, P., P.K. Atrey, and M. Kankanhalli. 2015. Multi-Camera Coordination and Control in Surveillance Systems: A Survey. *Proceedings of the ACM Transactions on Multimedia Computing, Communications, and Applications* 11(4): 57.
129. Nayak, M.G. 2004. *Music Synthesis for Home Videos*. PhD thesis.
130. Neo, S.-Y., J. Zhao, M.-Y. Kan, and T.-S. Chua. 2006. Video Retrieval using High Level Features: Exploiting Query Matching and Confidence-Based Weighting. In *Proceedings of the Springer International Conference on Image and Video Retrieval*, 143–152.
131. Ngo, C.-W., F. Wang, and T.-C. Pong. 2003. Structuring Lecture Videos for Distance Learning Applications. In *Proceedings of the IEEE International Symposium on Multimedia Software Engineering*, 215–222.
132. Nguyen, V.-A., J. Boyd-Graber, and P. Resnik. 2012. SITS: A Hierarchical Nonparametric Model using Speaker Identity for Topic Segmentation in Multiparty Conversations. In *Proceedings of the Annual Meeting of the Association for Computational Linguistics*, 78–87.
133. Nwana, A.O. and T. Chen. 2016. Who Ordered This?: Exploiting Implicit User Tag Order Preferences for Personalized Image Tagging. *In Proceedings of the arXiv preprint arXiv:1601.06439.*
134. Papagiannopoulou, C. and V. Mezaris. 2014. Concept-Based Image Clustering and Summarization of Event-related Image Collections. In *Proceedings of the Workshop on HuEvent at the ACM International Conference on Multimedia*, 23–28.
135. Park, M.H., J.H. Hong, and S.B. Cho. 2007. Location-Based Recommendation System using Bayesian User's Preference Model in Mobile Devices. In *Proceedings of the Springer Ubiquitous Intelligence and Computing*, 1130–1139.
136. Petkos, G., S. Papadopoulos, V. Mezaris, R. Troncy, P. Cimiano, T. Reuter, and Y. Kompatsiaris. 2014. Social Event Detection at MediaEval: a Three-Year Retrospect of Tasks and Results. In *Proceedings of the Workshop on Social Events in Web Multimedia at ACM International Conference on Multimedia Retrieval.*
137. Pevzner, L., and M.A. Hearst. 2002. A Critique and Improvement of an Evaluation Metric for Text Segmentation. *Proceedings of the Computational Linguistics* 28(1): 19–36.
138. Picard, R.W., and J. Klein. 2002. Computers that Recognise and Respond to User Emotion: Theoretical and Practical Implications. *Proceedings of the Interacting with Computers* 14(2): 141–169.
139. Picard, R.W., E. Vyzas, and J. Healey. 2001. Toward machine emotional intelligence: Analysis of affective physiological state. *Proceedings of the IEEE Transactions on Pattern Analysis and Machine Intelligence* 23(10): 1175–1191.
140. Poisson, S.D. and C.H. Schnuse. 1841. Recherches Sur La Pprobabilité Des Jugements En Mmatieré Criminelle Et En Matieré Civile. Meyer.
141. Poria, S., E. Cambria, R. Bajpai, and A. Hussain. 2017. A Review of Affective Computing: From Unimodal Analysis to Multimodal Fusion. *Proceedings of the Elsevier Information Fusion* 37: 98–125.
142. Poria, S., E. Cambria, and A. Gelbukh. 2016. Aspect Extraction for Opinion Mining with a Deep Convolutional Neural Network. *Proceedings of the Elsevier Knowledge-Based Systems* 108: 42–49.
143. Poria, S., E. Cambria, A. Gelbukh, F. Bisio, and A. Hussain. 2015. Sentiment Data Flow Analysis by Means of Dynamic Linguistic Patterns. *Proceedings of the IEEE Computational Intelligence Magazine* 10(4): 26–36.

144. Poria, S., E. Cambria, and A.F. Gelbukh. 2015. Deep Convolutional Neural Network Textual Features and Multiple Kernel Learning for Utterance-level Multimodal Sentiment Analysis. In *Proceedings of the EMNLP*, 2539–2544.
145. Poria, S., E. Cambria, D. Hazarika, N. Mazumder, A. Zadeh, and L.-P. Morency. 2017. Context-Dependent Sentiment Analysis in User-Generated Videos. *In Proceedings of the Association for Computational Linguistics.*
146. Poria, S., E. Cambria, D. Hazarika, and P. Vij. 2016. A Deeper Look into Sarcastic Tweets using Deep Convolutional Neural Networks. In *Proceedings of the International Conference on Computational Linguistics (COLING).*
147. Poria, S., E. Cambria, N. Howard, G.-B. Huang, and A. Hussain. 2016. Fusing Audio Visual and Textual Clues for Sentiment Analysis from Multimodal Content. *Proceedings of the Elsevier Neurocomputing* 174: 50–59.
148. Poria, S., E. Cambria, N. Howard, and A. Hussain. 2015. Enhanced SenticNet with Affective Labels for Concept-Based Opinion Mining: Extended Abstract. In *Proceedings of the International Joint Conference on Artificial Intelligence.*
149. Poria, S., E. Cambria, A. Hussain, and G.-B. Huang. 2015. Towards an Intelligent Framework for Multimodal Affective Data Analysis. *Proceedings of the Elsevier Neural Networks* 63: 104–116.
150. Poria, S., E. Cambria, L.-W. Ku, C. Gui, and A. Gelbukh. 2014. A Rule-Based Approach to Aspect Extraction from Product Reviews. In *Proceedings of the Second Workshop on Natural Language Processing for Social Media (SocialNLP)*, 28–37.
151. Poria, S., I. Chaturvedi, E. Cambria, and F. Bisio. 2016. Sentic LDA: Improving on LDA with Semantic Similarity for Aspect-Based Sentiment Analysis. In *Proceedings of the IEEE International Joint Conference on Neural Networks (IJCNN)*, 4465–4473.
152. Poria, S., I. Chaturvedi, E. Cambria, and A. Hussain. 2016. Convolutional MKL Based Multimodal Emotion Recognition and Sentiment Analysis. In *Proceedings of the IEEE International Conference on Data Mining (ICDM)*, 439–448.
153. Poria, S., A. Gelbukh, B. Agarwal, E. Cambria, and N. Howard. 2014. Sentic Demo: A Hybrid Concept-level Aspect-Based Sentiment Analysis Toolkit. In *Proceedings of the ESWC.*
154. Poria, S., A. Gelbukh, E. Cambria, D. Das, and S. Bandyopadhyay. 2012. Enriching SenticNet Polarity Scores Through Semi-Supervised Fuzzy Clustering. In *Proceedings of the IEEE International Conference on Data Mining Workshops (ICDMW)*, 709–716.
155. Poria, S., A. Gelbukh, E. Cambria, A. Hussain, and G.-B. Huang. 2014. EmoSenticSpace: A Novel Framework for Affective Common-sense Reasoning. *Proceedings of the Elsevier Knowledge-Based Systems* 69: 108–123.
156. Poria, S., A. Gelbukh, E. Cambria, P. Yang, A. Hussain, and T. Durrani. 2012. Merging SenticNet and WordNet-Affect Emotion Lists for Sentiment Analysis. *Proceedings of the IEEE International Conference on Signal Processing (ICSP)* 2: 1251–1255.
157. Poria, S., A. Gelbukh, A. Hussain, S. Bandyopadhyay, and N. Howard. 2013. Music Genre Classification: A Semi-Supervised Approach. In *Proceedings of the Springer Mexican Conference on Pattern Recognition*, 254–263.
158. Poria, S., N. Ofek, A. Gelbukh, A. Hussain, and L. Rokach. 2014. Dependency Tree-Based Rules for Concept-level Aspect-Based Sentiment Analysis. In *Proceedings of the Springer Semantic Web Evaluation Challenge*, 41–47.
159. Poria, S., H. Peng, A. Hussain, N. Howard, and E. Cambria. 2017. Ensemble Application of Convolutional Neural Networks and Multiple Kernel Learning for Multimodal Sentiment Analysis. *In Proceedings of the Elsevier Neurocomputing.*
160. Pye, D., N.J. Hollinghurst, T.J. Mills, and K.R. Wood. 1998. Audio-visual Segmentation for Content-Based Retrieval. In *Proceedings of the International Conference on Spoken Language Processing.*
161. Qiao, Z., P. Zhang, C. Zhou, Y. Cao, L. Guo, and Y. Zhang. 2014. *Event Recommendation in Event-Based Social Networks.*

162. Raad, E.J. and R. Chbeir. 2014. Foto2Events: From Photos to Event Discovery and Linking in Online Social Networks. In *Proceedings of the IEEE Big Data and Cloud Computing*, 508–515, .
163. Radsch, C.C.. 2013. The Revolutions will be Blogged: Cyberactivism and the 4th Estate in Egypt. Doctoral Disseration. American University.
164. Rae, A., B. Sigurbjörnssön, and R. van Zwol. 2010. Improving Tag Recommendation using Social Networks. In *Proceedings of the Adaptivity, Personalization and Fusion of Heterogeneous Information*, 92–99.
165. Rahmani, H., B. Piccart, D. Fierens, and H. Blockeel. 2010. Three Complementary Approaches to Context Aware Movie Recommendation. In *Proceedings of the ACM Workshop on Context-Aware Movie Recommendation*, 57–60.
166. Rattenbury, T., N. Good, and M. Naaman. 2007. Towards Automatic Extraction of Event and Place Semantics from Flickr Tags. In *Proceedings of the ACM Special Interest Group on Information Retrieval*.
167. Rawat, Y. and M. S. Kankanhalli. 2016. ConTagNet: Exploiting User Context for Image Tag Recommendation. In *Proceedings of the ACM International Conference on Multimedia*, 1102–1106.
168. Repp, S., A. Groß, and C. Meinel. 2008. Browsing within Lecture Videos Based on the Chain Index of Speech Transcription. *Proceedings of the IEEE Transactions on Learning Technologies* 1(3): 145–156.
169. Repp, S. and C. Meinel. 2006. Semantic Indexing for Recorded Educational Lecture Videos. In *Proceedings of the IEEE International Conference on Pervasive Computing and Communications Workshops*, 5.
170. Repp, S., J. Waitelonis, H. Sack, and C. Meinel. 2007. Segmentation and Annotation of Audiovisual Recordings Based on Automated Speech Recognition. In *Proceedings of the Springer Intelligent Data Engineering and Automated Learning*, 620–629.
171. Russell, J.A. 1980. A Circumplex Model of Affect. *Proceedings of the Journal of Personality and Social Psychology* 39: 1161–1178.
172. Sahidullah, M., and G. Saha. 2012. Design, Analysis and Experimental Evaluation of Block Based Transformation in MFCC Computation for Speaker Recognition. *Proceedings of the Speech Communication* 54: 543–565.
173. Salamon, J., J. Serra, and E. Gomez. 2013. Tonal Representations for Music Retrieval: From Version Identification to Query-by-Humming. In *Proceedings of the Springer International Journal of Multimedia Information Retrieval* 2(1): 45–58.
174. Schedl, M. and D. Schnitzer. 2014. Location-Aware Music Artist Recommendation. In *Proceedings of the Springer MultiMedia Modeling*, 205–213.
175. M. Schedl and F. Zhou. 2016. Fusing Web and Audio Predictors to Localize the Origin of Music Pieces for Geospatial Retrieval. In *Proceedings of the Springer European Conference on Information Retrieval*, 322–334.
176. Scherp, A., and V. Mezaris. 2014. Survey on Modeling and Indexing Events in Multimedia. *Proceedings of the Springer Multimedia Tools and Applications* 70(1): 7–23.
177. Scherp, A., V. Mezaris, B. Ionescu, and F. De Natale. 2014. HuEvent '14: Workshop on Human-Centered Event Understanding from Multimedia. In *Proceedings of the ACM International Conference on Multimedia*, 1253–1254.
178. Schmitz, P.. 2006. Inducing Ontology from Flickr Tags. In *Proceedings of the Collaborative Web Tagging Workshop at ACM World Wide Web Conference*, vol 50.
179. Schuller, B., C. Hage, D. Schuller, and G. Rigoll. 2010. Mister DJ, Cheer Me Up!: Musical and Textual Features for Automatic Mood Classification. *Proceedings of the Journal of New Music Research* 39(1): 13–34.
180. Shah, R.R., M. Hefeeda, R. Zimmermann, K. Harras, C.-H. Hsu, and Y. Yu. 2016. NEWSMAN: Uploading Videos over Adaptive Middleboxes to News Servers In Weak Network Infrastructures. In *Proceedings of the Springer International Conference on Multimedia Modeling*, 100–113.

181. Shah, R.R., A. Samanta, D. Gupta, Y. Yu, S. Tang, and R. Zimmermann. 2016. PROMPT: Personalized User Tag Recommendation for Social Media Photos Leveraging Multimodal Information. In *Proceedings of the ACM International Conference on Multimedia*, 486–492.
182. Shah, R.R., A.D. Shaikh, Y. Yu, W. Geng, R. Zimmermann, and G. Wu. 2015. EventBuilder: Real-time Multimedia Event Summarization by Visualizing Social Media. In *Proceedings of the ACM International Conference on Multimedia*, 185–188.
183. Shah, R.R., Y. Yu, A.D. Shaikh, S. Tang, and R. Zimmermann. 2014. ATLAS: Automatic Temporal Segmentation and Annotation of Lecture Videos Based on Modelling Transition Time. In *Proceedings of the ACM International Conference on Multimedia*, 209–212.
184. Shah, R.R., Y. Yu, A.D. Shaikh, and R. Zimmermann. 2015. TRACE: A Linguistic-Based Approach for Automatic Lecture Video Segmentation Leveraging Wikipedia Texts. In *Proceedings of the IEEE International Symposium on Multimedia*, 217–220.
185. Shah, R.R., Y. Yu, S. Tang, S. Satoh, A. Verma, and R. Zimmermann. 2016. Concept-Level Multimodal Ranking of Flickr Photo Tags via Recall Based Weighting. In *Proceedings of the MMCommon's Workshop at ACM International Conference on Multimedia*, 19–26.
186. Shah, R.R., Y. Yu, A. Verma, S. Tang, A.D. Shaikh, and R. Zimmermann. 2016. Leveraging Multimodal Information for Event Summarization and Concept-level Sentiment Analysis. In *Proceedings of the Elsevier Knowledge-Based Systems*, 102–109.
187. Shah, R.R., Y. Yu, and R. Zimmermann. 2014. ADVISOR: Personalized Video Soundtrack Recommendation by Late Fusion with Heuristic Rankings. In *Proceedings of the ACM International Conference on Multimedia*, 607–616.
188. Shah, R.R., Y. Yu, and R. Zimmermann. 2014. User Preference-Aware Music Video Generation Based on Modeling Scene Moods. In *Proceedings of the ACM International Conference on Multimedia Systems*, 156–159.
189. Shaikh, A.D., M. Jain, M. Rawat, R.R. Shah, and M. Kumar. 2013. Improving Accuracy of SMS Based FAQ Retrieval System. In *Proceedings of the Springer Multilingual Information Access in South Asian Languages*, 142–156.
190. Shaikh, A.D., R.R. Shah, and R. Shaikh. 2013. SMS Based FAQ Retrieval for Hindi, English and Malayalam. In *Proceedings of the ACM Forum on Information Retrieval Evaluation*, 9.
191. Shamma, D.A., R. Shaw, P.L. Shafton, and Y. Liu. 2007. Watch What I Watch: Using Community Activity to Understand Content. In *Proceedings of the ACM International Workshop on Multimedia Information Retrieval*, 275–284.
192. Shaw, B., J. Shea, S. Sinha, and A. Hogue. 2013. Learning to Rank for Spatiotemporal Search. In *Proceedings of the ACM International Conference on Web Search and Data Mining*, 717–726.
193. Sigurbjörnsson, B. and R. Van Zwol. 2008. Flickr Tag Recommendation Based on Collective Knowledge. In *Proceedings of the ACM World Wide Web Conference*, 327–336.
194. Snoek, C.G., M. Worring, and A.W. Smeulders. 2005. Early versus Late Fusion in Semantic Video Analysis. In *Proceedings of the ACM International Conference on Multimedia*, 399–402.
195. Snoek, C.G., M. Worring, J.C. Van Gemert, J.-M. Geusebroek, and A.W. Smeulders. 2006. The Challenge Problem for Automated Detection of 101 Semantic Concepts in Multimedia. In *Proceedings of the ACM International Conference on Multimedia*, 421–430.
196. Soleymani, M., J.J.M. Kierkels, G. Chanel, and T. Pun. 2009. A Bayesian Framework for Video Affective Representation. In *Proceedings of the IEEE International Conference on Affective Computing and Intelligent Interaction and Workshops*, 1–7.
197. Stober, S., and A. . Nürnberger. 2013. Adaptive Music Retrieval – A State of the Art. *Proceedings of the Springer Multimedia Tools and Applications* 65(3): 467–494.
198. Stoyanov, V., N. Gilbert, C. Cardie, and E. Riloff. 2009. Conundrums in Noun Phrase Coreference Resolution: Making Sense of the State-of-the-art. In *Proceedings of the ACL International Joint Conference on Natural Language Processing of the Asian Federation of Natural Language Processing*, 656–664.

199. Stupar, A. and S. Michel. 2011. Picasso: Automated Soundtrack Suggestion for Multimodal Data. In *Proceedings of the ACM Conference on Information and Knowledge Management*, 2589–2592.
200. Thayer, R.E. 1989. *The Biopsychology of Mood and Arousal*. New York: Oxford University Press.
201. Thomee, B., B. Elizalde, D.A. Shamma, K. Ni, G. Friedland, D. Poland, D. Borth, and L.-J. Li. 2016. YFCC100M: The New Data in Multimedia Research. *Proceedings of the Communications of the ACM* 59(2): 64–73.
202. Tirumala, A., F. Qin, J. Dugan, J. Ferguson, and K. Gibbs. 2005. Iperf: The TCP/UDP Bandwidth Measurement Tool. http://dast.nlanr.net/Projects/Iperf/
203. Torralba, A., R. Fergus, and W.T. Freeman. 2008. 80 Million Tiny Images: A Large Data set for Nonparametric Object and Scene Recognition. *Proceedings of the IEEE Transactions on Pattern Analysis and Machine Intelligence* 30(11): 1958–1970.
204. Toutanova, K., D. Klein, C.D. Manning, and Y. Singer. 2003. Feature-Rich Part-of-Speech Tagging with a Cyclic Dependency Network. In *Proceedings of the North American Chapter of the Association for Computational Linguistics on Human Language Technology*, 173–180.
205. Toutanova, K. and C.D. Manning. 2000. Enriching the Knowledge Sources used in a Maximum Entropy Part-of-Speech Tagger. In *Proceedings of the Annual Meeting of the Association for Computational Linguistics*, 63–70.
206. Utiyama, M. and H. Isahara. 2001. A Statistical Model for Domain-Independent Text Segmentation. In *Proceedings of the Annual Meeting on Association for Computational Linguistics*, 499–506.
207. Vishal, K., C. Jawahar, and V. Chari. 2015. Accurate Localization by Fusing Images and GPS Signals. In *Proceedings of the IEEE Computer Vision and Pattern Recognition Workshops*, 17–24.
208. Wang, C., F. Jing, L. Zhang, and H.-J. Zhang. 2008. Scalable Search-Based Image Annotation. *Proceedings of the Springer Multimedia Systems* 14(4): 205–220.
209. Wang, H.L., and L.F. Cheong. 2006. Affective Understanding in Film. *Proceedings of the IEEE Transactions on Circuits and Systems for Video Technology* 16(6): 689–704.
210. Wang, J., J. Zhou, H. Xu, T. Mei, X.-S. Hua, and S. Li. 2014. Image Tag Refinement by Regularized Latent Dirichlet Allocation. *Proceedings of the Elsevier Computer Vision and Image Understanding* 124: 61–70.
211. Wang, P., H. Wang, M. Liu, and W. Wang. 2010. An Algorithmic Approach to Event Summarization. In *Proceedings of the ACM Special Interest Group on Management of Data*, 183–194.
212. Wang, X., Y. Jia, R. Chen, and B. Zhou. 2015. Ranking User Tags in Micro-Blogging Website. In *Proceedings of the IEEE ICISCE*, 400–403.
213. Wang, X., L. Tang, H. Gao, and H. Liu. 2010. Discovering Overlapping Groups in Social Media. In *Proceedings of the IEEE International Conference on Data Mining*, 569–578.
214. Wang, Y. and M.S. Kankanhalli. 2015. Tweeting Cameras for Event Detection. In *Proceedings of the IW3C2 International Conference on World Wide Web*, 1231–1241.
215. Webster, A.A., C.T. Jones, M.H. Pinson, S.D. Voran, and S. Wolf. 1993. Objective Video Quality Assessment System Based on Human Perception. In *Proceedings of the IS&T/SPIE's Symposium on Electronic Imaging: Science and Technology*, 15–26. International Society for Optics and Photonics.
216. Wei, C.Y., N. Dimitrova, and S.-F. Chang. 2004. Color-Mood Analysis of Films Based on Syntactic and Psychological Models. In *Proceedings of the IEEE International Conference on Multimedia and Expo*, 831–834.
217. Whissel, C. 1989. The Dictionary of Affect in Language. In *Emotion: Theory, Research and Experience. Vol. 4. The Measurement of Emotions*, ed. R. Plutchik and H. Kellerman, 113–131. New York: Academic.

218. Wu, L., L. Yang, N. Yu, and X.-S. Hua. 2009. Learning to Tag. In *Proceedings of the ACM World Wide Web Conference*, 361–370.
219. Xiao, J., W. Zhou, X. Li, M. Wang, and Q. Tian. 2012. Image Tag Re-ranking by Coupled Probability Transition. In *Proceedings of the ACM International Conference on Multimedia*, 849–852.
220. Xie, D., B. Qian, Y. Peng, and T. Chen. 2009. A Model of Job Scheduling with Deadline for Video-on-Demand System. In *Proceedings of the IEEE International Conference on Web Information Systems and Mining*, 661–668.
221. Xu, M., L.-Y. Duan, C. Xu, M. Kankanhalli, and Q. Tian. 2003. Event Detection in Basketball Video using Multiple Modalities. *Proceedings of the IEEE Joint Conference of the Fourth International Conference on Information, Communications and Signal Processing, and Fourth Pacific Rim Conference on Multimedia* 3: 1526–1530.
222. Xu, M., N.C. Maddage, C. Xu, M. Kankanhalli, and Q. Tian. 2003. Creating Audio Keywords for Event Detection in Soccer Video. In *Proceedings of the IEEE International Conference on Multimedia and Expo*, 2:II–281.
223. Yamamoto, N., J. Ogata, and Y. Ariki. 2003. Topic Segmentation and Retrieval System for Lecture Videos Based on Spontaneous Speech Recognition. In *Proceedings of the INTERSPEECH*, 961–964.
224. Yang, H., M. Siebert, P. Luhne, H. Sack, and C. Meinel. 2011. Automatic Lecture Video Indexing using Video OCR Technology. In *Proceedings of the IEEE International Symposium on Multimedia*, 111–116.
225. Yang, Y.H., Y.C. Lin, Y.F. Su, and H.H. Chen. 2008. A Regression Approach to Music Emotion Recognition. *Proceedings of the IEEE Transactions on Audio, Speech, and Language Processing* 16(2): 448–457.
226. Ye, G., D. Liu, I.-H. Jhuo, and S.-F. Chang. 2012. Robust Late Fusion with Rank Minimization. In *Proceedings of the IEEE Conference on Computer Vision and Pattern Recognition*, 3021–3028.
227. Ye, Q., Q. Huang, W. Gao, and D. Zhao. 2005. Fast and Robust Text Detection in Images and Video Frames. *Proceedings of the Elsevier Image and Vision Computing* 23(6): 565–576.
228. Yin, Y., Z. Shen, L. Zhang, and R. Zimmermann. 2015. Spatial temporal Tag Mining for Automatic Geospatial Video Annotation. *Proceedings of the ACM Transactions on Multimedia Computing, Communications, and Applications* 11(2): 29.
229. Yoon, S. and V. Pavlovic. 2014. Sentiment Flow for Video Interestingness Prediction. In *Proceedings of the Workshop on HuEvent at the ACM International Conference on Multimedia*, 29–34.
230. Yu, Y., K. Joe, V. Oria, F. Moerchen, J.S. Downie, and L. Chen. 2009. Multiversion Music Search using Acoustic Feature Union and Exact Soft Mapping. *Proceedings of the World Scientific International Journal of Semantic Computing* 3(02): 209–234.
231. Yu, Y., Z. Shen, and R. Zimmermann. 2012. Automatic Music Soundtrack Generation for Outdoor Videos from Contextual Sensor Information. In *Proceedings of the ACM International Conference on Multimedia*, 1377–1378.
232. Zaharieva, M., M. Zeppelzauer, and C. Breiteneder. 2013. Automated Social Event Detection in Large Photo Collections. In *Proceedings of the ACM International Conference on Multimedia Retrieval*, 167–174.
233. Zhang, J., X. Liu, L. Zhuo, and C. Wang. 2015. Social Images Tag Ranking Based on Visual Words in Compressed Domain. *Proceedings of the Elsevier Neurocomputing* 153: 278–285.
234. Zhang, J., S. Wang, and Q. Huang. 2015. Location-Based Parallel Tag Completion for Geo-tagged Social Image Retrieval. In *Proceedings of the ACM International Conference on Multimedia Retrieval*, 355–362.
235. Zhang, M., J. Wong, W. Tavanapong, J. Oh, and P. de Groen. 2004. Media Uploading Systems with Hard Deadlines. In *Proceedings of the Citeseer International Conference on Internet and Multimedia Systems and Applications*, 305–310.

236. Zhang, M., J. Wong, W. Tavanapong, J. Oh, and P. de Groen. 2008. Deadline-constrained Media Uploading Systems. *Proceedings of the Springer Multimedia Tools and Applications* 38(1): 51–74.
237. Zhang, W., J. Lin, X. Chen, Q. Huang, and Y. Liu. 2006. Video Shot Detection using Hidden Markov Models with Complementary Features. *Proceedings of the IEEE International Conference on Innovative Computing, Information and Control* 3: 593–596.
238. Zheng, L., V. Noroozi, and P.S. Yu. 2017. Joint Deep Modeling of Users and Items using Reviews for Recommendation. In *Proceedings of the ACM International Conference on Web Search and Data Mining*, 425–434.
239. Zhou, X.S. and T.S. Huang. 2000. CBIR: from Low-level Features to High-level Semantics. In *Proceedings of the International Society for Optics and Photonics Electronic Imaging*, 426–431.
240. Zhuang, J. and S.C. Hoi. 2011. A Two-view Learning Approach for Image Tag Ranking. In *Proceedings of the ACM International Conference on Web Search and Data Mining*, 625–634.
241. Zimmermann, R. and Y. Yu. 2013. Social Interactions over Geographic-aware Multimedia Systems. In *Proceedings of the ACM International Conference on Multimedia*, 1115–1116.
242. Shah, R.R. 2016. Multimodal-based Multimedia Analysis, Retrieval, and Services in Support of Social Media Applications. In *Proceedings of the ACM International Conference on Multimedia*, 1425–1429.
243. Shah, R.R. 2016. Multimodal Analysis of User-Generated Content in Support of Social Media Applications. In *Proceedings of the ACM International Conference in Multimedia Retrieval*, 423–426.
244. Yin, Y., R.R. Shah, and R. Zimmermann. 2016. A General Feature-based Map Matching Framework with Trajectory Simplification. In *Proceedings of the ACM SIGSPATIAL International Workshop on GeoStreaming*, 7.

Index

R. Shah, R. Zimmermann, *Multimodal Analysis of User-Generated Multimedia Content*, Socio-Affective Computing 6, DOI 10.1007/978-3-319-61807-4

Y

Z